# Gravity's Fatal Attraction

## *Black Holes in the Universe*

Mitchell Begelman

Martin Rees

**SCIENTIFIC**
**AMERICAN**
**LIBRARY**
A division of HPHLP
New York

Begelman, Mitchell C.
Gravity's fatal attraction: black holes in the Universe / by Mitchell C. Begelman and Martin J. Rees.
p. cm.—(Scientific American Library series, ISSN 1040-3213: no. 58)
Includes bibliographical references and index.
ISBN 0-7187-5074-0 (hardcover) ISBN 0-7167-6029-0 (pbk)
1. Black holes (Astronomy) 2. Quasars. I. Rees, Martin J., 1942- . II. Title. III. Series.
QB843.B55B44 1995
523.8′875—dc20 95–42959
CIP

ISSN 1040-3213


Printed in the United States of America

Scientific American Library
A division of HPHLP
New York

Distributed by W. H. Freeman and Company
41 Madison Avenue, New York, NY 10010
Houndmills, Basingstoke RG21 6XS, England

First printing 1998

# Contents

# Preface

As early as 1967 black holes were creeping into public consciousness. On the television program *Star Trek,* Captain Kirk and the crew of the starship *Enterprise* were caught in the gravitational field of an "uncharted black star" and were propelled backward in time. Astronomers then had no clue whether black holes were real or just theoreticians' constructs. Certainly no serious evidence of their reality existed, and few would have guessed that they would soon become the object of intense astronomical study. When we entered the field in the late 1960s and early 1970s, the black hole was still a novel concept, studied by specialists in Einstein's general theory of relativity. And that theory itself (though already several decades old) had only tentative empirical support. Gravity, one of the fundamental forces of nature, was still poorly understood.

Thanks to a technological revolution in observational astronomy—new detectors and mirror designs in optical telescopes, radio telescopes that offer far sharper images than even the best optical instruments, and observations made from space, revealing the sky at infrared, ultraviolet, X-ray, and gamma-ray wavelengths—we are now confident that there are millions of black holes in every galaxy. Each of these holes is the remnant of an ordinary star several times more massive than the Sun. More remarkably, giant black holes, weighing as much as millions (or, dare we say, even billions and billions) of suns, lurk in the centers of most galaxies. The most energetic phenomena in the Universe—quasars, and jets a million light-years long erupting from the centers of galaxies—are powered by black holes. The same phenomena, in miniature, are energized by the smaller holes within our own Galaxy.

This book describes the extraordinary ways in which black holes make their presence known. We discuss the designs and accidents through which they were discovered, and how far we have come toward understanding their relationship to other structures in the cosmos. Every advance in technology has disclosed an assortment of dazzling, unexpected phenomena. Some of these phenomena, like "gravitational lenses," are well understood and are being co-opted as tools in the search for black holes and other forms of "dark matter"; others, such as gamma-ray bursts, remain a mystery; in between lies a range of phenomena for which understanding is substantial but incomplete. And new questions for the future are constantly arising. Can gravitational waves from merging black holes be detected? How much could black holes contribute to the "dark matter" in the Universe? Could microscopic holes exist—the size of an atomic nucleus but the weight of a mountain? As the Universe evolves, what is the ultimate fate of the matter "swallowed" by black holes?

As theorists in a profession heavily dominated (and rightly so) by observers, we demand empirical proof of the existence of black holes. Is there a "smoking gun" that removes *all* doubt of their existence, or of their ubiquity at the level we claim? The answer to the last question is still being debated by some, but, as we aim to show, the accumulating evidence is painting the skeptics into an increasingly tight corner.

This book went to press in the autumn of 1995. The timing was lucky, because important developments earlier that year proved crucial to our theme. In particular, two discoveries—one by radio astronomers, the other by X-ray astronomers—greatly bolstered the evidence for huge black holes in the centers of galaxies. Had we finished the book sooner, we would have had to present a more tentative case. The evidence is now compelling that black holes are an important feature of our cosmic environment. We are beginning to understand the exotic ways they manifest themselves, and how astronomical observations can probe their immediate surroundings.

These cosmic "fireworks," spectacular as they are, may ultimately prove to be of greatest value as stepping-stones to even more profound knowledge. Shrouded from view, deep inside black holes, lurk mysteries that will not be understood without a unification of Einstein's theory of gravity with the other great "pillar" of twentieth-century physics, the quantum theory. Such new insights, when they are achieved, may change our view of the nature of time, and of our entire Universe. The strong new evidence that black holes indeed exist strengthens the motivation for this fundamental quest.

Research in astrophysics is, for us, an intensely interactive activity. Many observers have generously discussed their latest work with us; we have also benefited from other theorists who have special knowledge and expertise that we lack. We especially acknowledge all we have learned from Roger Blandford, a frequent collaborator during the last twenty years. We are fortunate to have, among our immediate colleagues, leading authorities such as Andrew Fabian, Stephen Hawking, Donald Lynden-Bell, and Richard McCray. We are grateful to them, and to our many other collaborators. Among our many colleagues who took the time to prepare illustrations—often of results "hot off the press," or even prepublication—we especially thank John Bally, Joshua Barnes, Adam Burrows, David Devine, Robert Hjellming, Keith Horne, Jean-Paul Kneib, Juan Marcaide, James Moran, Michael Norman, Michael Nowak, C. R. O'Dell, Richard Perley, Kevin Rauch, Max Ruffert, Ralph Sutherland, Robert Wagoner, and Ann Wehrle.

We have been especially fortunate in our publisher and editors—they have completely spared us the frustrations and delays that befall some authors. Jonathan Cobb's encouragement motivated us to set aside other projects for long enough to get the book finished. Susan Moran and Nancy Brooks have helped to make our text clearer than it otherwise would have been. And we are grateful to Travis Amos for tracking down so many attractive and appropriate pictures—his efforts have much enhanced our text by giving the book real visual appeal. We thank Colin Norman, Joseph Silk, and Virginia Trimble for offering helpful comments on our draft text. And, as always, we are grateful to our respective staffs—especially Judith Moss in Cambridge and Elaine Verdill and her colleagues in Boulder—for keeping the three-way exchange of text and pictures running smoothly.

October 1995

When we sent the hardcover edition of *Gravity's Fatal Attraction* to press late in 1995, we congratulated ourselves on our timing. That year had been a banner one for black hole research; little did we suspect that the pace of discovery would actually quicken. New X-ray data are promising to provide a new, precision probe of the geometry of space just outside black holes, and even give us a handle on measuring black hole spins. The existence of a massive black hole at the center of the Milky Way has now been confirmed beyond reasonable doubt. And one of the great mysteries of astrophysics—the nature of gamma-ray bursts—is one important step closer to solution. We have taken advantage of this paperback edition to bring the story of black holes back up to date, however briefly.

December 1997

*Chapter* 1

# Gravity Triumphant

*In the spiral galaxy NGC 4258, there is compelling evidence for a central black hole weighing as much as 36 million suns. Radio astronomers have discovered a disk about a light-year across (shown here in an artist's impression) composed of orbiting gas that is probably swirling down into a black hole. Similar black holes lurk in the center of many galaxies, in some cases generating far more power than all the stars in the galaxy and producing jets of energy squirting outward at almost the speed of light.*

Gravity is the one truly universal force. No substance, no kind of particle, not even light itself, is free of its grasp. It was Isaac Newton who, over three hundred years ago, realized that the force holding us to the ground, and governing a cannon ball's trajectory, is the same force that holds the Moon in its orbit around the Earth. This force, which later came to be called gravity, is the mutual attraction among all bodies. Newton showed how the motion of each planet in the Solar System is the combined outcome of the gravitational pull of the Sun and all the other planets, each contributing according to its mass and distance from the others. Out of Newton's conceptually simple prescriptions came the calculations that have guided spacecraft to all the planets, told us a year in advance that Comet Shoemaker-Levy would hit Jupiter, and enabled us to determine the mass of the Milky Way Galaxy. On cosmic scales gravity dominates over every other kind of force. Every significant level of structure in the Universe—stars, clusters of stars, galaxies, and clusters of galaxies—is maintained by the force of gravity.

Nowhere is gravity stronger than near the objects we call black holes. The effect of gravity on the surface of a body intensifies if the mass of the body is increased, or its size is decreased. Gravity's strength can be characterized by how fast a rocket must be fired to escape from the body. For the Earth, the escape speed is 40,000 kilometers per hour; for the Sun it is nearly a hundred times greater. This tendency for the escape speed to become ever larger for more massive, or more compact, bodies raises an obvious conceptual problem. The gravitational pull of the Sun dominates the Solar System, yet its escape speed is only $\frac{1}{500}$ the highest speed possible, the speed of light. What would be the effect of gravity around a body for which the escape speed was as high as that of light?

The Reverend John Michell, an underappreciated polymath of eighteenth-century science, puzzled over this question in 1784. Arguing on the basis of Newton's laws and assuming that light behaves like particles of matter (the view favored by Newton and his contemporaries), Michell wrote a paper in which he noted, "If the semi-diameter of a sphere of the same density as the Sun were to exceed that of the Sun in the proportion of five hundred to one, and supposing light to be attracted by the same force in proportion to its *vis inertiae* [inertial mass] with other bodies, all light emitted from such a body would be made to return towards it, by its own proper gravity." With remarkable foresight, Michell was suggesting, as did the French astronomer and mathematician Pierre-Simon Laplace a decade later, that the most massive objects in the Universe might be undetectable by their direct radiation but still manifest gravitational effects on material near them. Nearly two hundred years later, astronomers discovered objects for which gravity is as strong as Michell envisaged—these powerfully gravitating bodies are neutron stars and black holes.

Neutron stars, discovered in 1968, are objects slightly heavier than the Sun but extending only 10 to 20 kilometers across, with an escape speed about half that of light. While light can escape from a neutron star, its path is strongly curved by gravity, just like the trajectory of a rocket boosting a spacecraft into interplanetary space from the Earth. The extreme strength of gravity around a neutron star leads to consequences far removed from common-sense experience. Because light itself follows a strongly curved trajectory, an observer's perspective would be severely distorted near a neutron star. One would be able to see much farther to the horizon, and from high above the surface, more than half the entire star would be visible. So powerful is gravity that a pen dropped from a height of one meter above a neutron star would release as much energy as a ton of dynamite.

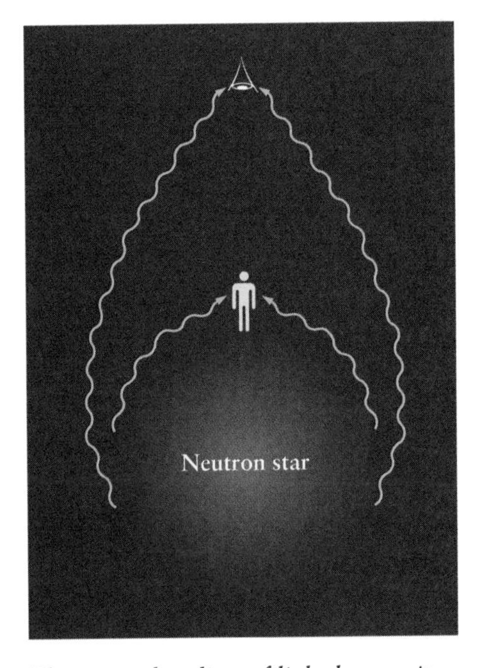

*The severe bending of light by gravity means that an observer on the surface of a neutron star can see farther to the horizon. An observer high above the star could see much more than half the surface.*

Conditions near a neutron star are not quite so extreme as Michell described for his hypothetical light-retaining body. But if a neutron star were a few times smaller, or just slightly heavier, it would trap all the light in its vicinity. It would then be a black hole. Black holes are not merely stars whose surface is hidden from view by the bending of light. In 1905, Einstein's special theory of relativity established that the speed of light sets a limit to the speed at which any kind of matter or signal propagates, so that the kind of object envisaged by Michell would be completely cut off from the outside Universe. Nothing that ventures inside it would ever be able to escape. As it turns out, in a black hole gravity so overwhelms other forces that matter is crushed virtually to a point. These bizarre and counterintuitive properties led many researchers to suppose that, somehow, matter could never become quite so condensed that it produced a black hole. But we now know that black holes are not mere theoretical constructs; they exist in profusion and account for many of the most spectacular astronomical discoveries of recent times.

This book describes how black holes were found and what their existence implies for cosmic evolution and our understanding of gravity. The first candidate black holes were discovered soon after neutron stars. Since then, evidence has mounted that millions of black holes exist in our Milky Way Galaxy, each with a mass roughly 10 times that of the Sun. And in the centers of most galaxies, there almost certainly exist monster black holes weighing as much as 100 million suns, similar in mass and radius to the objects conjectured by Michell.

## General Relativity and Black Holes

Neither Newtonian gravity nor Newton's laws of motion are applicable to situations where the gravitational fields are as strong as those conjectured by Michell and Laplace. On the Earth, as elsewhere in the Solar System, Newton's theory of gravitation affords an excellent approximation. But where gravity is so strong that it can accelerate matter to nearly the speed of light—approximately 300,000 kilometers per second—the Newtonian description no longer works. We cannot formulate a consistent picture of a black hole without deeper insights into gravity than Newton's theory offers. Fortunately, long before black holes were discovered, Einstein's general theory of relativity had provided such insights and laid a firm theoretical foundation for our understanding of black holes. Einstein's theory has been verified by increasingly precise experimental tests, and its equations have solutions that represent situations where gravity is overwhelmingly strong, as it is in black holes.

The papers that established Albert Einstein as the greatest physicist since Newton are all contained in just two volumes of the journal *Annalen der Physik,* published in the years 1905 and 1916. The old backnumbers of scientific journals, relegated to remote stockrooms in academic libraries, are usually seldom consulted, but these two particular volumes are rare and valued collector's items. In 1905, the 26-year-old Einstein enunciated his "special theory of relativity," which (among other breakthroughs) emphasized the special significance of the speed of light and postulated the fundamental interconvertibility or "equivalence" between mass and energy popularly paraphrased in the expression $E = mc^2$. He also proposed that light is quantized into "packets" of energy and formulated the statistical theory of Brownian motion, the frenetic dance of dust particles suspended in a liquid or gas. These contributions alone would cause Einstein to be ranked among the half-dozen great pioneers of twentieth-century physics.

But it is his gravitational theory, "general" relativity, developed eleven years later, that puts Einstein in a class by himself. Had he written none of his 1905 papers, it would not have been long before the same concepts would have been put forward by some of his distinguished contemporaries: the ideas were "in the air." Well-known inconsistencies in earlier theories and puzzling experimental results would, in any case, have focused interest on these problems. General relativity, however, was not a response to any particular observational enigma. Motivated by more than the desire to explain concrete observations, Einstein sought simplicity and unity. Whereas Newton conceived gravity as a force instantaneously transmitted between bodies—a view clearly inconsistent with the speed limit on the propagation of signals—Einstein postulated that gravitational fields are manifestations of the curvature of space itself. Masses do not "exert" a gravitational pull, deflecting bodies from a straight path. Rather, their presence distorts the space within and around them. According to general relativity, bodies moving through space follow the straightest path possible through an amalgam of space and time called "spacetime." But when space is distorted, these paths become curved and accelerating trajectories that we might interpret as the reaction to a force. In the words of the noted relativist John Archibald Wheeler, "Space tells matter how to move; matter tells space how to curve."

As Einstein himself said when he announced his new work, "Scarcely anyone who has understood the theory can escape from its magic." The physicist Hermann Weyl described it as "the greatest example of the power of speculative thought," and Max Born called it "the greatest feat of human thinking about Nature." Had it not been for Einstein, an equally comprehensive theory of gravity might not have arrived until decades later, and might well have been approached by quite a different route. Einstein's creativity thus put a uniquely individual and long-lasting imprint on modern physics. Most scientists hope that their work will last, but Einstein is unique in the degree to which his work retains its individual identity.

Indeed, general relativity was proposed so far in advance of any real application that it remained, for forty years after its discovery, an austere intellectual monument—a somewhat sterile topic isolated from the mainstream of physics and astronomy—whose practitioners, according to Thomas Gold, were "magnificent cultural ornaments." This remoteness from the mainstream stands in glaring contrast to its more recent status as one of the liveliest frontiers of fundamental research. Its reputation for being an exceptionally hard subject was sometimes exaggerated, not least by

*Albert Einstein in 1905, aged 26. In that year he proposed the special theory of relativity, deduced the quantum nature of light, and explained Brownian motion of dust particles.*

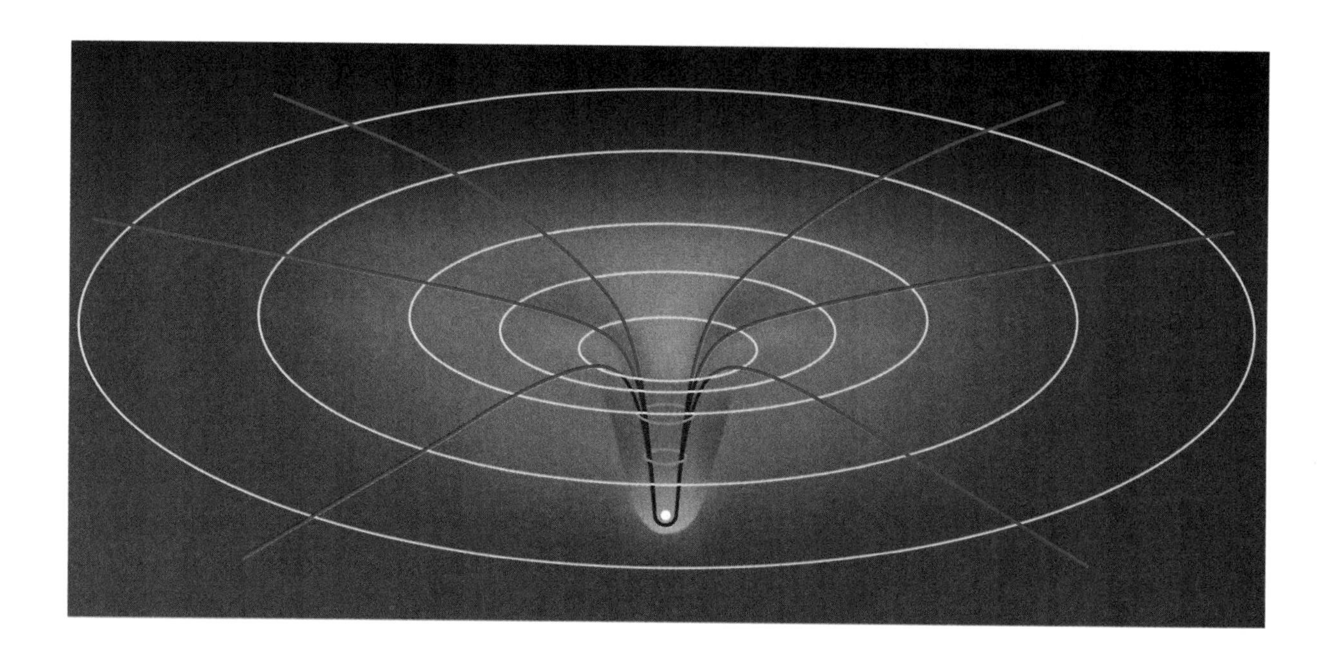

*According to the general theory of relativity, a gravitating body (white dot) stretches and distorts the space around it, much like a lead weight resting on a rubber membrane. Particles following the straightest possible paths in curved spacetime appear to be responding to an attractive force. If the gravitating object represented were a black hole, the depression would be infinitely deep.*

its early practitioners. (When Arthur Eddington was asked if it were true that just three people in the world understood general relativity, he is rumored to have wondered who the third might be.) The theory is now taught almost routinely to physics students, along with quantum physics, electromagnetism, and the rest of the physics canon.

To be sure, general relativity also had its share of early experimental successes. It accounted for long-recognized anomalies in Mercury's orbit, and was famously confirmed by the results of the 1919 solar eclipse expedition led by the British astrophysicist Arthur Eddington. Eddington's observations revealed that light rays passing close to the Sun were deflected by just the amount predicted by Einstein's theory. (A Newtonian argument, as might have been used by Michell, would have predicted half the observed amount.) For slowly moving bodies in weak gravitational fields, the predictions of general relativity agree almost exactly with those of Newton's laws—as they must, given that the latter have withstood three hundred years of testing. In the Solar System, the predicted departures from Newtonian theory amount to only about one part in a million. But experiments, including the tracking of space probes and higher-precision versions of Eddington's light-deflection measurements, are now precise enough to determine these tiny differences within a better than 1 percent margin of error. General relativity has been verified in every single case and, moreover, has proved more accurate than other, similar theories.

The recent revival of interest in general relativity has been inspired not only by the high-precision measurements whose results have favored Einstein's theory over its rivals, but also—to a greater extent—by dramatic advances in astronomy and space research. In neutron stars and black holes, the distinctive features of general relativity take center stage; they are not merely trifling modifications to Newtonian theory. These objects offer us the opportunity to test Einstein's theories in new ways. The other arena where relativity is crucial is cosmology—the study of the entire expanding Universe. During the last thirty years, cosmology has evolved from a subject about which we knew little more than Hubble's law of universal expansion (which states that distant galaxies appear to recede from us at a speed that is proportional to their distance) to a science that allows us to infer cosmic history reliably back to one second after the big bang, and to make compelling conjectures about still earlier eras. The entire fate of the Universe depends on the relativistic curvature of space induced by everything in it. And black holes may have played a special role in the early Universe, as we shall show in our final chapter.

From a relativistic point of view, the strength of a gravitational field can be characterized by the ratio of the velocity required to escape from the field (the "escape speed" often discussed in the context of space exploration) to the speed of light, all squared. When this ratio is very small, as it is near the Sun (1/250,000), Newtonian theory provides a near-perfect description of gravity. But when this ratio approaches 1, as it does close to a neutron star or black hole, general relativity predicts very bizarre and dramatic effects. We have already mentioned the bending of light rays, which becomes extreme as they traverse the curved space associated with a strong gravitational field. Perhaps even stranger is the "time dilation" effect—the apparent slowing down of time in the presence of gravity. If space were marked out by a set of fixed clocks, those located where gravity is strong would seem to a distant observer (located where gravity is weak) to run slow; conversely, distant clocks would seem to run fast when viewed by observers near the central mass. No matter how much the rate at which time passes varies from place to place, at a given spot time would pass at the same rate for *every* process or event. If we could send astronauts to the surface of a neutron star, and listen to them through a radio or view them through a telescope, their speech and movements would appear to us to be about 25 percent slower than normal. As far as they were concerned, however, they would be operating at normal speed and our messages would seem speeded up. Any oscillation of the kind that generates an electromagnetic wave would behave just like the ticks of a clock. To a distant observer watching such an oscillation near the surface of a large mass, the interval from one oscillation to the next would appear longer, and so would the wavelength of the corresponding electromagnetic wave. Because in the

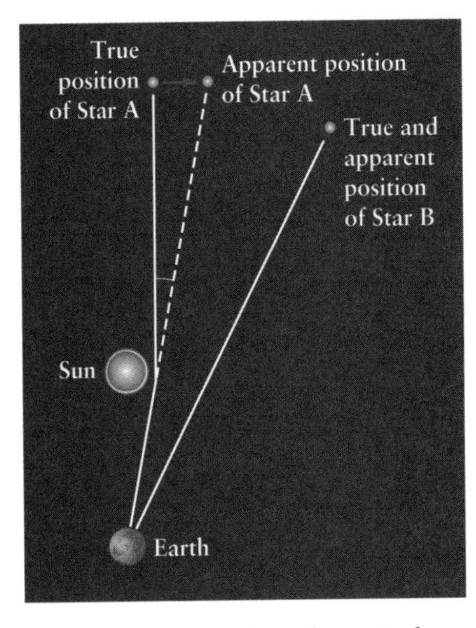


*During the 1919 solar eclipse, Arthur Eddington observed the deflection of starlight by the Sun's gravity. The effect is tiny; according to Einstein's theory the maximum angle of deflection would be 1.75 seconds of arc ($\frac{1}{2000}$ of a degree) for a light ray grazing the Sun's surface.*

optical spectrum red light is longer in wavelength than the other colors, the lengthening of wavelength is called a redshift. Thus light emitted close to the central mass would display a "gravitational redshift" when observed farther out; in other words, its observed wavelength is longer than the wavelength when it was emitted.

Black holes exhibit the effects of "strong" gravity in its purest, and most extreme, form. The possible existence of objects that have collapsed to such small dimensions that neither light nor any other signal can escape from them is predicted by most theories of gravity, not merely by Einstein's general relativity. But it is within the context of general relativity that the most complete understanding of black holes and their properties has been developed.

The first theoretical description of a black hole within the framework of general relativity was given in 1916 by Karl Schwarzschild, who calculated the distortions of space outside a spherically symmetrical body of a given mass. An observer, sitting at a fixed distance from the center of the sphere, would measure the same curvature of space (and therefore the same gravitational strength) regardless of the body's radius. However, if the body were smaller than a certain size, it would not be seen, even though its gravity would still be felt. This boundary between visibility and invisibility, or "horizon," corresponds to the minimum radius from which light or anything else can escape to an external observer, or, almost equivalently, the radius at which the gravitational redshift is infinite and the escape speed is equal to the speed of light. Objects (and light) can cross the horizon in the inward direction, but nothing can come out. Any spherical body smaller than this radius is a "Schwarzschild black hole." (Schwarzschild's equations are discussed in the box on pages 10–11.)

*Karl Schwarzschild in his academic robe of the University of Göttingen, Germany. By the winter of 1915–16, he had voluntarily taken leave from his position as director of Potsdam Astrophysical Observatory for active duty in World War I. He was dying of a rare disease contracted on the eastern front when he discovered the Schwarzschild radius.*

The horizon of a Schwarzschild black hole lies at a radius of 3 $M$ kilometers, where $M$ is the mass of the black hole measured in solar masses (units of mass each equivalent to the mass of our Sun). Coincidentally, this is exactly the same radius that Michell and Laplace would have calculated using the incorrect Newtonian theory. Whereas the gravitational deflection of light is a very small effect in the Solar System and in most astronomical contexts, light is severely bent in the strongly curved space close to the Schwarzschild radius. Indeed, light is even more strongly perturbed than around a neutron star. If our astronauts could hover just outside the Schwarzschild radius, they would need to aim their radio beam almost directly outward to keep it from being dragged back and eventually swallowed by the black hole. If they ventured within this radius, they would be unable to send any signals whatever to the outside world.

Although the region enclosed within the horizon is shrouded from an external observer's view, the explorers would initially find nothing unusual as they passed through the critical radius. Nevertheless, they would then have entered a region from which, however hard they fired their rockets, they could never escape. They would be inexorably drawn to a central point—the "singularity"—where tidal forces (the difference in the gravitational acceleration from one side of their spaceship to the other) would become infinite, and they and their ship would be crushed out of existence. As measured by their own clock, the trip from the horizon to the singularity would have taken just about as long as the time required for a light beam to travel the length of the Schwarzschild radius. But an external observer would not even see the astronauts cross the Schwarzschild radius: as they approached the horizon, their clock would appear to the distant observer to run slower and slower, and any signal that they sent would become more and more redshifted. Photons—packets of electromagnetic energy carrying the signal—would arrive less and less frequently, even as they weakened individually, until the signal practically ceased. Soon it would become undetectable, even by the most sensitive of receivers. As the redshift continues growing, it doubles in the length of time light takes to travel along the Schwarzschild radius. For a black hole with the mass of the Sun, this time is so short that the astronauts would disappear and their signals fade in less than a millisecond. Thus, one cannot from a safe distance observe anything about the extreme conditions near the central singularity. To learn about the interior of a black hole, one must feel a Faustian urge sufficiently strong to make one venture inside the horizon despite the inevitable destruction this implies!

Schwarzschild's theory of the horizon was at first regarded as a mathematical curiosity, but surely not a representation of a real object; then, in 1939, J. Robert Oppenheimer and Hartland Snyder suggested that the implosion of a dying star might actually create a black hole. Their conjecture generated surprisingly little interest. Physicists were critical of the highly idealized nature of the calculation through which Oppenheimer and Snyder showed what would happen to a collapsing star. In order to handle the mathematics without access to electronic computers (which had not yet been invented), they were forced to assume that the collapsing star was perfectly spherical and nonrotating, without the slightest blemish or imperfection. This, of course, was an outrageously unrealistic assumption. Most physicists who reflected on the result had the "gut feeling" that any slight deviation from perfect symmetry would be grossly amplified during the collapse. Different parts of the infalling star would "miss" one another, no horizon would form, and the result would be some indeterminate mess but certainly not a black hole. Another reason for the early dismissal of black holes was the troubling presence of a singularity, a place where (according to the equations) some physical quantity becomes infinite. Nature is simply not supposed to behave this way, so a singularity is usually interpreted as evidence that physicists, not nature, made a mistake!

## The Schwarzschild Metric and the Curvature of Space

According to Newton, the gravitational force outside a spherical object is inversely proportional to the square of the distance from the object's center. From this finding, it follows that orbits around such an object must be ellipses. In 1916, Karl Schwarzschild determined how gravity would behave outside a spherical object according to Einstein's theory. The results are expressed as the *Schwarzschild metric,* describing the curvature of space outside a spherically symmetric mass.

Where gravity is weak and the orbital motion is very much slower than the speed of light, the predictions of Schwarzschild's metric agree almost exactly with Newton's. The Earth moves around the Sun at about one ten-thousandth the speed of light, and Newton's theory is an excellent approximation for all orbits in the Solar System. According to general relativity, however, orbits around a central mass don't close up as exact ellipses, as Newton's theory predicts. Whereas Newtonian theory predicts that the orientation of an elliptical orbit remains fixed in space, Einstein's theory predicts that the direction of the "long axis" of the ellipse gradually rotates. The expected effect has been detected in Mercury's orbit (though the "Newtonian" perturbations due to the other planets are much stronger). Another consequence of the Schwarzschild metric is that light moving through the Solar System is deflected twice as much as it would be if you simply applied Newton's laws to a "particle" moving at the speed of light. This effect, though still very small, has been measured accurately.

Einstein's theory has remarkable consequences when gravity is strong—wherever, in Newton's theory, the orbital speeds would be a good fraction of the speed of light. Here the underlying concept that particles take the "straightest" path through curved space—which offers a deeper insight into gravity even when it is weak—becomes indispensable.

To understand the concept of "curvature," it is helpful to think first of how we could check whether an ordinary two-dimensional surface was curved or flat. Suppose you drew two circles of latitude around the North Pole of the Earth, one slightly larger than the other, and measured their circumferences by crawling round them. If the Earth were flat, the differ-

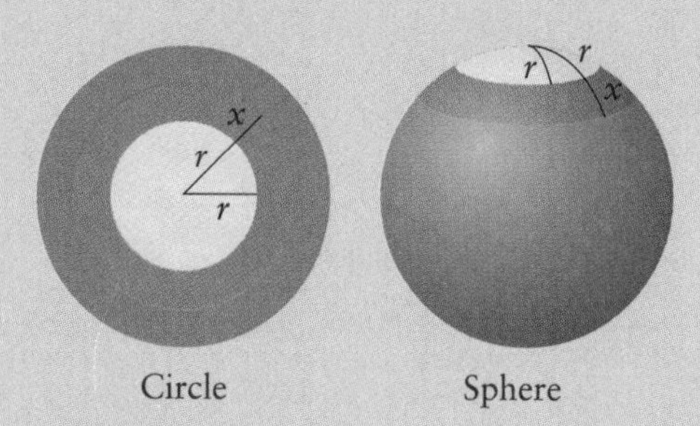


*Concentric circles drawn on a plane differ in circumference by an amount equal to $2\pi$ times their difference in radius. If we drew concentric circles on a sphere and measured their radii along the surface of the sphere, the difference in circumference would be smaller than $2\pi$ times their difference in radius. This discrepancy arises because the sphere is curved, whereas the plane is flat.*

ence would be $2\pi x$, when $x$ is the difference between the radii. However, on the Earth, where $x$ is measured along a meridian of longitude, the difference would be less than $2\pi x$. Measurements of this kind can tell two-dimensional beings whether the surface they live on is curved.

The curvature of three-dimensional space around a point mass can be understood in the same way: the circumferences of circles aren't $2\pi r$, where $r$ is the radial distance. Schwarzschild's equations tell us what the curvature is—how the circumference differs from $2\pi r$, and how it depends on radius.

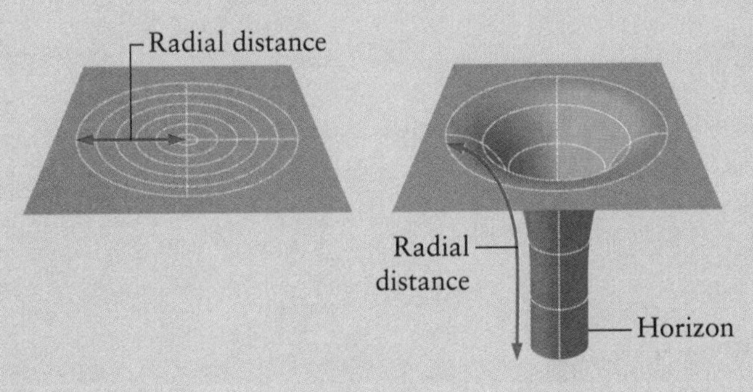


*Close to a Schwarzschild black hole, space is severely stretched in the radial direction compared to the perpendicular directions. Because of the curvature of space, an object falling inward never seems to get any closer to the center of the hole than the radius of the horizon. This three-dimensional effect can be represented in two dimensions by a so-called embedding diagram that resembles the mouth of a trumpet.*

According to these equations, a bizarre thing happens to the circumference when the radius reaches $3M$ kilometers, where $M$ is the central mass in solar masses. Instead of continuing to decrease with decreasing radius, as we would expect, the circumference levels off and becomes independent of radius! This means that, to a distant observer, an object can keep falling and falling toward the center of gravity and never appear to get any closer. In other words, its inward motion seems to "freeze out" when it reaches $3M$ kilometers. The reader has probably noted that this radius is precisely the radius of the black hole's horizon; indeed, what we have just described is an interpretation of the horizon in terms of the curvature of space.

Where does the numerical value of the horizon's radius come from? Using Schwarzschild's equations, a physicist might write the "Schwarzschild radius" in the form $2GM/c^2$. Here, $M$ is the central mass, $c$ is the speed of light, and $G$ is the "universal constant of gravitation" from Newton's theory.

Schwarzschild's equations imply that gravity, in effect, increases with decreasing distance more steeply than the inverse square law of Newton—it becomes more than four times stronger when the distance halves. A particle close to the central mass needs more energy than in Newtonian theory in order to stay in orbit. When an object is orbiting at less than three times the Schwarzschild radius, this effect is so drastic that a slight nudge to a body would send it spiraling toward the center. (In Newtonian theory, conservation of angular momentum would prevent this from happening.) Indeed, at 1.5 Schwarzschild radii, a body would have to move at the speed of light to stay in orbit.

*Trajectories of light rays become severely curved close to the horizon of a black hole. As the horizon is approached, a light beam must be aimed within an increasingly narrow cone in order to avoid being dragged into the hole. At 1.5 Schwarzschild radii a carefully aimed light beam can orbit the black hole indefinitely. At the horizon, the escape cone closes up—no light rays emitted from this radius can avoid being sucked into the black hole. Radiation from just outside the Schwarzschild radius would reach a distant observer with a large redshift, and a clock near the hole would appear to run slow.*

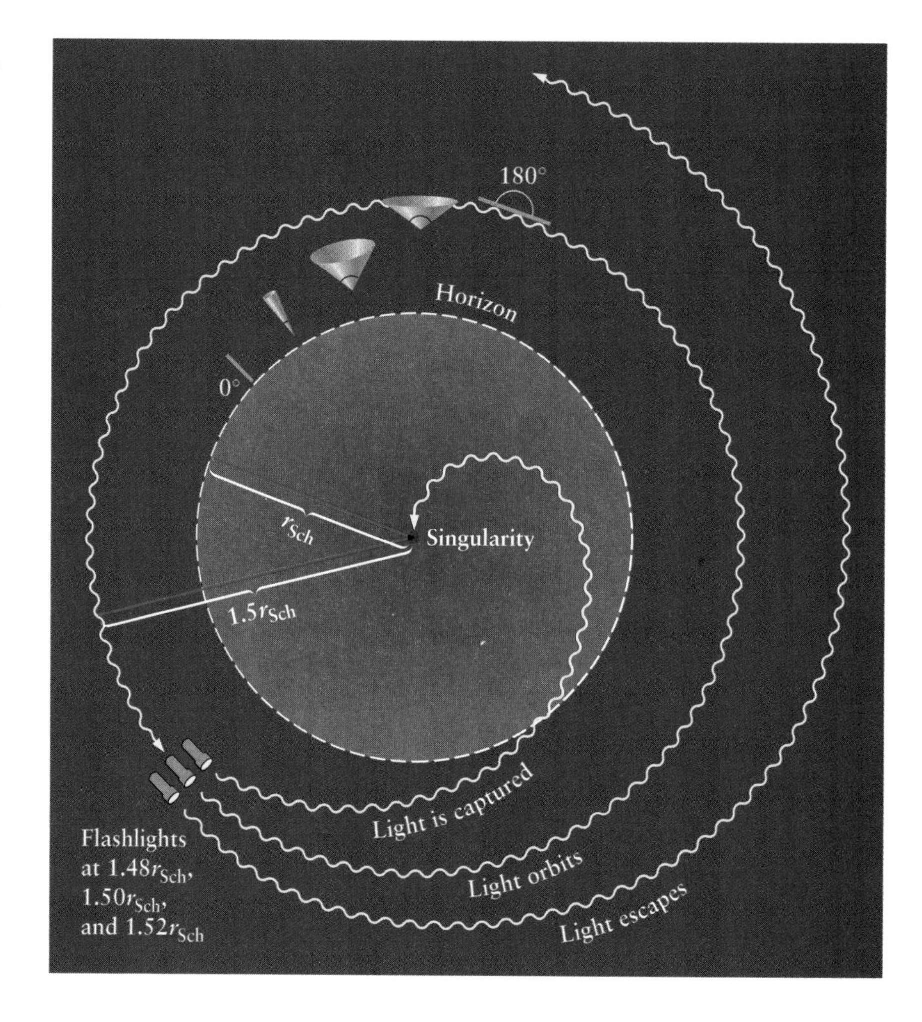


We now understand that the equations of general relativity do indeed predict that black holes must contain a singular point (inside the horizon, and thus hidden from our view) where gravitation is infinitely strong. And physicists generally agree that general relativity's prediction of a singularity signals the theory's inability to cope with situations where gravitational effects vary over too small a distance. It is hoped these problems will be remedied in the future, once the theory of gravity is combined successfully with quantum theory, which treats phenomena occurring on tiny scales. But it is not thought that the modifications of such a "unified theory" will have much impact on our understanding of the black holes discovered in nature. Indeed, the real black hole singularity—the one at the center—

was *not* the one that disturbed physicists between 1940 and 1960. They were worried about what *seemed* to be a singularity at the radius of the horizon, where time apparently came to a standstill. We now know that the latter is a singularity in Schwarzschild's *description* of the physics, and not in the physics itself.

Addressing the problem of asymmetry in the collapse proved to be more subtle. Its final resolution, in the late 1960s and early 1970s, was the key piece of theoretical progress in general relativity that finally convinced many astrophysicists that black holes were objects that might actually exist. When an exactly spherical body collapses, even Newtonian theory predicts that everything crashes together at the center—but nobody worried about singularities in Newtonian theory, because any deviation from symmetry would prevent the collapse focusing onto a single point. However, Stephen Hawking and Roger Penrose showed, in the late 1960s, that Einstein's theory always leads to a singularity surrounded by a horizon, even when the black hole forms as the consequence of a realistic and irregular collapse. Moreover, the resulting black hole, as viewed from outside, has remarkably simple and standardized properties. As a black hole forms, any residual asymmetry in its structure is "radiated away" in a burst of gravitational waves (explained in Chapter 8), and the black hole quickly settles down to a stationary state. In this state, the hole reveals itself to us only by its external gravitational field, which is characterized by only two quantities, the hole's mass and a quantity called angular momentum that is related to the black hole's rate of spin. The reduction of a black hole to two simple properties is colloquially described as the theorem that "black holes have no hair." (In principle, a black hole could also have a net electric charge, but it is unlikely that any realistic collapse could produce a black hole with significant charge because that charge would be neutralized by attracting an equal and opposite charge from its surroundings.)

One more notable piece had earlier been added to the theoretical picture of a black hole. In 1963, the New Zealander Roy Kerr had discovered an exact mathematical description of black holes that is more general than Schwarzschild's solution because it represents the gravitational field around a collapsed object that is rotating, whereas Schwarzschild's equations apply only to a black hole that is not. Thus the so-called Kerr solution, originally regarded (even by Kerr himself) as a mathematical curiosity, has now acquired paramount importance because it describes the distortions of space and time around *any* realistic black hole.

*The mathematician Roy Kerr, of the University of Canterbury, New Zealand, who in 1963 discovered the equations that are now believed to describe all black holes in nature.*

The realization that the singularity lay *inside* the horizon and not *at* the horizon triggered a drastic shift in the way people thought about collapsed objects and introduced a powerful new metaphor to everyday speech and thought. Before 1968, the hypothetical collapsed remnants of stars were generally called "frozen stars." Physicists had a mental image in which the collapse appeared to slow down and "freeze" at the Schwarzschild radius, even though it was already known that the light from the collapsed object would rapidly become so redshifted that the object's image would quickly disappear from view. It was John Archibald Wheeler who first coined the term "black hole," and the name immediately caught on. Here was something infinitely more mysterious, and perhaps much more sinister as well—a place where *any* form of matter or energy could enter, lose its identity, and be lost forever to the Universe. Is it any wonder that such a concept would resonate immediately with contemporary anxieties, as in the following quote from the June 16, 1980, issue of *Time* magazine, which appears in the *Oxford English Dictionary* definition of "black hole": "To the 1.7 million people added to the jobless rolls in April and May, the U.S. economy may well seem to have . . . been sucked into a black hole." Even if black holes did not exist in nature, they could well have been invented for their metaphorical significance.

## Gravitational Equilibrium and Collapse

Black holes represent the ultimate triumph of gravity over all other forces. Gravity is always attractive and, given the chance, will draw everything it can grasp toward a common center. Why, then, is our view of the Universe dominated by structures—stars, clusters of stars, galaxies, clusters of galaxies, and, indeed, the Universe itself—in which gravity seems to be held in check?

The answer lies in the concept of *gravitational equilibrium*. The basic structures in our cosmic environment all engage in a balancing act between gravitational attraction and the dispersive effects of pressure or particle motions. The entire observable Universe displays a similar competition: its galaxies are flying apart from one another, yet the expansion of the Universe is being slowed, and may eventually be brought to a halt, by the combined gravitational pull of its inventory of mass and energy. Collapse can be staved off as long as a balance is maintained. If particle motions become too large, a system can fly apart, as the entire Universe did some 10 to 20 billion years ago during the big bang. But the systems that form black holes are those in which gravity gets the upper hand, with no recovery possible.

Gravity is not the decisive force in *every* structure, only in the large-scale structures studied by astronomers and cosmologists. It is amazingly weak on the atomic scale: within a single atom, in which negatively charged electrons are electrically bound to a positive nucleus, gravitational forces are weaker than electrical forces by almost 40 powers of 10. In a hydrogen molecule, consisting of two protons neutralized by two (much lighter) electrons, the gravitational binding between the two protons is feebler than their electrical coupling by a factor of $10^{-36}$. Gravity is almost undetectable between laboratory-scale bodies as well. In a rock, a person, an asteroid, or even a small planet, the electrical forces between atoms and molecules dominate over their mutual gravitation attraction. But gravitation becomes progressively more significant for larger objects: its relative importance rises a hundredfold for each thousandfold increase of mass.

Why does gravity become powerful only on large scales? In any macroscopic object the positive and negative charges are present in such nearly equal quantities that the electrical forces tend to cancel out. But there is no such cancellation for gravity. Everything has, as it were, the same sign of gravitational charge and attracts everything else. Gravity is thus the controlling force on sufficiently large scales. How large? Only for masses as large as those of planets can gravity compete with other forces. Indeed, the Solar System's largest planet, Jupiter—made up of $10^{54}$ atoms ($2 \times 10^{27}$ kilograms, or one-thousandth the mass of the Sun)—is just about the most massive body whose gravity can be balanced by the kind of interatomic forces that operate in solids or liquids. A cold body much larger than Jupiter would actually have a smaller radius because gravity would be strong enough to compress it to an extraordinarily high density.

To comprehend how gravitational equilibrium is *lost,* leading to the formation of black holes, we need to understand first how it is *maintained.* A satellite in orbit around the Earth is continuously falling, but it never actually hits the Earth's surface because its sideways motion carries it just far enough around the circumference to always "miss." Likewise, the motions of all the stars in a galaxy or all the atoms in a star prevent them from being drawn together by gravity to a common center.

The Sun—a typical star—contains a thousand times more mass than Jupiter. If it were cold, gravity would compress it to a million times the density of an ordinary solid: it would be a "white dwarf" star, about the same size as the Earth but 330,000 times more massive. But the Sun is not a white dwarf, because its center has a temperature of about 15 million degrees Celsius, thousands of times hotter even than its glowing surface. At these high temperatures the atomic nuclei inside the Sun are moving randomly at speeds of hundreds of kilometers per second. It is the pressure of this hot interior—caused by the motions of the fast-moving electrons and nuclei—that counteracts the effect of gravity in all stars like the Sun.

Individual stars are the "atoms" of a galaxy. Galaxies are in some respects simpler systems than their constituent stars. The stars in a galaxy are so widely dispersed, compared to their sizes, that they are unlikely to hit each other; so, to a good approximation, their orbits can be traced independently of one another. Their motions are determined solely by

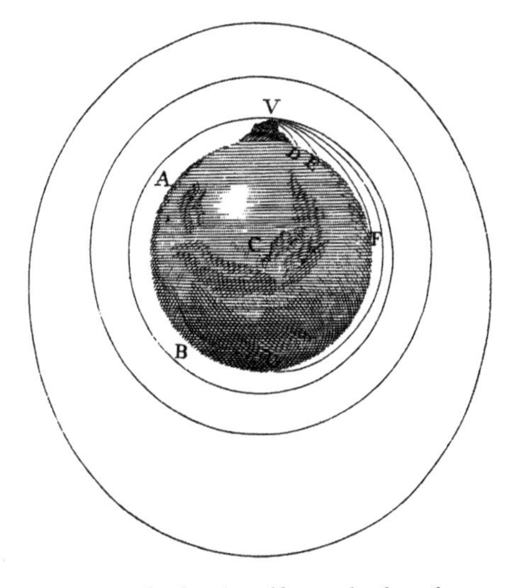


*A projectile fired sufficiently fast from a mountaintop would go into orbit because its trajectory would curve downward no more sharply than the Earth's surface. This illustration appeared in* The System of the World, *a "popular" version of the third volume of Newton's* Principia.

gravity, whereas inside an individual star other physical processes are crucial in determining what happens.

There are about 100 billion stars in our Milky Way Galaxy, which is itself just one of the billions of galaxies within range of modern telescopes. A galaxy is held in equilibrium by a balance between gravity, which tends to pull the stars inward, and the stellar motions, which, if gravity didn't act, would cause the galaxy to fly apart. Disk galaxies, of which our Milky Way is an example, contain a flat "platter" of stars all moving in nearly circular orbits. The star motions are quite regular. In elliptical galaxies the motions are more random, each star tracing out a complicated path under the gravitational influence of all the others. Ellipticals look like big fuzzy balls of stars, sometimes flattened or a bit lopsided. Orbiting around the center of our Galaxy (and others like it) are several hundred *globular clusters* of stars: these are small-scale analogues of elliptical galaxies, generally containing between 100,000 and a few million stars in random orbits.

The amount of motion needed to oppose gravitational collapse can be predicted if one knows how the matter is distributed in space. Imagine that we could dismantle the globular cluster 47 Tucanae, taking each of the million or so stars and carting it off to deep space, so far away from the other stars that their mutual gravitational forces are negligible. To overcome the gravity holding the cluster together we would have to expend a considerable amount of energy, equal to the *gravitational binding energy* of the cluster. The energy required to pull apart a cluster, or any other gravitating object, depends only on the initial distribution of matter and not on the exact procedure used to demolish it. By virtue of their motion, moving stars have the form of energy called *kinetic energy;* it is this energy of motion that counteracts the pull of gravity. In the Newtonian theory of gravity, which is applicable to all systems where the typical motions are much slower than the speed of light, the amount of kinetic energy required for gravitational equilibrium turns out to be exactly equal to the gravitational binding energy. This means, for example, that if we gave the planets in the Solar System an additional amount of kinetic energy equal to what they already have, they would just be able to break free of the Sun.

Exactly the same balance of energies holds for entire galaxies and clusters of galaxies. This very general result is often called the *virial theorem.* It also applies to stars, except that in the case of a star like the Sun we refer to the kinetic energy as "thermal energy" because a hot gas consists of randomly moving atoms whose energies are proportional to the temperature. Gravitational equilibrium always requires a balance between the gravitational binding energy and the internal energy opposing gravity.

Gravitational equilibrium does not always last forever. Maintaining the equilibrium in the Sun, for example, requires a central power source to replenish the heat that leaks out through the Sun's surface and is lost to

*47 Tucanae is one of several hundred globular clusters in orbit around the center of the Milky Way Galaxy. The random motions of its stars balance their mutual gravitational attraction, preventing the cluster from collapsing.*

space. Fusion of hydrogen into helium has kept the Sun shining for 4.5 billion years, but this power source is not inexhaustible. The Solar core will run out of hydrogen in another 5 billion years; more massive stars, which shine more brightly, have shorter lifetimes.

When a star faces an energy crisis, gravity compresses its core further. Some other power source may then take over—for instance, fusion of helium into carbon and other elements further up the periodic table. But, at last, all the nuclear fuel that can be tapped will run out. The Sun will cool down to become a white dwarf. Heavier stars collapse further, transforming themselves into more exotic objects.

Even systems composed of stars may not be able to maintain their equilibrium indefinitely. Globular clusters exhibit a gradual form of evolution, which can be observed by comparing the structures of clusters with different masses and ages. A star inside a globular cluster not only responds to the average, "smoothed out" gravitational field of all the other stars but also feels discrete tugs from its neighbors. The tendency of all these individual encounters is to equalize the average kinetic energies of stars in all parts of the cluster. This sounds as if it should help to preserve the equilibrium, but in fact it has just the opposite effect. The equilibrium is endangered because the gravitational field is weaker in the outer parts of the cluster than it is near the center. If all stars had similar kinetic energies, on average, then some of the stars near the outer edge would be moving fast enough to escape. Thus, globular clusters and other stellar systems can never actually reach a permanent equilibrium; in a sense, they are always slowly *evaporating*.

Let us see how far we can carry the metaphor of evaporation. The loss of the fastest-moving stars from the outer parts of the cluster leaves the remaining stars, on average, with less energy. Thus, it would seem that "stellar evaporation" cools the cluster down, just as evaporating water cools the skin. The random motion of stars in the cluster is analogous to the temperature, so we might expect them to move more slowly as the cluster evaporates. However, what actually happens is that gravity pulls the remaining stars more tightly together, so that they end up moving faster than before. The cluster does not cool down as it loses energy—it actually *heats up!*

This unusual (and counterintuitive) property is a general feature of gravitating systems: when they *lose* energy, they get *hotter.* If, for instance, the Sun's nuclear energy supply were turned off, its core would gradually deflate as heat continued to leak out. But as the core shrinks, it has to become even hotter to support itself against the (now stronger) gravitational force. Something similar happens in a star cluster when stars "evaporate" from it. If we consider the structure of a star cluster in more detail, we see that what actually happens is that the central star concentration increases in a runaway fashion. This *gravothermal catastrophe,* first predicted by Vadim Antonov in 1962, can lead to globular clusters with unusually compact, tightly bound cores. More important, it illustrates a general tendency of gravitating systems to develop compact nuclei with very intense gravitational fields.

Globular clusters have been evolving since their formation 10 to 15 billion years ago. In that time, about 20 percent of them have developed dense cores. These cores do not, however, form an exotic object such as a black hole or a neutron star in the center of the cluster because the number of stars remaining in the core decreases too rapidly as the cluster evolves. But, aided by additional processes such as stellar collisions and the infall of gas, the dense clusters of stars in the nuclei of galaxies *can* evolve into black holes. Some objects, such as stars, can lose their gravitational equilibrium all at once, suffering a catastrophic collapse or explosion: this is what happens when a massive star runs out of the fuel needed to power thermonuclear reactions. In some cases the object is able to find a new equilibrium, as when the core of a moderately massive star collapses to form a neutron star. But in other cases no equilibrium is possible. The general theory of relativity permits one and only one outcome in this situation—collapse to a black hole.

## The Search for Black Holes

Most theories of gravity, not just general relativity, predict that black holes can form, although with slightly different properties depending on the theory. A strong motivation for searching for black holes is that they represent objects where gravity has overwhelmed all other forces, allowing one to test theories of gravitation under the most extreme conditions. But there are other reasons for studying black holes. Stellar-mass black holes interest the astrophysicist because they are in a sense the "ghosts" of dead stars and tell us something about the last stages of stellar evolution. Surprisingly, they give themselves away through the pyrotechnics of flickering X-ray sources. The supermassive black holes in galactic nuclei may be the "trash heaps" of galaxies, or they may be the "seeds" of galaxy formation in the early Universe. In either case, the quasars and jets of gas they produce are the most energetic phenomena known. All these phenomena, from the predictions of general relativity to cosmic jets, offer the astrophysicist a rich source of insights about black holes, and they are the topics of the following chapters.

Black holes are objects that have collapsed, cutting themselves off from the rest of the Universe but leaving a gravitational imprint frozen in the space they have left. A black hole has, in effect, a "surface" resembling a semipermeable membrane. Things can fall in, but the interior is shrouded from view because not even light can escape. No matter how complicated the ancestral object was, once a black hole forms, its appearance from outside has only a few simple characteristics. The thought that complicated and "messy" astrophysical processes could lead to such a simple and "clean" endpoint has endless philosophical appeal for both astronomers and physicists.

To the physicist, gravitational collapse is important also because the singularity that develops inside the hole must be a region where the laws of classical gravitation break down and one needs (just as one does in discussing the initial instants after the big bang) to extend general relativity into the quantum regime to understand what is going on. Thus, black holes are at the frontier of modern physics. Many people have claimed that the paradoxes associated with the singularity within a black hole are as fundamental and far-reaching in their implications as the puzzles connected with blackbody radiation and the stability of bound electron orbits in atoms that confronted physicists at the beginning of the twentieth century and triggered the development of the quantum theory.

Black holes also have an important bearing on our general concepts of space and time because near them spacetime behaves in peculiar and highly "non-intuitive" ways. For instance, time would "stand still" for an

*"It's black, and it looks like a hole. I'd say it's a black hole."*

observer who, managing to hover or orbit just outside the horizon, could then see the whole future of the external Universe in what, to him, was quite a short period. Stranger things might happen if one ventured inside the horizon. It is therefore hardly surprising that black holes have captured the imagination of the public in a way that few physical concepts have matched. Even though we cannot yet predict what would happen very close to the singularity itself, that limitation need not deter us from the study of black holes. The "singularity" is hidden within the black hole and cannot directly affect the external Universe. For this reason, our uncertainty about the "interior" of black holes doesn't reduce our confidence in predicting their astrophysical consequences. As an analogy, there are many mysteries deep inside the atomic nucleus, but this realization doesn't prevent physicists from calculating the properties of atoms measured by spectroscopists and chemists.

Do black holes exist, and are they accessible targets for study by astronomers? Nearly all astrophysicists would now answer in the affirmative. Evidently some massive stars do leave behind black hole remnants, and the nuclei of many galaxies harbor single, supermassive black holes containing millions of solar masses. How these black holes may have formed, and how they make their presence knowable to us, is the subject of the tale that follows.

Chapter 2

# Stars and Their Fates

*The stars of the "Trapezium," in the Orion Nebula, illuminate nearby clumps of gas and dust where stars are forming. Cometlike trails radiating from the star-forming clouds surrounding the brightest star consist of evaporating gas swept back by the intense stellar wind and ultraviolet radiation. The Trapezium stars themselves are only a few million years old.*

Ordinary stars like the Sun cannot last forever. The Sun is held in equilibrium by a balance between gravitation and the pressure in its hot interior. Take away that pressure and the Sun would go into free-fall, halving its size in less than an hour. If, on the other hand, gravity were magically switched off, the hot interior would just as suddenly explode and disperse. To us, the Sun looks like a steady beacon in the sky, and in fact its appearance has hardly changed for more than 4 billion years. But the heat of its interior is continuously ebbing. The rays of sunlight you catch on a beach or in a park are plain enough evidence that the Sun is losing energy. And although this heat is now being replenished by nuclear reactions at the Sun's center, these reactions cannot continue indefinitely at their present rate. Eventually the Sun, like any other star, will use up its nuclear fuel, and its finely tuned gravitational equilibrium will fail.

The need for a central power source to replenish the energy lost by a star was recognized in the nineteenth century. If there were no power source at the center, the heat loss would cause the Sun to deflate, at a rate first calculated by Lord Kelvin. The deflation would be far slower than free-fall, because the energy takes a long time to leak out of the Sun's interior—indeed, the process would take about 10 million years. Kelvin knew, however, that 10 million years was much shorter than the timescale that biologists and geologists had estimated for the age of the Earth and its life forms. Some energy source within the Sun must prolong its life, and Kelvin readily showed that chemical energy would be utterly inadequate. To reconcile his theory with the Earth's estimated age would require, as Kelvin wrote, "some unknown source of energy laid down in the storehouse of creation." Not until the 1930s, when the energetic potential of nuclear fusion was recognized, was there a solution to Kelvin's paradox. The Sun's center is hot enough for hydrogen to fuse into helium at just the rate needed to compensate losses; moreover, the energy release is so high that the supply of hydrogen could keep the Sun shining for several billions of years.

What happens after a star's nuclear fuel is exhausted? This question concerns not only the far distant future. Stars of similar mass to the Sun, but which formed several billion years earlier when our Milky Way was young, have already faced this crisis. Moreover, stars much heavier than the Sun are so much brighter that they use up their fuel more quickly—the most massive ones in only a few million years. Thus, the remnants of dead stars are all around us. These remnants are all much more compact and have much stronger gravity than their precursors.

As their fuel is exhausted, the centers of stars evolve through a sequence of ever more tightly bound gravitational equilibria, punctuated by episodes of contraction. The Sun will become a white dwarf, a slowly cooling cinder about the size of the Earth, and can remain in this state indefinitely. But stars of much greater mass contract further—they end their lives as neutron stars or black holes. In this chapter, we show how and why these end states are attained and set the stage for discussing, in Chapter 3, the dramatic and unexpected ways in which these seemingly inert remnants have revealed themselves.

## Stellar Evolution

Since the 1930s, substantial progress has been made in understanding the structures and life cycles of stars, particularly stars like the Sun. Such stars start their lives by condensing gravitationally from interstellar clouds. At first they contract more or less as Kelvin envisaged, but this process stops when the central temperature becomes high enough for fusion to start in the core. The star then settles down to the so-called main sequence, the longest-lived epoch in a star's life. The main sequence phase lasts until most of the hydrogen in the core (about 10 percent of the star's total hydrogen) has been incinerated. During this phase, the power output, size, and surface temperature of the star are all related uniquely to the mass, and to one another.

The most obvious way to discriminate among stars, to a casual observer, is by noting their surface temperatures: temperature determines color (or "spectral type," in astronomers' jargon). Look up on any dark night and you will see that the brighter stars show a distinct range of colors from obvious reds to subtle blue-whites. Though filtered by the

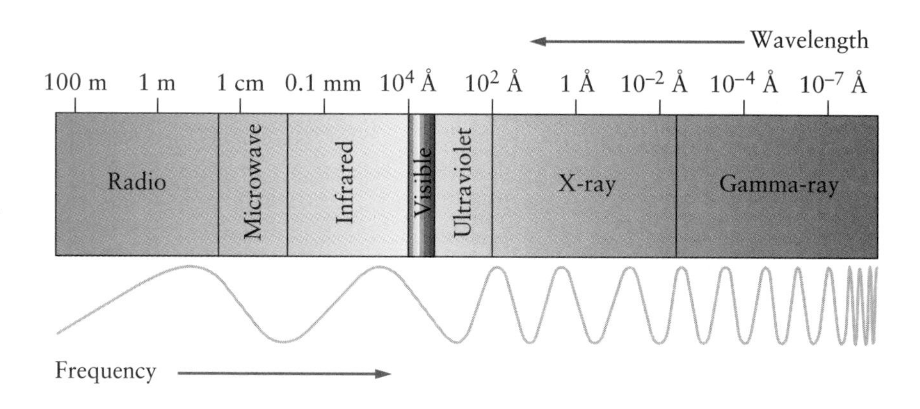


*The electromagnetic spectrum. The longest wavelengths and lowest frequencies correspond to photons carrying the smallest amounts of energy. Wavelengths are usually measured in angstroms (1 Å = $10^{-10}$ m), frequency in Hertz (1 Hz = 1 cycle per second).*

response of our eyes, these perceived color differences are very real. The color of light is associated with the wavelength of the corresponding electromagnetic waves. Bluer light has shorter wavelengths; the wavelengths of redder light are longer. Photons, the discrete packets into which light is parsed, each contain a fixed amount of energy that *also* depends on the color. Photons of blue light carry more energy than do photons of red light. Since temperature is generally a measure of energy per particle and photons are effectively "particles" of light, it is not surprising that hotter stars are bluer and cooler stars are redder. Simple as this explanation may sound, its ramifications pervade the foundations of modern physics. Why heated objects produce a unique spectrum of radiation that depends only on their temperature was one of the principal puzzles facing physicists at the turn of the twentieth century. Its resolution (due, in part, to Einstein) led to the realization that light behaves both as a wave and as a particle, and thence directly to the quantum theory.

More massive stars are hotter and bigger, but, most important, they are much more luminous. Hence more massive stars burn up their supply of fuel much more rapidly than less massive stars, even though the fuel supply is bigger to start with. The Sun, and stars like it, have enough nuclear fuel to keep shining for 10 billion years. But a 20-solar-mass star, for example, is nearly 10,000 times as luminous as the Sun, and lives only a few million years.

Paradoxically, stars at first become *more luminous* as they use up their store of nuclear fuel. This curious property is related to the fact that gravitating systems "heat up as they cool down," as we saw when we considered the fate of a globular cluster in the last chapter. As nuclear fusion proceeds in the core of the Sun, more and more hydrogen is converted to helium. As the fuel is exhausted, the production of nuclear energy becomes less efficient, and the core shrinks under its own weight. But with this contraction, the gravitational force felt by the atoms in the core increases and the remaining hydrogen atoms speed up. As atomic motion increases, so do the rates of nuclear reactions and, consequently, the power output. Since nuclear reactions can last only as long as the fuel supply, this increase of luminosity cannot last indefinitely.

The changes in the Sun will not be very noticeable for another 5 billion years or so. Thereafter, the solar orb will swell to the enormous dimensions of a red giant star, possibly vaporizing the Earth in the process. But do not imagine that the Sun will be simply dispersing. Deep within the bloated envelope of the red giant Sun, the nuclear furnace will have shrunk to the size of the Earth. During its red giant phase, lasting nearly a billion years, the star burns hydrogen in a shell around the now-helium core. A briefer stage follows, in which helium itself is burnt, after which the star settles down to a quiet demise as a white dwarf.

The story of stellar evolution sketched by astrophysicists is an elaborate one; their theories show that the evolution of a star is determined by the interplay of complicated nuclear reaction chains with subtle aspects of radiation and gravitational physics. How can we check these theories? We cannot perform experiments on stars, we can merely observe them. Moreover, stars are so long-lived compared with astronomers that we only have, in effect, a single snapshot of each one. But as compensation, billions are

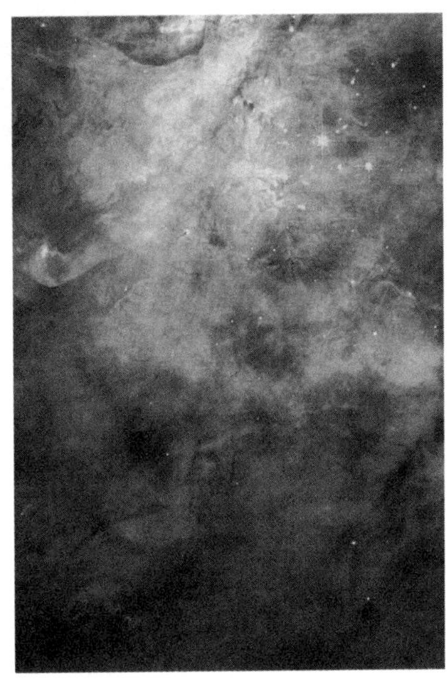

*The Orion Nebula is a vast stellar nursery, some 25 light-years across. Located about 1400 light-years from us, the fan-shaped nebula is illuminated and ionized mainly by the ultraviolet light from the brightest star in the Trapezium, the centrally located grouping of four bright stars. The detailed image at right, by the Hubble Space Telescope, shows shock fronts criss-crossing the nebula and dense clouds where stars are forming.*

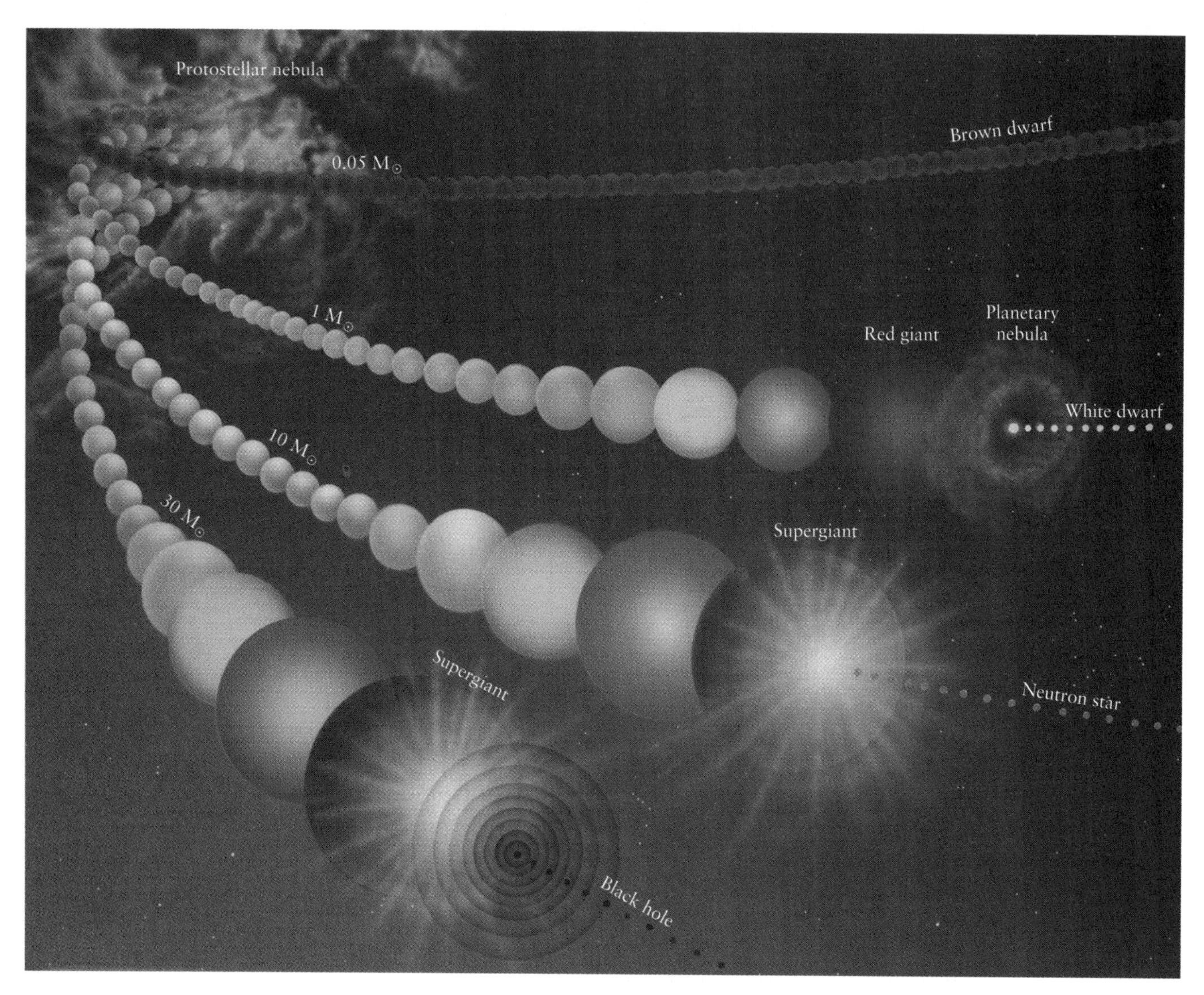


*Stars evolve in both structure and appearance as they use up their nuclear fuel. As the nuclear energy supply dwindles, gravity tends to draw the central regions of the star into an ever more tightly bound core. Even during the red giant phase, the bloated stellar envelope conceals a superdense core not much bigger than the Earth. As illustrated here, a star's ultimate fate depends on its initial mass, the most massive stars ending up as neutron stars or black holes and the less massive stars settling down as white dwarfs. Bodies of very low mass never become hot enough for thermonuclear fusion to start, and exist as brown dwarfs. The demise of a massive star is accompanied by an explosion known as a supernova; stars like the Sun return some of their mass to interstellar space by ejecting shells of gas that form a planetary nebula.*

available for study, and we can check our theories of how stars form and evolve by observing stars of different ages, just as you could infer the life cycle of a tree, even if you had never seen one before, by one day's observation in a forest. The British astronomer William Herschel expressed this point in the eighteenth century: "Is it not almost the same thing, whether we live successively to witness the germination, blooming . . . and wither of a plant, or whether a vast number of specimens, selected from every stage through which the plant passes, be brought at once to our view?" Of special interest in this regard are places like the Great Nebula in Orion, where even now new stars, and probably new solar systems, are condensing within glowing gas clouds. Astronomers can also test their theories by studying the stellar populations in globular clusters—swarms of up to a million stars, held together by their mutual gravity, containing stars of different sizes that can be presumed to have formed at the same time.

## White Dwarfs

Gravity does not always win in the end; the Sun's fate as a white dwarf promises to be a stable one. White dwarfs are a kind of star that can remain in gravitational equilibrium indefinitely, without any need of nuclear reactions or other power supply. Like ordinary stars, white dwarfs are supported against gravitational collapse by the random motions of the particles of which they are composed. But, unlike those in a normal star, these motions are *not* related to the temperature of the gas in the interior. So, as white dwarfs radiate away their internal thermal energy and cool down, they do not lose pressure support and contract.

The kind of pressure that supports a white dwarf is called *degeneracy pressure,* and it arises from a quantum mechanical effect called the *Pauli exclusion principle:* two identical particles cannot have the same position and momentum. The principle applies to the basic building blocks of atoms—protons, electrons, and neutrons. Such particles are collectively called fermions, after the great Italian physicist Enrico Fermi. What they all have in common is that they are endowed with precisely one-half (or, more precisely, a half-integer) unit of quantum mechanical angular momentum, or "spin." The spin is an intrinsic property of each kind of particle, and there is nothing one can do to change it. Despite the esoteric sound of the concept, "half-integer spin" has a very concrete effect on the collective behavior of groups of fermions: it prevents them from being packed together too closely. More precisely, it prevents any two of them from occupying the same "quantum state," meaning that they cannot both

*In 1930, Subrahmanyan Chandrasekhar calculated that white dwarfs more massive than 1.4 suns would collapse under their own weight, paving the way for the theoretical prediction of neutron stars and black holes. In an extraordinarily long and prolific career (mainly at the University of Chicago) he wrote major books on many topics in astrophysics, including* The Mathematical Theory of Black Holes, *published in 1983. He published a detailed commentary on Newton's* Principia *shortly before his death in 1995.*

lie very close together in space *and* have similar velocities. When fermions are packed closely together, they cannot all be slow moving. The exclusion principle forces most of them to have high velocities. The more you compressed them, the faster they would move. These motions give rise to a pressure that resists compression, over and above any pressure the particles might exert as a result of their thermal energy.

When most of the particle motion in a gas is the result of this quantum-mechanical resistance to compression, the gas is called *degenerate*. The random motions associated with degeneracy create a pressure that can support a star against gravitational collapse, just as the more familiar thermal motions create the pressure that supports the Sun. A star can become a white dwarf when its constituent electrons become degenerate; this happens when the star is compressed to a density about a million times that of water. White dwarfs are roughly the size of the Earth, but with a gravitational field about a million times larger!

There is, however, an upper limit on the mass of a white dwarf. If you examined a sequence of white dwarfs of increasing mass, you would find that the degeneracy-induced random speed of the electrons increases with the mass of the dwarf, approaching the speed of light for masses just above that of the Sun. In 1930, the young theorist Subrahmanyan Chandrasekhar (known to his colleagues as Chandra) pondered this problem during the long voyage from India to England, where he had enrolled as a graduate student, and reached an astonishing conclusion. He found that no gravitational equilibrium could exist for white dwarfs with masses above about 1.4 solar masses. At higher masses, the degeneracy pressure could never balance gravity, no matter how much you compressed the star. Chandra realized that forcing the motion of the electrons to speeds close to the speed of light would weaken the resiliency of the degenerate gas, its ability to resist compression; consequently, a white dwarf with a mass above the "Chandrasekhar limit" would go into gravitational free-fall and collapse in about one second.

The implications of Chandrasekhar's discovery were disturbing. What would eventually happen to stars of mass greater than 1.4 solar masses once they had exhausted their nuclear energy? Some suggested that a massive star in the course of its evolution would lose most of its mass, so that no matter how massive it started out it would eventually settle down as a white dwarf of mass below the Chandrasekhar limit. We now know that massive stars do lose a lot of mass during their lifetimes, through powerful stellar winds, and in this way, stars moderately more massive than 1.4 solar masses probably end up as white dwarfs. But there is no evidence to suggest that stars of much greater mass can avoid Chandra's mass limit. Another possibility is that the outer layers may be ejected explosively in a supernova event at the end of the star's lifetime, leaving a neutron star. But theory sets a limit to the mass of a neutron star. The exact limit is a bit uncertain, but it certainly lies below 3 solar masses. The third possibility is unchecked collapse to a black hole.

In 1930 neutron stars had not yet been conceived, and so unpalatable was the third possibility that many scientists at the time simply refused to accept Chandrasekhar's prediction at face value. The famous theorist Arthur Eddington led a public attack, rejecting the fundamental physical basis of Chandra's arguments. "Various accidents may intervene to save the star, but I want more protection than that," he proclaimed at a meeting of the Royal Astronomical Society in London in 1935. "I think there ought to be a law of Nature to prevent a star from behaving in this absurd way!" Eddington was ultimately proved wrong, but Chandra, stung by the ferocity of the attack, left British academia permanently for the more welcoming environment of the University of Chicago. Chandra went on to a brilliant career as a pioneer in many other areas of astrophysics, but he was so discouraged by the white dwarf episode that it was decades before he could bring himself to tackle the problem of gravitational collapse once again—although he did so in a characteristically exhaustive fashion in his treatise on the mathematical theory of black holes, published in 1983.

## Massive Stars and Supernovae

Stars heavier than the Sun evolve in an increasingly complicated and dramatic way. Most end up (just as the Sun will) as white dwarfs composed mainly of carbon and oxygen (although there may be a thin layer of hydrogen and/or helium remaining on the surface). Their interiors never become hot enough to trigger the nuclear fusion of carbon and oxygen into heavier nuclei. Whereas electrical or gravitational forces can act over arbitrarily large distances, the forces that cause nuclear reactions operate only over distances less than about $10^{-13}$ centimeters, or about one hundred-thousandth the size of an atom. Therefore, to trigger a nuclear fusion reaction it is first necessary to bring the reactants very close together. (However, one need not bring them as close as $10^{-13}$ centimeters, because a quantum mechanical effect called tunneling makes the particles behave as if they were smeared out over a larger region in space.) Although a nuclear fusion reaction requires close encounters between nuclei, the protons in a nucleus electrically repel the protons in other nuclei (since like charges repel). This electrical repulsion tries to prevent the particles from getting too close together. Heavier atomic nuclei contain a larger number of protons and therefore repel each other electrically with more force. High relative velocity—or, equivalently, high temperature—is needed to overcome the repulsion.

Massive stars are powered in later life by a sequence of nuclear reactions involving heavier and heavier elements. As each nuclear fuel is exhausted—hydrogen fused into helium, then helium into carbon and oxygen, etc.—the inner part of the star contracts, becoming even hotter, until a new set of fusion reactions can take the atoms further up the periodic table. Computer models of stellar evolution indicate that the Sun will probably burn much of its core to carbon and oxygen before it settles down as a white dwarf. But for stars more massive than 5 to 8 suns (the borderline is uncertain), this process would proceed all the way up to iron. At every stage up to this point, the creation of heavier atomic nuclei releases energy that staves off gravitational collapse. But there are no nuclear reactions that can release energy from iron; iron is the end of the nuclear road for a star.

What happens next is one of the most spectacular events known in astronomy. A sufficiently massive star (probably any star above about 8 solar masses) develops an iron core that weighs more than 1.4 suns. In other words, the core is too massive to reach equilibrium as a white dwarf—it exceeds Chandrasekhar's limit. Since there are no nuclear reactions that can extract energy from iron, the supply of fuel is shut off and the core suffers sudden and catastrophic collapse, reaching the density of an atomic nucleus (about a million billion times the density of water!) in a fraction of a second. During this collapse, the nuclear reactions that have built up the iron core are undone—the iron nuclei are broken apart into smaller and smaller pieces until a soup of subatomic particles is all that remains.

If a sequence of nuclear or chemical reactions has released a certain amount of energy, then reversing those reactions requires the introduction of an equivalent amount of energy. Where does this energy come from? Gravity, of course! The collapse of the core releases an amount of energy equal to about 10 *times* the energy output of the star during its entire nuclear burning lifetime. To put this another way: over the lifetime of a massive star, gravity, not nuclear reactions, provides most of the energy, but it does so almost entirely within the last second of the star's life.

The dismantling of the nuclear fusion products does not stop with the breaking apart of iron into subatomic particles. The density in the collapsing core becomes so great that the protons and electrons among these particles are fused together to form neutrons, electrically neutral subatomic particles with about the same mass as protons. Normally, neutrons are stable only when they are bound to protons inside atomic nuclei. An isolated neutron will decay into a proton, an electron, and a third particle called an antineutrino in about 10 minutes. But in the dense core, the existing electrons are already so tightly packed that the Pauli exclusion principle does

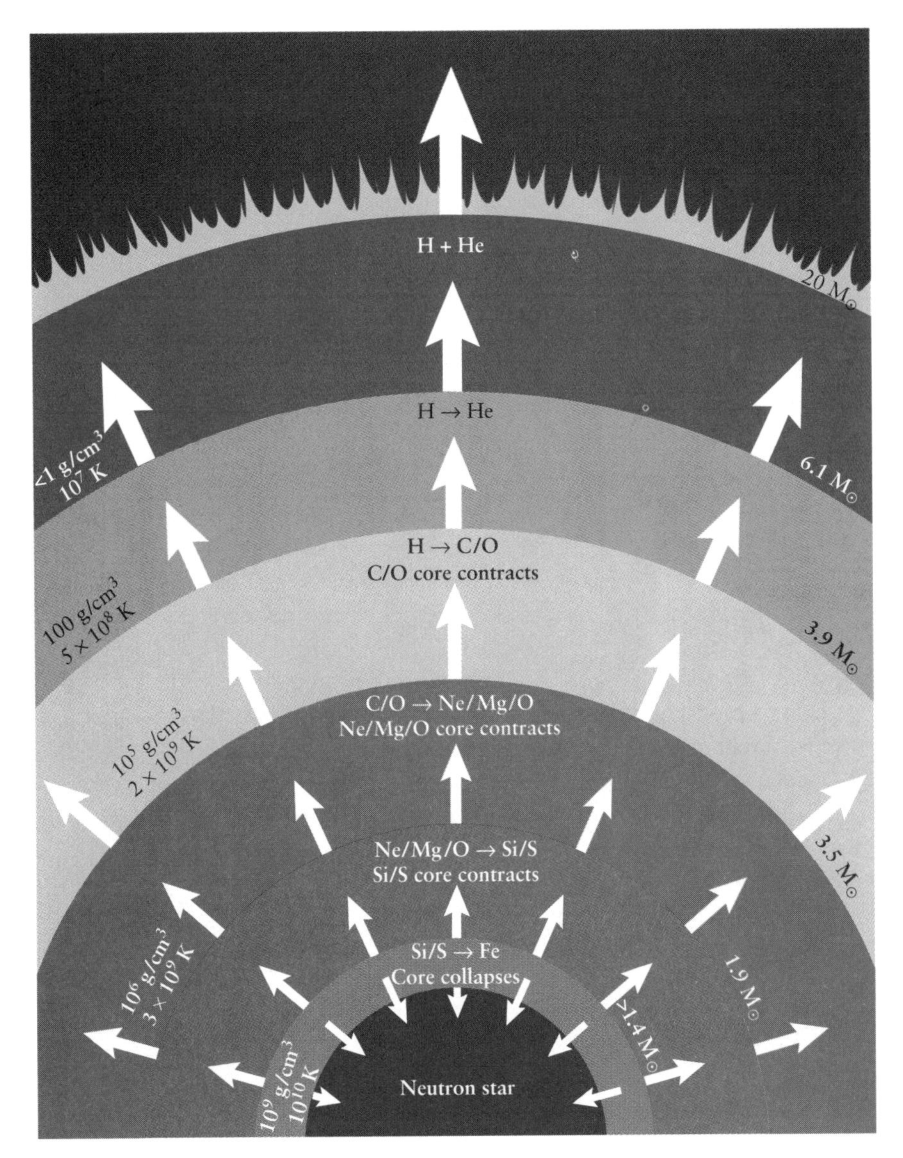


*The interior of a massive star goes through a series of nuclear-burning stages in which increasingly heavy elements are created until the core consists of pure iron. Outside the core the star has an "onionskin" structure in which the chemical composition of each layer is determined by the maximum temperature it reached during the star's life. No nuclear energy can be extracted from iron, and if the core is more massive than 1.4 suns it will collapse, forming a neutron star or black hole. In this illustration showing the structure of a 20-solar-mass star, the collapse of the core blows off the outer layers, producing a supernova explosion.*

not allow any room for additional electrons to be created by neutron decay. The core of the collapsing star becomes a sort of gigantic atomic nucleus containing $10^{57}$ neutrons, and will (if it is not too massive) evolve into a neutron star. The force that supports a neutron star against collapse is the degeneracy pressure set up among the neutrons, which also belong to the fermion class of particles.

If gravity has released 10 times more energy than needed to dismantle the atomic nuclei, where does the surplus 90 percent of the energy go?

*The Crab Nebula is the remnant of the supernova explosion observed in A.D. 1054. The reddish filaments are shredded debris of the star's envelope; the diffuse bluish light comes from electrons gyrating at close to the speed of light in a magnetic field that pervades the nebula.*

About 90 percent of that surplus is carried away by subatomic particles called neutrinos, which are released during the formation of the neutrons. Neutrinos are particularly elusive particles that hardly interact with ordinary matter; most of them pass through the envelope of the dying star without doing much damage. The other 10 percent, however, triggers an explosion that blows off the outer layers of the star. These explosions are called supernovae, rare events that occur on average only once or twice per century in our Galaxy.

One such event created the Crab Nebula in the constellation Taurus. Although the light from this explosion reached the Earth in the year A.D. 1054, we fortunately have records of the supernova's appearance from the Orient, particularly China. In the fifth month of the first year of the Chih-Ho reign, the Emperor received a visit from Yang Wei-Te, the Chief Calendrical Computer, equivalent perhaps to the British Astronomer Royal. Yang is recorded to have said to the Emperor, "Prostrating myself before your Majesty, I have observed the appearance of a guest star. On the star there was a slightly iridescent yellow color. Respectfully I have prognosticated, and the result said, the guest star does not infringe upon Aldebaran. This shows that a Plentiful One is Lord, and the country has a Great Worthy. I request that this prognostication be given to the Bureau of Historiography to be preserved." Now, nearly a thousand years later, we see the expanding debris from the explosion witnessed by Yang Wei-Te. Supernova explosions signify a violent end point of stellar evolution. Astronomers now distinguish several different types of supernovae, but the most common mechanism (the one that gave rise to the Crab Nebula) comes about when a star much heavier than the Sun, too massive to leave behind a white dwarf, exhausts its nuclear fuel. The star then faces an energy crisis. Its core catastrophically implodes, and the outer layers are blown off.

Supernovae are bright enough to be observed at great distances. Thus, although they are rare events in any one galaxy, by watching the behaviors of many supernovae discovered in other galaxies, astronomers have pieced together a systematic description of how their spectra and luminosities change with time following the initial explosion. But the large distances of typical supernovae have handicapped observers, especially when the supernova fades out of view after a few months. Thus the discovery, on February 23, 1987, of a "nearby" supernova was a wonderful stroke of luck for astronomers. Supernova 1987A exploded a mere 170,000 light-years away in the Large Magellanic Cloud, a satellite galaxy of the Milky Way. Supernova 1987A is one of the few supernovae for which the star was observed *before* it exploded. The doomed star did not exactly fit theoretical expectations, being blue rather than red, but this anomaly (which is now consistent with revised theory) is minor compared with several impressive verifications of our general theoretical picture of supernovae. The star was massive—about 20 solar masses—and its mass fits the picture of a core collapse. But more important, two deep underground neutrino detectors—Kamiokande II in Japan and the IMB detector in Cleveland, Ohio—each captured a handful of neutrinos produced during the collapse and "neutronization" of the stellar core. Having traveled all the way through the stellar envelope and intervening space, these neutrinos reached the Earth in a burst lasting less than a minute, arriving several hours before optical telescopes could observe the shock wave erupt through the surface of the star. (However, the data from the neutrino detectors were not analyzed until after the supernova had been discovered by the traditional optical means.)

The remarkable observation of neutrinos provides the first direct confirmation of the core collapse theory for supernovae. In fact, it was the first detection of *any* neutrinos from an astronomical object other than the Sun. Neutrinos from supernovae are not rare, but they are difficult to detect because they hardly interact with any ordinary form of matter, including the matter in a neutrino detector. Just as most of the neutrinos produced during a supernova escape even the dense core of the collapsing star, so did virtually all of the neutrinos reaching Earth from SN 1987A pass through unnoticed, neutrino detectors and all. In fact, about 100 trillion neutrinos from 1987A passed through each person on the planet. Neutrino detectors put up a considerably larger obstacle than a person, and caught only 10 apiece. Some astronomical wag has estimated that a

*"Before" and "after" pictures of Supernova 1987A in the Large Magellanic Cloud. The supernova is at the top right-hand extremity of the galaxy. Using more detailed images, astronomers have determined exactly which star exploded.*

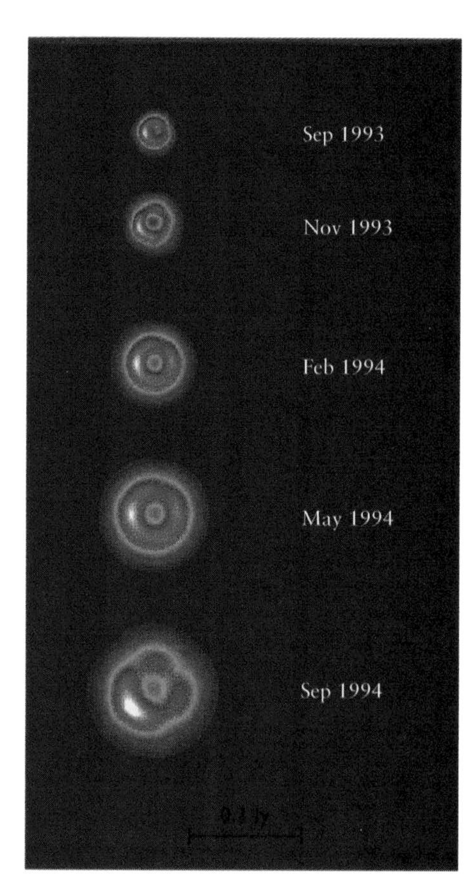


*When a star explodes as a supernova, its outer layers are blown off at up to 20,000 kilometers per second. Its brightness flares up within a few days, and then gradually fades. To optical astronomers, a young supernova looks like a point of light. However, the techniques of very-long-baseline interferometry (see Chapter 6) allow radio astronomers to obtain sharp enough images to reveal the actual expansion. This series shows Supernova 1993J, in the galaxy M81, during the first two years after the explosion. These images are from the observations of Juan Marcaide and collaborators.*

half-dozen or so of the 5 billion Earthlings might have had a 1987A neutrino actually hit a molecule inside an eyeball, perhaps creating a noticeable flash. Please do not contact us if you believe you had such an experience!

Most of what we know about Supernova 1987A comes not from neutrino detectors but from spectral analysis of the supernova's radiation, as explained in the box on pages 37–39.

The detection of spectral lines of cobalt in the aftermath of Supernova 1987A was an important confirmation of supernova theory. In the moments before and during the explosion, the stellar core becomes so hot, and the nuclear reactions so rapid, that some unstable nuclei heavier than iron are synthesized. Among these are cobalt and nickel. The gradual decline in a supernova's light, lasting several months, is attributed to the radioactive decay of unstable forms (isotopes) of cobalt and nickel, produced behind the violent shock at the base of the expanding envelope. Supernova 1987A gave us the first direct evidence that cobalt was indeed present in the expected amounts. Unprecedented monitoring of the expanding remnant, at wavelengths ranging from radio waves to gamma rays, has allowed us to "reconstruct" the structure of the stellar envelope and the dynamics of the explosion. And observers are still watching carefully for the first signs of the expected neutron star peeking through.

It may seem strange that the catastrophic collapse of the inner part of a star should lead to the violent explosion of the outer part. Indeed, it has proved quite difficult to produce a satisfactory theory to explain the details of supernova explosions. There seem to be a number of effects that conspire to throw off the star's envelope at high speed. One is the so-called core-bounce mechanism. When the core approaches the density of an atomic nucleus, the now-degenerate neutrons become very "stiff," resisting further compression. This sudden increase in pressure is enough to halt the collapse. But the core is contracting so rapidly that instead of screeching to a halt it overshoots and bounces back a little. The idea of the core-bounce mechanism is that the bouncing core runs into the still-infalling envelope, giving it a kick that sends it flying outward. Unfortunately, computer models of this process indicate that it falls slightly short, and that the shock wave driving the explosion "stalls" inside the envelope. A second contributory mechanism is needed to finish the job, and this is thought to come from the large outpouring of neutrinos from the very hot bouncing core. Although neutrinos barely interact with matter, there are so many of them that they are thought to give the envelope an extra "push," re-energizing the shock wave and finally blowing off the envelope.

While we seem to have the basic theoretical picture in place, we have not yet reached the point where we can simulate a supernova explosion inside a computer. The first calculations assumed (for computational

## Spectroscopy

An atom consists of a positively charged nucleus surrounded by orbiting electrons. Electrons on smaller, closer-in orbits have lower energy than electrons on more distant orbits. An electron can lose energy and drop into a lower orbit by emitting radiation, but if it radiated according to the nineteenth-century "classical" theory developed by James Clerk Maxwell, it would spiral right into the nucleus. Modern-day quantum theory states that this is impossible. According to quantum theory, electrons cannot be concentrated arbitrarily close to the nucleus; instead, there is a well-defined orbit of lowest energy called the ground state. Higher-energy orbits, or "excited" states, are also discrete. So an electron that has become electrically bound to a nucleus, instead of spiraling in gradually, changes its energy in discrete steps and reaches a definite lowest state where it can radiate no more.

Radiation can be viewed either as a wave or as consisting of "quanta," or particles, called photons. A

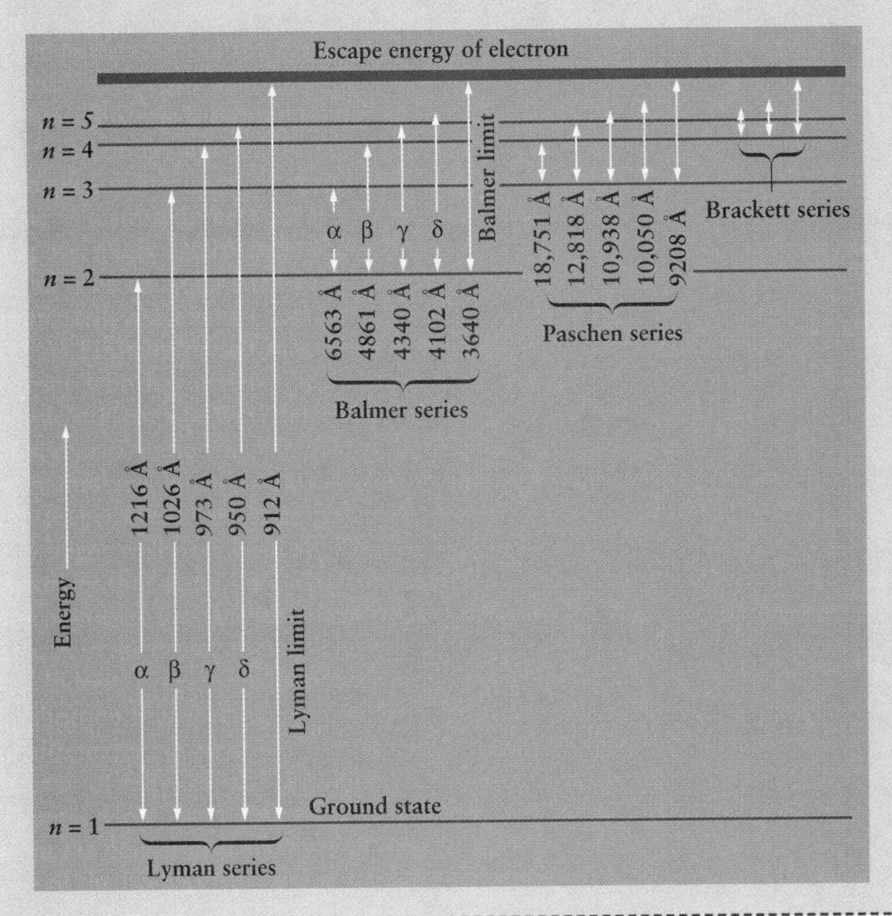


*A diagrammatic representation of the permitted energy levels of an electron in a hydrogen atom. Transitions from lower to upper levels absorb energy in discrete amounts, while those from upper to lower levels release energy in the same amounts. The energy released or absorbed in a transition is proportional to the length of the arrow connecting the levels. There is a definite relationship between the energy of a "quantum" of radiation (or "photon") and its wavelength. The spectrum of the light from hydrogen consequently displays a set of lines at well-defined wavelengths. Other atoms produce equally distinctive (though generally more complicated) sets of spectral lines.*

photon carries an energy equal to the frequency (equivalently, the speed of light divided by the wavelength) of the light it represents multiplied by Planck's constant. It requires bigger "jumps" of energy to emit blue light than red light, because blue light is carried in more energetic quanta. Ultraviolet and X-ray emissions require more energy still.

The simplest atom is hydrogen, with just one electron; its energy states correspond to a simple ladder. In very hot gases, as are found in stars and galactic nuclei, collisions and other processes can dislodge electrons from atoms, creating what is called a "plasma" composed of atomic nuclei and free electrons. Whenever an electron is recaptured by a nucleus and initially lodges in an excited state, one or more photons (each with the frequency characteristic of the particular energy jump) will be emitted as it cascades down to the ground state.

A hot gas of hydrogen emits a series of discrete frequencies, which are in simple mathematical ratios. The strongest emission of all, Lyman alpha, corresponds to a transition from energy state $n = 2$ to the ground state $n = 1$. Other transitions ending at $n = 1$ (3 to 1, 4 to 1, 5 to 1, etc.) are known as Lyman beta, gamma, and so on; there is another set of frequencies corresponding to transitions that end up at $n = 2$ rather than $n = 1$; these are the Balmer series. Other series end at $n = 3$, $n = 4$, and so on.

The variation of radiation intensity with wavelength is called the *spectrum*. A *spectrograph* displays a spectrum by spreading the wavelengths in a band across a piece of film or other detector, from short wavelengths at one end to long wavelengths at the other. At the wavelengths where atomic transitions are producing a lot of emission, the spectrograph shows sharp, bright stripes called *emission lines*. Atoms of each element in the periodic table emit radiation with a characteristic set of discrete frequencies, producing a distinctive set of lines in the spectrum that are the "fingerprint" of that particular element. We can also tell whether an atom has lost some of its electrons—that is, become *ionized*—by measuring the pattern of spectral lines. By measuring the *relative* intensities of different emission lines, we can often deduce the temperature and density of the gas.

The spectrum emitted by a hot, dilute gas consists mainly of emission lines. But a dense or opaque gas, or a heated liquid or solid (such as the filament of an incandescent light bulb), emits radiation at all wavelengths. This kind of continuous spectrum is called *blackbody radiation.* The variation of intensity with wavelength depends only on the temperature of the radiating material, and not at all on its composition. Inside a star, for instance, the radiation would have an almost exact blackbody spectrum.

A pure blackbody spectrum has no features that can reveal the nature of the material emitting the radiation. We can nevertheless learn about the composition of stars by measuring their spectra. The Sun, for instance, has a surface temperature of about 5800 degrees Kelvin. But the surface is not completely "sharp," and successive layers near the surface have slightly different temperatures. When a hot continuous spectrum passes through cooler gas along the line of sight, the inverse of emission—absorption—can occur. An electron can be excited to a higher level by photons of the right energy. This process happens in the outer layers of stars, where a cooler gaseous atmosphere overlays the hot gas of the stellar disk. The light from the star is therefore depleted at the frequencies that have been absorbed. Spectra then show dark rather than bright lines at the wavelengths characteristic of the atoms and ions in the star's outer layers.

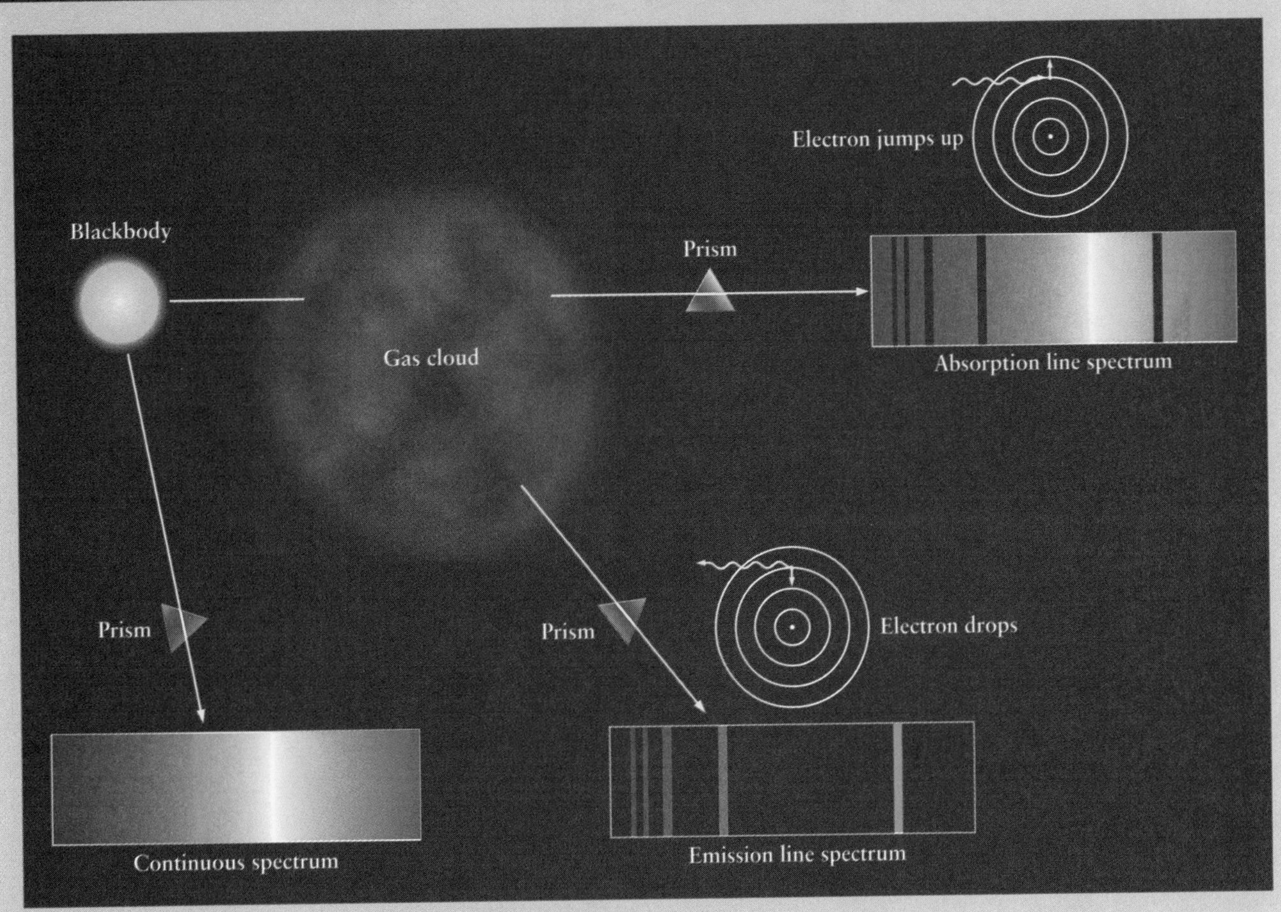


*Spectrographs can be used to probe the compositions, temperatures, and densities of astronomical bodies. A cloud of hot, rarefied gas produces a spectrum of emission lines, whose wavelengths reveal the cloud's chemical composition and level of ionization. The relative intensities of different lines can give indirect information about the temperature and density of the radiating gas. A dense or opaque gas produces a continuous spectrum, which carries no information about the composition of the source but gives us a direct measurement of its temperature. If a source of continuous radiation is viewed through a cooler intervening cloud of gas, one sees absorption lines superimposed on the continuous spectrum.*

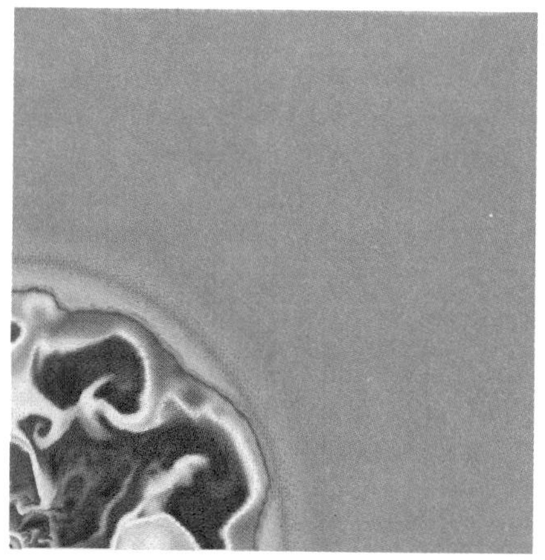
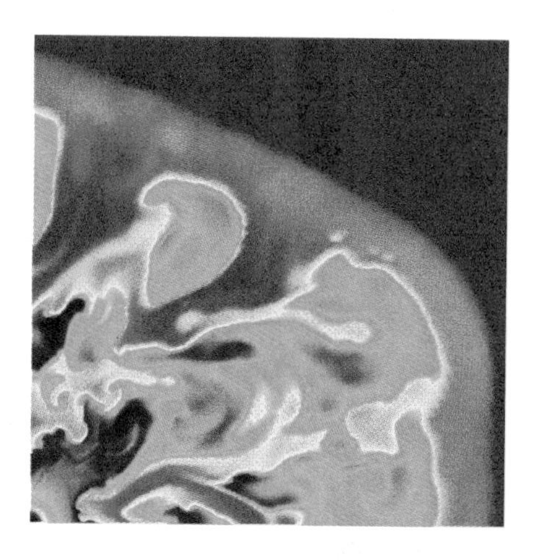

*Snapshots from a computer simulation of a supernova explosion, by Adam Burrows, John Hayes, and Bruce Fryxell of the University of Arizona. Convection—the violent mixing that occurs as tongues of hot gas from the deep interior penetrate the cooler outer layers—plays a crucial role in triggering the explosion. Supercomputers have only recently become capable of simulating dynamical processes at this level of detail. Dramatic as they appear, these simulations are still a highly simplified representation of reality.*

simplicity) that the star was exactly spherical, but this proved to be too much of a simplification. Supernova explosions are extremely messy, turbulent events. They are sensitive to many details, such as asymmetries of the collapsing core and envelope, chemical inhomogeneities, and the effects of rotation. Supercomputers have only recently become capable of handling these complications: they can now cope with rotation, and even with convection, complex internal motions that are crucial to the explosion, but it will still be some time before most of the complications are understood. Nevertheless, Supernova 1987A has been a great ego-builder, showing supernova theorists that they have probably been on the right track.

Supernovae may seem like very distant and remote events, but they are crucial for our very existence. Throughout our Milky Way, gas is being converted into stars, and material from stars recycled back to gas, enriched in newly synthesized chemical elements through supernova explosions. Complex chemical elements are built from hydrogen via the nuclear reactions that provide the power source in the cores of ordinary stars. Although the elements in the core are destroyed during the collapse just before a supernova explosion, plenty remain in the star's outer shells to be scattered through space by the force of the explosion, and the heaviest elements are created in the explosion itself as the shock wave ignites nuclear reactions in the star's outer envelope. Only by studying the births of stars, and their explosive deaths, can we tackle such an everyday question as where the atoms we are made of came from.

The abundances of the elements are quite well known: they can be determined in the Solar System by direct measurement and inferred spectroscopically in stars, nebulae, and even in other galaxies. The relative abundances display many similarities from place to place, and the pattern is now reasonably well understood. Most elements heavier than helium are made in stars, and blown off either in stellar winds or by a final explosion. We understand why oxygen is common, but gold and uranium are rare. All the carbon, oxygen, and iron on the Earth could have been manufactured in stars that exhausted their fuel supply before the Sun formed. The Solar System then condensed from gas contaminated by debris ejected from these early stars.

Pooh-Bah (in Gilbert and Sullivan's *Mikado*) traced his ancestry back to a "protoplasmic primordial atomic globule," but the astronomer goes back further still. Each atom on Earth can be traced back to nuclear reactions inside stars that died before the Solar System formed. An oxygen atom, forged in the core of a massive star and ejected when the star explodes as a supernova, may spend hundreds of millions of years wandering in interstellar space. It may then be once again in a stellar interior, where it can be transmuted into still heavier elements. Or it may find itself out on the boundary of a new solar system, incorporated into the fabric of a planet, and maybe eventually into a living cell. We are literally the ashes of long-dead stars.

## Neutron Stars

The ejecta of supernova explosions contain all the familiar chemical elements of which we are made, but the remnant left behind could not be more exotic or alien to our everyday experiences. Sixty years ago, the astronomers Walter Baade and Fritz Zwicky made a very bold prognostication, one more physical (though no less speculative) than Yang Wei-Te's. In a 1934 paper they wrote, "With all reserve we advance the view that a supernova represents the transition of an ordinary star into a neutron star, consisting mainly of neutrons." They conjectured that a supernova explosion was driven by the gravitational energy impulsively released when the star's core collapsed, and that a tiny cinder should remain: a star squeezed to a size of 10 kilometers, and having an internal density comparable to that of an atomic nucleus, $10^{15}$ times higher than an ordinary solid and a billion times higher even than a white dwarf.

Although theorists, among them J. Robert Oppenheimer and George Volkoff, explored the physics of neutron stars in the late 1930s, their existence remained conjectural until 1968. Jocelyn Bell, then a graduate student at Cambridge University's Mullard Radio Astronomy Observatory, and her thesis advisor, Anthony Hewish, were surveying the sky with a new radio telescope when they found a source emitting regular pulses of radio energy every 1.3 seconds. They soon found some others, each "ticking" with a distinctive regular period. The nature of these sources—later

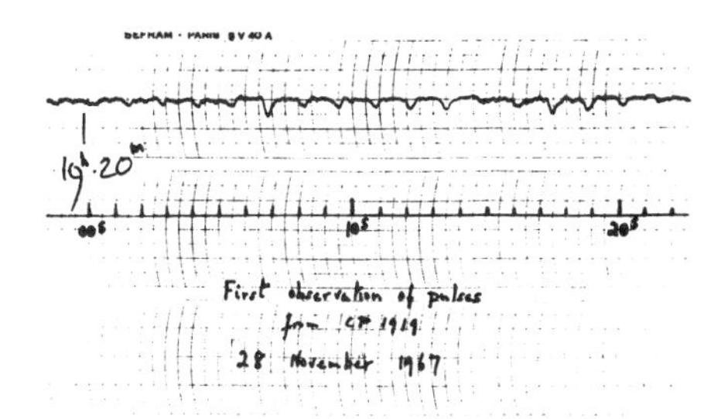

*Strip-chart data taken by Jocelyn Bell in 1967 record the first detection of a pulsar.*

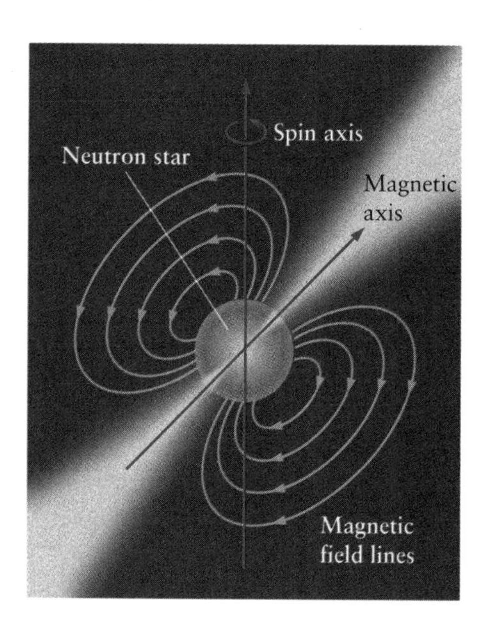


*Neutron stars have ultra-strong magnetic fields. If the stars are spinning, these fields generate strong electric currents, as in a dynamo, and accelerate electrons. These electrons emit an intense beam of radiation from each magnetic pole. Because the magnetic field is misaligned with the star's spin axis, these beams sweep around the sky once per revolution.*

called pulsars—was baffling. They clearly had to be very compact in order to vary so rapidly. One idea, favored by Franco Pacini and Thomas Gold, was that the sources were spinning neutron stars.

Only a year later, John Cocke, Michael Disney, and Don Taylor at the University of Arizona discovered that the ordinary-looking little star in the middle of the Crab Nebula was actually flashing on and off 30 times a second. Nothing much larger than a neutron star could pulse or spin this frequently—even a white dwarf would be too large. So the discovery of the pulsar in the Crab Nebula clinched the case for pulsars being spinning neutron stars whose strong magnetic fields produce a kind of lighthouse beam of radiation that swings through our line of sight once every revolution. Since then, several hundred pulsars have been discovered in the Milky Way Galaxy. Most of them were formed within the last 10 million years and are rather close by; there is every reason to believe that there are at least 100 million neutron stars in our Galaxy, one for practically every supernova that has occurred.

The discovery of a fast-spinning pulsar solved the puzzle of why the Crab Nebula still glows so brightly, more than nine hundred years after the explosion that created it. The glow of a typical nebula, like the Orion Nebula, comes from turbulent shocked gas and from atoms bathed in ultraviolet radiation from hot young stars. The nebula pervading the Pleiades star cluster in Taurus glows blue because tiny dust particles scatter the light of the stars. But the diffuse blue light from the Crab Nebula has a different origin. It does not come from ordinary hot gas; rather, it is synchrotron radiation, emitted by electrons or their antimatter counterparts, positrons, moving at very close to the speed of light through a magnetic field. It is straightforward to calculate how quickly these ultra-energetic particles (with energies a million times higher than their mass-energy equivalent, $mc^2$) will radiate their energy away—they radiate so quickly that they cannot have retained their energy since the explosion occurred in

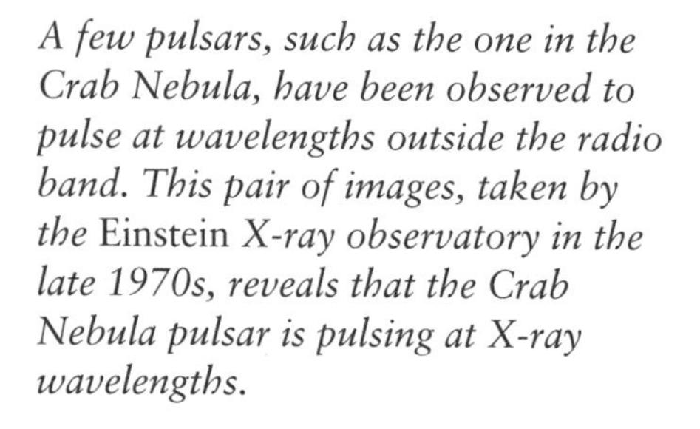

*A few pulsars, such as the one in the Crab Nebula, have been observed to pulse at wavelengths outside the radio band. This pair of images, taken by the* Einstein *X-ray observatory in the late 1970s, reveals that the Crab Nebula pulsar is pulsing at X-ray wavelengths.*

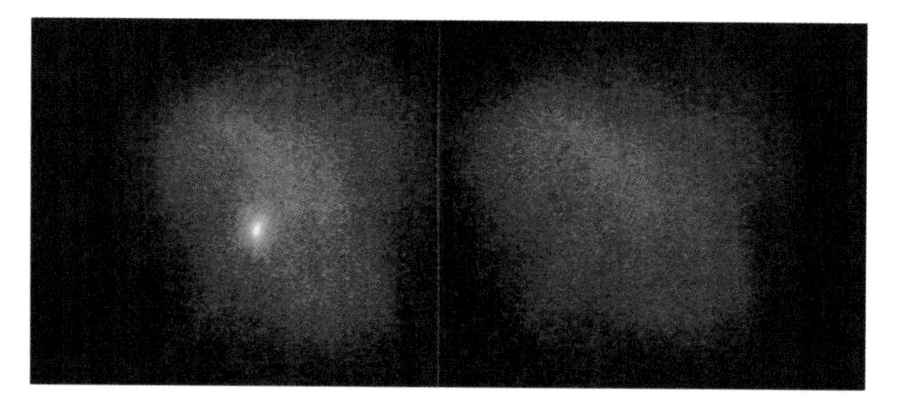

A.D. 1054. The tiny spinning star must be acting like a particle accelerator, continuously regenerating the fast electrons that produce the blue light. In effect, the star transfers some of its rotational energy to the electrons. The timing of the pulses shows that the star is slowing down, and the rate at which it is losing its rotational energy is more than enough to keep the nebula shining.

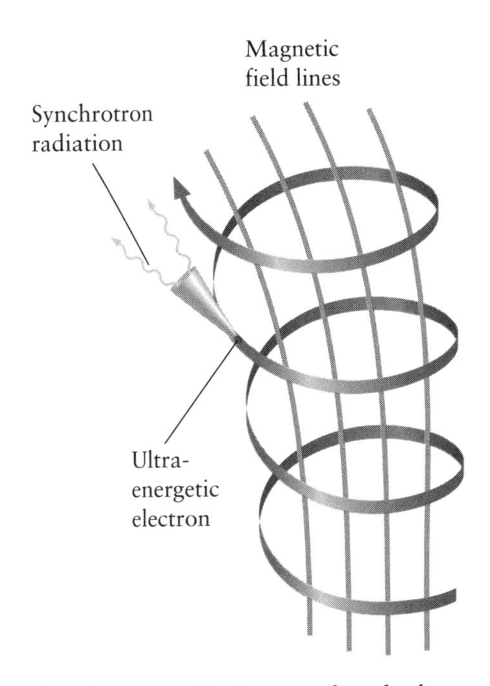


*Synchroton radiation, produced when electrons move through a magnetic field at close to the speed of light, is responsible for the bluish glow of the Crab Nebula. The electrons trace out helical trajectories centered on the magnetic lines of force. Synchrotron emission from each electron is concentrated in a narrow cone around its direction of motion.*

Through neutron stars, the cosmos offers us a laboratory where we can study how material behaves under conditions far more extreme than can be produced on Earth. Basically, a neutron star is an analogue of a white dwarf in which degeneracy pressure among neutrons, rather than electrons, provides the force that opposes gravity. We cannot determine the upper limit for the mass of a neutron star by simple analogy with the Chandrasekhar-limit argument for white dwarfs because nuclear forces figure importantly in the gravitational equilibrium and our understanding of the behavior of nuclear matter at such high densities is incomplete. Still, we can be quite confident, from general theoretical considerations, that neutron stars cannot exist with masses exceeding 3 solar masses. At least theory does tell us quite a bit about what the interior of a neutron star should be like. It should have a solid crust, and a core that is a liquid, indeed a so-called superfluid having almost no viscosity.

How can we test whether these properties really characterize the inside of a neutron star? One cannot, of course, go drilling below the surface of a neutron star, as a geologist can drill below the surface of the Earth. At first sight, it might seem hopeless to check out ideas about the inside of a neutron star by looking at one thousands of light-years away, but amazingly enough we can.

Pulsars can be timed with microsecond precision. Though they are on average slowing down, there are occasional "glitches," when they suddenly speed up by a tiny amount. There is one straightforward reason to expect glitches. A spinning object is flattened as it rotates by inertial forces (often called "centrifugal force," although some physics teachers excoriate that term), creating an equatorial bulge. As a pulsar slows down, these forces decrease and the bulge tries to shrink. If the neutron star were a fluid, the equatorial bulge would diminish continuously. But if the star has a solid crust, a continuous readjustment is impossible. Stress will build up, and when it exceeds a certain limit the crust will crack. Thus the readjustment takes place in jerks, causing sudden increases in the spin rate, the size and frequency of which tell us how thick and rigid the crust is. These changes in spin rate can be measured with such precision that a decrease of a few millionths of a meter in equatorial radius would show up clearly in the timing data on pulsars thousands of light-years from us.

The behavior of glitches has been studied in detail, and it is now believed that the most common ones result from a different effect: slippage between the solid crust and the fluid core. The drag that slows down a pulsar's rotation is applied to the crust and does not directly slow down the liquid core. A differential spin rate therefore develops between crust and core. The slowdown of the crust is communicated inward by a frictional force. This force doesn't act smoothly, however: instead, it operates in jerks rather like a car's worn-out clutch and causes the crust to spin up whenever the friction suddenly increases. The precise study of how the spin rate changes is just one of the ways we can learn about the insides of neutron stars—this kind of astrogeology is a feasible and serious subject.

A slice through a neutron star would in some respects look rather like a slice through the Earth, with a crust, a liquid interior, and maybe a solid core. The theory is detailed enough to predict that the star should deviate from a pure neutron composition at various depths. The outer crust would consist mainly of iron, but at higher densities the main constituents would be progressively more exotic neutron-rich nuclei such as nuclei of nickel, germanium, and krypton, arranged in a crystalline lattice. Many of these nuclei would be unstable in the laboratory, as are isolated neutrons: they undergo beta decay, emitting an electron and antineutrino. In a neutron star, however, this decay is inhibited according to the Pauli exclusion principle, because there are no unoccupied quantum states into which the electrons can go.

Digging still deeper below the crust one reaches densities approaching $10^{14}$ grams per cubic centimeter. Under these conditions the matter is mainly in the form of free neutrons. Moreover, the neutrons display the properties of a superfluid analogous to those displayed by terrestrial helium at temperatures very close to absolute zero. Superfluids have very unusual properties, including negligible internal viscosity. In fact, a vortex once set spinning within a superfluid will spin indefinitely. Within this fluid formed of free neutrons, a small fraction of the material remains in the form of protons and electrons. The proton component (and perhaps the electron component as well) would behave as an electrical superconductor, that is, a material with zero electrical resistance. The pressure exerted by "nuclear matter" at these densities is not exactly known; there is consequently some uncertainty in the internal structure and radius of a neutron star. Right in the core, where the density may attain values of several times $10^{14}$ grams per cubic centimeter, the character of the material is even more uncertain, and perhaps stranger still. One possibility, still highly controversial, is that the mix of elementary particles in a neutron star's core could condense into "quark nuggets"—a solid made from crumbled bits of protons and neutrons.

Neutron stars also appear to resemble the Earth in that they possess magnetic fields, but the magnitudes of these fields are vastly larger. The surface magnetic strength of a pulsar is typically several trillion times stronger than that of the Earth and a million times stronger than the strongest laboratory fields. For comparison, the magnets you might be using to attach notes to your refrigerator are about 100 to 1000 times stronger than the Earth's field, or about 10 billion times weaker than a pulsar's. Pulsar magnetic fields are so strong that they would severely distort the structures of atoms exposed to them, stretching them into cigarlike shapes.

Processes occurring in the magnetic field outside a neutron star are believed to generate the "lighthouse beam" of radio waves that we see from pulsars. These fields also supply the coupling that allows the pulsar to operate as a highly efficient particle accelerator, converting its spin energy into the very energetic electrons and positrons that keep the Crab Nebula shining. The radiation intensities produced by pulsars are extreme, surpassing those in terrestrial laser beams. According to one theory, the interaction of this radiation with the intense magnetic field gives rise to the beamed emission: in such an environment, numerous "electron–positron pairs"—literally, pairs of particles consisting of an ordinary electron and its antimatter counterpart—are created *from the vacuum of space* by a

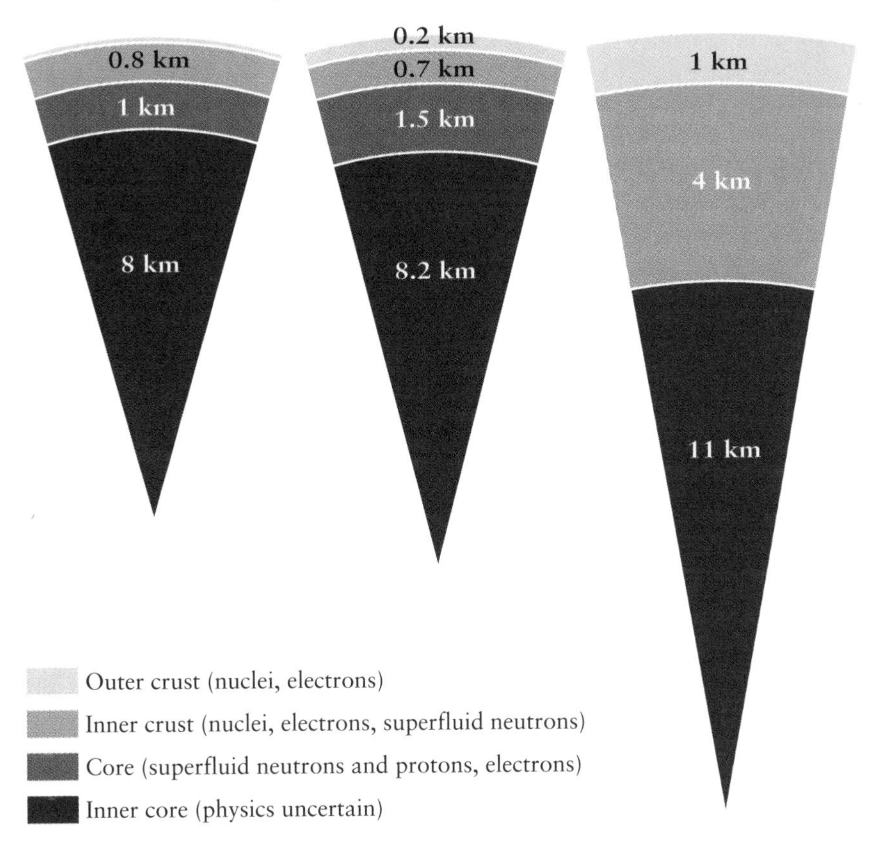


*The internal structure of a neutron star is shown in cross section for each of three theoretical models. Matter near the core is more than 100 trillion times denser than water; its density exceeds even that of an atomic nucleus. The star depicted here has a mass 1.4 times that of the Sun, yet its radius is only somewhere between 10 and 16 kilometers, the size of a typical city. The models differ because they make different assumptions about the pressure of superdense nuclear matter, about which little is known.*

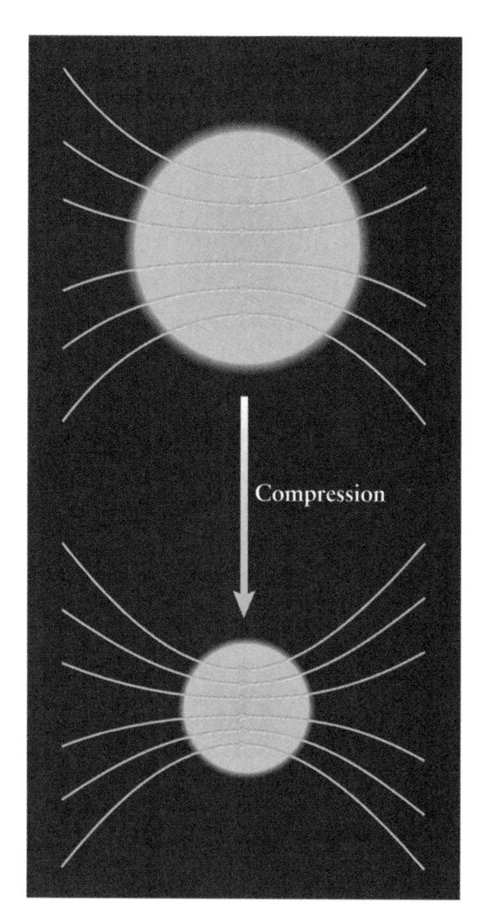


*Lines of magnetic force threading an ionized gas, such as a stellar interior, follow the gas motions. When a stellar core collapses to form a neutron star, the magnetic field lines running through it are compressed. Since the density of field lines determines the strength of the field, this process can explain how neutron stars can acquire the enormous magnetic fields inferred to exist in pulsars.*

process never observable in terrestrial laboratories. These particles could subsequently produce the radiation that we observe.

Neutron stars are extreme, too, in the strength of their gravitational fields: the force of gravity at the surface is $10^{12}$ times that on Earth. Their surfaces would be smooth: no mountain could stand more than a millimeter high. But so strong is gravity that the energy expended in climbing a millimeter mountain on a neutron star would still exceed that needed to escape entirely from the Earth's gravitational field. Because no structure would escape being squashed on the surface of a neutron star, the example of the mountain climber is even more far-fetched than it sounds. One could only consider observing a neutron star from a free-falling or orbiting platform. But, even there conditions would be uncomfortable because of the tidal effects—the difference between the gravitational pull on head and feet.

A rocket would have to attain about half the speed of light in order to escape from the surface of a neutron star, compared to about 11 kilometers per second ($\frac{1}{27,000}$ the speed of light) to escape from the Earth. The gravitational binding energy—the energy that would have to be supplied in order to disperse the star—is 10 to 20 percent of the so-called rest-mass energy—the energy that would be released by converting all the matter in the star to energy, according to $E = mc^2$. This is a truly enormous amount of energy. The most efficient nuclear reactions produce less than 1 percent of $mc^2$, while chemical reactions typically convert only a few billionths of a percent. Indeed, it is from gravitation that the magnetic and rotational energies of a pulsar are derived. Just as a skater spins faster by drawing her outstretched arms to her side, an imploding stellar core increases its rotation rate, and its rotational energy, as it contracts. This faster rotation is a consequence of the conservation of angular momentum. And just as the skater must expend energy drawing her arms in (against the centrifugal force, which tries to keep them outstretched), so the gravitational force does work on the star that goes into the increased rotational energy.

In an analogous way, an ordinary stellar magnetic field can be amplified by a factor of 10 billion or more by being squeezed during the compression to a neutron star. Lines of magnetic force in an ionized gas, such as makes up the interior of a star, behave as if they are "frozen-in," meaning that they are dragged along wherever the gas goes. Since the strength of a magnetic field is determined by the density of field lines, the magnetic field in a collapsing stellar core can increase by a huge factor. Here, the work done by gravity against the "magnetic pressure" force is what builds up the magnetic energy. Thus, gravitation not only energizes the supernova explosion itself but also leaves a "dead" star with more energy than nuclear reactions could generate over its entire previous lifetime. Gravity is the ultimate source of the energy that powers the Crab Nebula.

Before leaving the Crab Nebula, we offer an historical speculation. The nebula's central pulsar can be seen through any large optical telescope, but the pulse repetition rate is so high that the eye responds to it as a steady source. But had the pulsar been spinning, say, 10 times a second rather than 30 times, the remarkable properties of the little star in the Crab Nebula could have been discovered sixty years ago, long before the development of radio astronomy. How would the course of twentieth-century physics have been changed if superdense matter had been detected in the 1920s, before neutrons were discovered on Earth? One cannot guess, except that astronomy's importance for fundamental physics would surely have been recognized far sooner.

## Black Hole Stellar Remnants

Gravitational collapse does not appear to be a fate that stars work actively to avoid. Stars can shed a great deal of mass during their evolution—those of up to 5 (or even 8) solar masses can end up as white dwarfs. But the discovery of neutron stars in supernova remnants proves that still heavier stars do not shed enough gas to bring themselves safely below the Chandrasekhar limit. The upper mass limit for a neutron star is 2 or 3 solar masses, not very much larger than the limit for a white dwarf. No theoretical arguments suggest that stellar evolution should permit the existence of remnants too massive to become white dwarfs while forbidding those too massive to become neutron stars.

Stellar remnants more massive than about 2 or 3 solar masses therefore seem fated to undergo complete gravitational collapse to black holes, because no form of internal pressure can maintain them in equilibrium once their nuclear energy sources are exhausted. These gravitationally collapsed bodies may not be quite so common as neutron stars, but there may still be millions of them in our Galaxy. Some types of supernovae may form black holes rather than neutron stars; or, some massive stars may collapse without producing any conspicuous supernova-type explosion.

Where are these alleged black hole remnants? Can we identify them? Theoretical arguments are nourishing up to a point, but we must demand empirical evidence before we can replace conjecture with confidence. Fortunately, the development of a new kind of observational astronomy in the 1960s—X-ray astronomy—led straight to some very convincing evidence for stellar-mass black holes in our own Galactic neighborhood.

*Chapter* 3

# Black Holes in Our Backyard

*An artist's conception of a binary star system with an accretion disk. A stream of gas from the surface of the red star is being drawn away by the gravitational pull of its compact companion, shown here in white. Because of the orbital motion of the binary, the gas cannot fall directly onto the compact object, so it forms a disk through which it spirals in slowly. A "hot spot" forms where the stream of gas hits the outer edge of the disk. When the compact object is a neutron star or black hole, this mode of mass transfer can produce a powerful source of X-rays.*

How would you go about searching for the black hole remnant of a massive star in our Galaxy? When this first became a serious question among astronomers early in the 1960s, surprisingly few of the experts on black holes bothered to ask it. The reasons, in part, were "cultural." C. P. Snow had written about "two cultures"—scientific and humanistic—diverging in language and habits of thought, and becoming less and less able to communicate with each other. But cultural divergences had occurred *within* the physical sciences as well. Black holes, with their elegant mathematical properties, were primarily the intellectual toys of a highly specialized group of relativists—that is, scientists specializing in the subtleties of general relativity. In the tradition of Einstein, relativists liked to rely on pure thought and mathematical elegance to deduce deep truths about the nature of the Universe. On the other hand, astrophysics had a strong empirical tradition. Theory was important, but it was (and remains) observation that drove progress in the field.

The discovery of black holes in nature required synergism between astronomers and relativists; fortunately, just such an alliance had developed by the mid-1960s. Perhaps the "send-off" was the First Texas Symposium on Relativistic Astrophysics, which was held in Dallas in December 1963 in the aftermath of the discovery of the extraordinarily luminous and distant objects known as quasars. There the astrophysicist Thomas Gold (who in 1968 became the first to identify pulsars as rotating magnetized neutron stars) heralded the collaboration in his famous after-dinner speech, suggesting "that the relativists with their sophisticated work are not only magnificent cultural ornaments but might actually be useful to science! Everyone is pleased: the relativists who feel that they are being appreciated and are experts in a field [i.e., relativistic astrophysics] they hardly knew existed; the astrophysicists for having enlarged their domain, their empire, by the annexation of another subject—general relativity. It is all very pleasing, so let us hope that it is right. What a shame it would be if we had to go and dismiss all the relativists again."

Perhaps, too, the prospects for detecting black holes had been largely ignored because success seemed unlikely. But the possibility of failure did not deter the Soviet theorist Yakov Zel'dovich, the American Edwin Salpeter, and a few others from boldly speculating on possible observational signatures. Once formed, black holes are essentially "passive," and the best hope of locating them lies in discerning their gravitational effects on neighboring stars or gas. Zel'dovich's first proposal was to search for binary star systems in which an ordinary star is orbiting an invisible, yet massive, companion. By applying Newton's laws (which would still hold quite accurately far outside the horizon of the black hole) to the measured speed and orbital period of the normal star, one can sometimes estimate the mass of the unseen companion. By the 1960s this technique had already been long established as the most direct method for estimating stellar masses. However, it is usually fraught with uncertainties, and does not yield unique values for the masses of both stars. At best one could hope to obtain a lower limit on the mass of the unseen star. While the method could work in principle, to this day Zel'dovich's initial suggestion of searching for an invisible star has not yielded any plausible black hole candidates.

*Yakov Zel'dovich (right) and Igor Novikov. After a versatile earlier career, Zel'dovich became an outstanding researcher in relativity and cosmology. With Igor Novikov and other younger colleagues in Moscow, he was, from 1960 onward, responsible for many of the key ideas in relativistic astrophysics. Zel'dovich died in 1987; Novikov now heads a research group in Copenhagen, Denmark.*

But Zel'dovich and Salpeter also came up with other ideas. They noted that a black hole moving through space would absorb matter from the tenuous interstellar medium—an atmosphere of dust and gas that fills the volume between the stars in our Galaxy. This taking in of matter is the process called accretion, of which we shall hear much more later in this chapter. What Zel'dovich and Salpeter realized was that a black hole accreting interstellar gas might conceivably be quite luminous. As the gas falls deep into the gravitational field of the black hole (but before it crosses the horizon, of course), it will be strongly compressed and heated. In principle, the hot infalling gas could radiate into space as much as a few percent of the energy locked up in its mass, according to Einstein's relation $E = mc^2$. If reality matched theory, a black hole accreting a given amount of matter could produce more than 10 times as much energy as the most efficient thermonuclear incineration of the same material.

In a synthesis of these two proposals, Zel'dovich and his former student Igor Novikov proposed in 1966 that the best place to look for compact stellar remnants—either neutron stars or black holes—was in close binary systems, where the black hole or neutron star could capture and accrete matter from its stellar companion. Shock waves in the falling gas

*In the early days of space astronomy, observations were made from "sounding rockets," which gathered only a few minutes' worth of useful data. These pictures show the launch and "recovery" of an experiment to study the cosmic X-ray background—a diffuse X-ray glow coming from all points in the sky. The experimenter, Andrew Fabian, obtained his Ph.D. on the basis of this work, and has gone on to become a leading X-ray astronomer.*

would heat the gas to temperatures on the order of $10^8$ degrees Kelvin; at these temperatures the gas would emit radiation predominantly in the X-ray band. Rapid, random flickering of the X-rays, on timescales of a few milliseconds, was later recognized as an additional indication that the emission was being produced in an extremely compact region with the dimensions of a black hole or neutron star.

These signatures are exactly what led astronomers in the 1970s to the discovery of stellar-mass black holes, along with close binaries in which the compact accreting object is a neutron star rather than a black hole. Insightful as it proved to be, this lucky guess was not taken in an observational vacuum. Even the most brilliant theoretical astronomers have seldom foreseen the startling variety of real phenomena that appear in the sky. At the time that Zel'dovich and Novikov made their suggestion, the journals and conference circuits were already abuzz with the amazing discoveries being made using an entirely new kind of observatory, one that gave us a view of the Universe through "X-ray eyes." It is hard to know the extent to which their suggestion was stimulated by the debate then going on over the nature of the first X-ray "star" to be discovered, Scorpius X-1.

## X-Rays from the Cosmos

Physicists at the turn of the twentieth century knew well that heated gases give off radiation over a wide range of wavelengths, and that extremely hot gases emit X-ray and even gamma-ray radiation. But until the middle of the century, astronomers' view of the sky was restricted to the narrow band of wavelengths centered on visible light. If we had to identify one stimulus for the revolution in our conception of the Universe during the past fifty years, it would have to be the opening up of the electromagnetic spectrum to observational astronomy, first in the radio band (as we will discuss in Chapter 6), then in the X-rays, and more recently, in the infrared, ultraviolet, and gamma-ray bands. Rather than just confirming long-held theories, each advance into a new wavelength regime revealed entirely unexpected phenomena.

When we use an X-ray telescope to pinpoint and measure the sources of X-rays on the sky, we now know that we are studying the most energetic features of the Universe—the hottest gases, the most intense gravitational fields, the most energetic explosions. X-ray observations provided the first strong evidence that black holes actually exist. But observing the sky at X-ray wavelengths was far from easy. Unlike visible light, X-rays cannot penetrate very far into the Earth's atmosphere. To observe them, we must send detectors above the atmosphere, using rockets and satellites. X-ray astronomy was truly the first "new astronomy" of the space age.

Our first glimpses of the X-ray sky came from detectors carried on board sounding rockets, small boosters originally designed to make measurements of the atmosphere. Each suborbital flight yielded only about 4 minutes of useful data, enough to detect X-rays from the Sun, as some astronomers expected. But in 1962, a rocket flown by the American Science and Engineering (AS&E) group, led by Riccardo Giacconi, discovered an apparently discrete source of X-rays from beyond the Solar System in the direction of the constellation Scorpius. This became known as the source Scorpius (or "Sco") X-1. Now we know that Sco X-1 contains a neutron star orbiting a normal star, and that the X-rays are produced close to the surface of the neutron star. A subsequent rocket flight by Herbert Friedman's group at the U.S. Naval Research Laboratory (NRL), using improved apparatus, provided a more accurate location for Sco X-1 and also detected X-rays from the direction of the Crab Nebula, the remnant of Yang Wei-Te's supernova of A.D. 1054.

The discovery of Sco X-1 was a great surprise. Astronomers had been pessimistic about the prospect of detecting X-rays from beyond the Solar System, since the X-ray emission from known stellar objects was expected to be too weak to be picked up with any feasible instruments. Sco X-1, emitting thousands of times more energy in the X-rays than in visible light, was clearly an entirely new and unsuspected type of source. The same AS&E rocket flight also revealed an unexpectedly strong X-ray glow in all directions in the sky, now known as the *X-ray background*. The discover-

*Riccardo Giacconi's experiments using sounding rockets led to the discovery of the first cosmic X-ray source (Sco X-1) and the cosmic X-ray background. He directed the pioneering* Uhuru *satellite project, followed by the much more ambitious* Einstein *observatory (shown here). Giacconi was the first Director of NASA's Space Telescope Science Institute. Now, as Director-General of the European Southern Observatory, he is overseeing the construction of the world's largest optical telescope.*

ies from these experiments—each based on just a few minutes of data—provided the initial impetus to the entire subject of X-ray astronomy.

X-ray astronomy took a major leap forward when detectors were flown on satellites. Now it was possible to observe faint sources for extended periods of time and to make surveys of the whole sky, neither of which was feasible during brief rocket flights. The first such payload was conceived, built, and operated by Giacconi and his colleagues at AS&E. It flew on NASA's small *Uhuru* X-ray satellite, launched in 1970. (Launched from a platform off the coast of Kenya on the seventh anniversary of that country's independence, the satellite was named with the Swahili word for "freedom.") *Uhuru* detected numerous extragalactic X-ray sources, including "active" galaxies emitting intense concentrations of energy from their centers and clusters of galaxies. But *Uhuru*'s remarkable success stemmed from its unexpected discovery of a whole new class of sources, the binary X-ray sources, of which Sco X-1 is now believed to be a member.

X-ray binaries have a wide range of properties. They include sources like Hercules X-1 and Centaurus X-3, which pulsate regularly every few seconds, and sources like Cygnus X-1, which seem to flicker at random on timescales down to a few milliseconds. What is the difference between these two types of sources? For reasons that will become clear shortly, Cygnus X-1 has been identified as the first black hole candidate and is still one of the firmest such candidates. On the other hand, the regularly pulsing sources in binary systems are interpreted as spinning neutron stars, sometimes called *X-ray pulsars.* The masses of several of these periodic compact sources, of which about two dozen are now known, have been determined from the Newtonian dynamics of the binary system. All have masses between 1 and 2 times the mass of the Sun, below the likely mass limit for neutron stars.

Although these neutron stars pulsate with a steady frequency at X-ray wavelengths, the radiation mechanism is apparently very different from that of *radio pulsars,* like the one in the Crab Nebula. The intense X-rays are generated by infalling gas, captured from the companion star, impacting on the neutron star's surface at one-third to one-half the speed of light. The kinetic energy of the gas is converted to thermal energy on impact, and radiated at temperatures of roughly 10 to 100 million degrees Kelvin, giving predominantly X-rays. The explanation for the regular pulsations is that the gas does not fall uniformly over the surface of the neutron star, but is channeled onto the magnetic polar caps by the extremely strong field. As the neutron star spins, the X-rays from the pair of hot spots come in and out of view, producing regular pulses. Some radio pulsars, including the Crab, also pulse in X-rays, but with not nearly the same intensity.

It was an accident of timing that neutron stars were discovered by radio astronomers (in their manifestation as pulsars) rather than by X-ray astronomers. The X-ray astronomers had been expecting to discover neutron stars, because at the time people thought that the surfaces of neutron stars might be very hot. (Isolated, nonpulsing neutron stars have now been detected, but they are much cooler than the early calculations had suggested.) In the earliest days of rocket experiments, Friedman's group at NRL observed the Crab Nebula during the few minutes it was covered on the sky by the Moon. Since the position of the Moon's limb was known very accurately, this lunar occultation was a unique opportunity to probe the structure of the Crab X-ray source with unprecedented angular resolution. The purpose was to test whether the X-rays came from a point source such as a hot neutron star or from the extended nebula around it. The result—that the Crab X-ray source was extended—was corroborated by later work. But had Friedman's experiment been slightly more sensitive, he would have found that 10 percent of the Crab's X-rays do indeed come from a central point source, and thus he would have discovered the X-ray emission of the Crab pulsar. He did not have an accurate clock in his detector, however, and would not have discovered that the source was flashing 30 times every second. But the X-rays certainly would have tipped off investigators that this was a likely detection of a neutron star. One wonders how much longer it would have taken before the pulsations were discovered. (Data recorded during a 1964 balloon experiment by Richard Haymes of Rice University actually contained evidence of regular

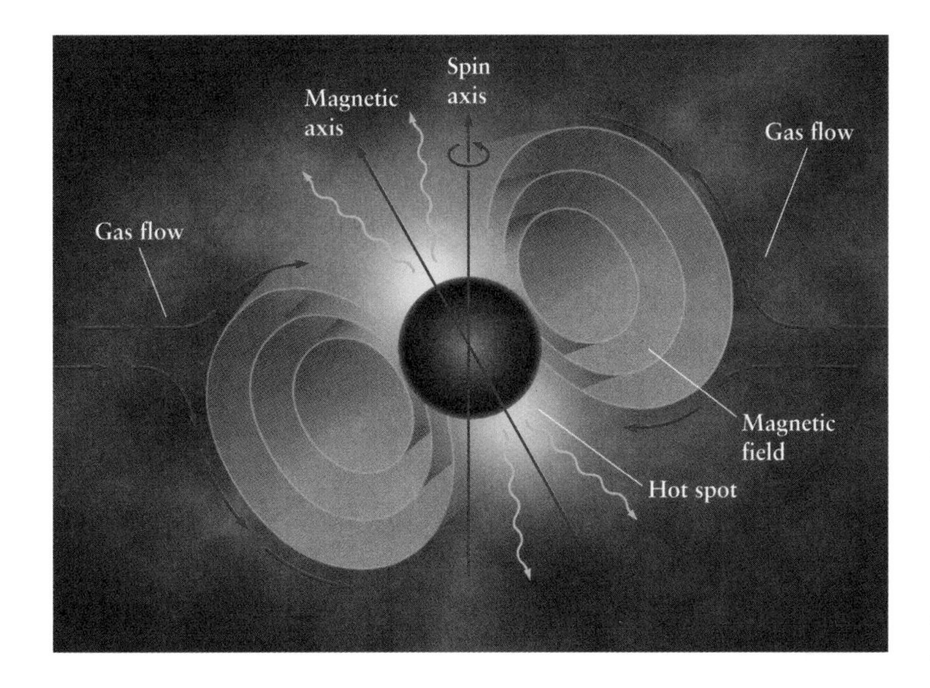


*In an X-ray pulsar, the neutron star's enormous magnetic field diverts the flow of infalling gas and channels it onto the magnetic polar caps. The impact of the gas on the star's surface creates a pair of X-ray hotspots. The star's rotation produces the appearance of periodic X-ray pulsations.*

pulsations in the X-rays from the Crab, but this was only discovered when the data were reanalyzed several years later.)

On the other hand, had Jocelyn Bell been less perceptive in 1967, neutron stars might have escaped notice by radio astronomers, and their discovery would have been deferred until the launch of the *Uhuru* satellite, three years later. The X-ray sources Her X-1 and Cen X-3 discovered by *Uhuru* in 1971 would have been quickly interpreted as neutron stars. It was thus bad luck for the X-ray researchers that radio astronomers won the credit for discovering neutron stars—X-ray astronomers had been seeking them actively in the 1960s, and did in fact discover them by a quite different route three years later. However, even though X-ray astronomers lost the glory of discovering neutron stars, X-ray data have subsequently yielded important information about neutron star physics, complementary to what can be learned from radio observations.

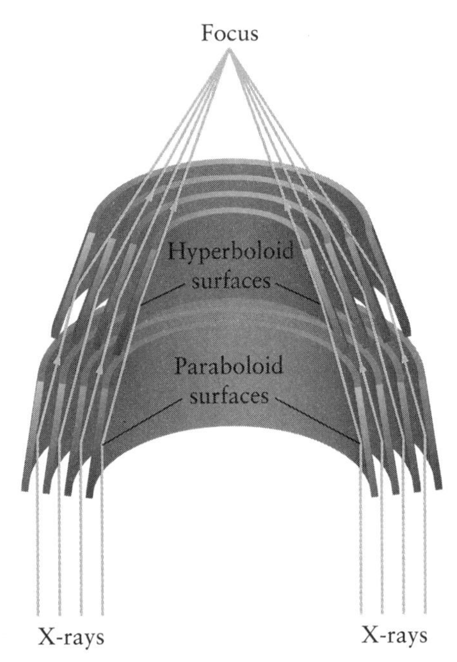


*X-rays cannot be focused by conventional mirror arrangements like those used in optical telescopes. In X-ray telescopes like the one flown on the* Einstein *satellite, X-rays are focused into an image by reflecting off the insides of tapered metallic tubes at very shallow angles, or "grazing incidence." To achieve a large total "collecting area" and therefore high sensitivity, several of these reflecting surfaces are nested inside one another. This mirror system gathers the X-rays entering the telescope through an annulus.*

Whereas regular pulsations are a sure giveaway that an X-ray–emitting object is a neutron star, such pulsations are not always present. Many sources, like Sco X-1, are suspected of harboring neutron stars even though they show no sign of regular pulsations. Why, then, are we convinced that the X-ray source in Cygnus X-1 is a black hole and not a neutron star? Distinguishing a black hole from a neutron star in such a system is not trivial, since the gravitational field near a neutron star is nearly as strong as that near a black hole. The effects on infalling gas are broadly similar; therefore, any differences in the radiation emitted in the two cases are likely to be subtle. There are probably some differences in the spectrum of the radiation as well as in the nature and strength of the random flickering. After all, neutron stars have real surfaces whereas black holes have horizons, which represent the complete absence of a physical surface. The magnetic fields and rotation (however slow, or masked by an opaque screen of gas) of neutron stars would also imprint features on the radiation that would be absent in a black hole. But no one has succeeded in predicting the differences purely from theoretical considerations. In fact, having used other methods to select black hole "candidates," astronomers have been trying to determine the spectral signatures of black holes empirically, with mixed success so far.

Fortunately, there is a decisive discriminant that can at least rule out the presence of a neutron star from some X-ray binary systems. If one can show that the compact object has a mass greater than 3 solar masses, then one can exclude a neutron star. In lieu of any other plausible candidates for a massive compact object capable of producing flickering X-rays, these systems are thought to contain black hole remnants of stars.

The spectacular and highly publicized early successes of X-ray astronomy led to a rapid development of the field. Following *Uhuru,* several small X-ray satellites were launched during the 1970s, signaling the entry of other nations into a field that had hitherto been predominantly a U.S. enterprise. These projects consolidated the discoveries made with *Uhuru,* and discovered additional kinds of X-ray binaries. The American *Einstein* satellite, launched in 1978, carried the first X-ray telescope combining high angular resolution with a large collecting area—an innovation, made possible by the telescope's nested mirror configuration, that offered much higher sensitivity. The results from *Einstein* had broad ramifications for many areas of astronomy, including the search for black holes in both X-ray binaries and the nuclei of galaxies. Major projects in X-ray astronomy have by now been carried out by all spacefaring nations, and more are being planned for the late 1990s and beyond. Progress has led from simple detectors in sounding rockets to major telescopes that offer images as sharp as those made by big optical telescopes. The imaging and spectroscopic capabilities of modern X-ray telescopes and their phenomenal sensitivity constitute a prime scientific success of the space program.

## Mass Transfer and Accretion in Binaries

In all X-ray binaries, matter is being transferred from a normal star to a compact object, whether neutron star or black hole. The infall of gas onto a star or other gravitating system is called *accretion.* Astronomers first considered the possibility of a star's accreting matter when they realized

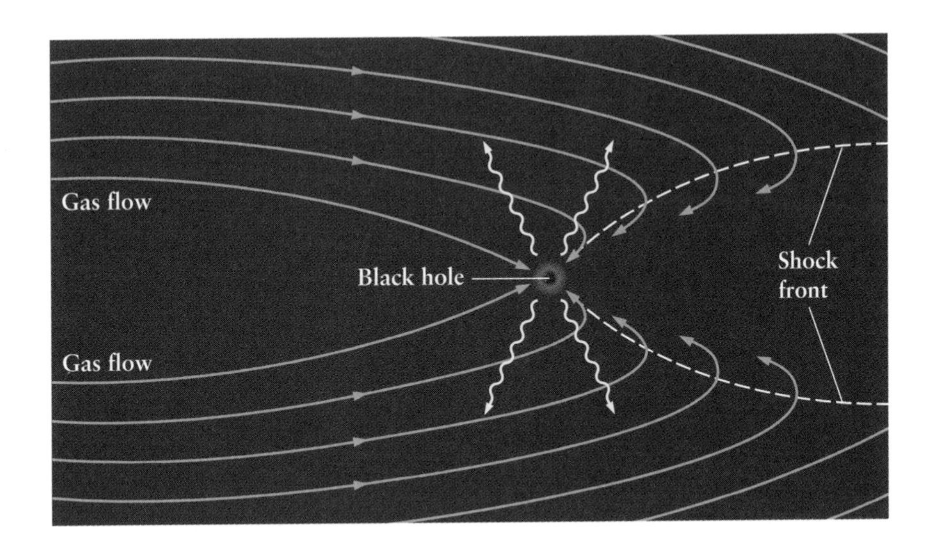


*This diagram depicts how accretion can occur onto a black hole moving through diffuse gas. In the hole's frame of reference, the gas is flowing past as though in a wind. The hole's gravity causes the flow to converge into a "wake," and a shock front forms "downstream." The shocked gas can lose energy and fall into the black hole. This process was first discussed (with application to normal stars rather than to black holes) by Hermann Bondi, Fred Hoyle, and Raymond Lyttleton.*

that the space between the stars is not a perfect vacuum but is pervaded by a tenuous gas (the interstellar medium).

Astronomers' early interest in the process had nothing to do with the emission of X-rays (or any other form of energy) by a star, but rather, with examining whether a star could grow appreciably in mass by accretion, or at least undergo an alteration of the chemical composition of its surface. (The latter is still thought to be a viable mechanism for explaining certain chemical abundance anomalies in white dwarfs.) In 1926, Eddington considered the rate at which a star would sweep up matter, taking into account the enhancement due to an effect called gravitational "focusing." The rate at which gas particles would hit a nongravitating sphere is simply determined by the surface area of the sphere and the speed and density of the particles. But gravity has the effect of focusing the trajectories of the gas particles toward the center of mass, so that some particles that would have missed the nongravitating sphere will hit the gravitating one. The accretion rate would therefore be larger by a factor that depends on the mass. The Cambridge astrophysicists Fred Hoyle and Raymond Lyttleton took Eddington's argument further by showing that the accretion rate would be greatly enhanced if particles were colliding close to the star. This idea was extended and formalized by Hermann Bondi in 1952, and most astronomers believe that the Bondi–Hoyle–Lyttleton theory gives a fair picture of accretion by a star moving through a gaseous background.

The idea that mass can be transferred between stars in a binary system also predates the discovery of binary X-ray sources. The star Algol (from *al ghoul,* Arabic for "the demon") in the constellation Perseus is an eclipsing binary with a special place in the history of stellar evolution studies. Every 2.9 days the smaller (but brighter) blue star is hidden from view by a

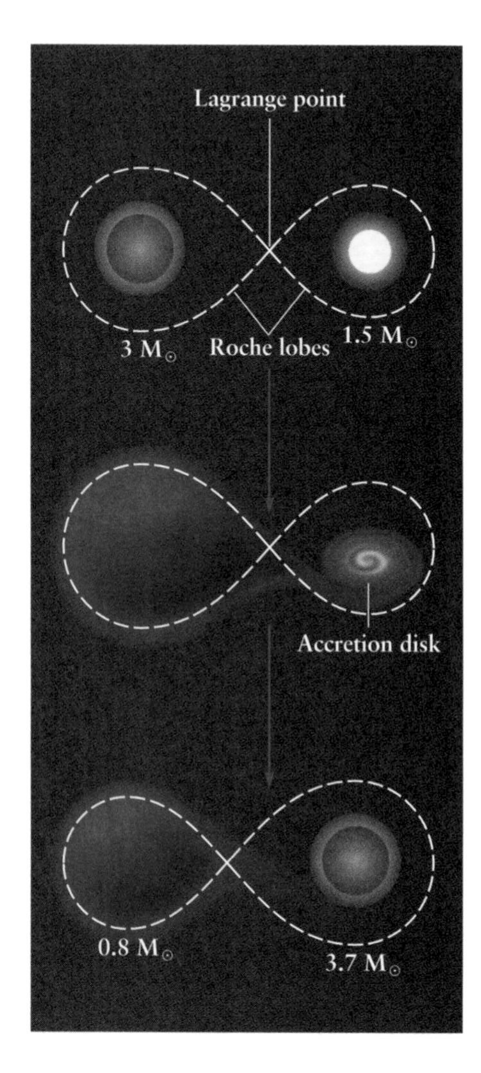


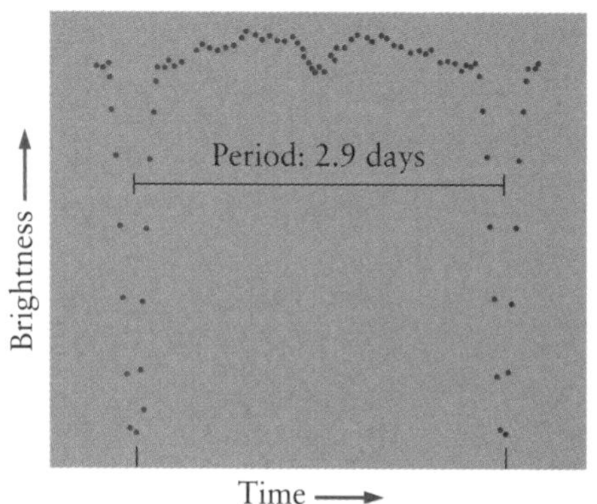


*The evolutionary history of the Algol binary explains the "Algol paradox." As the initially more massive star swelled into a red giant, it overflowed its Roche lobe and dumped so much matter onto its main sequence companion that the relative masses were reversed. According to stellar evolution theory, if a binary contains a red giant and a normal star, the red giant should be the more massive star. Because of mass transfer, this rule is violated in Algol—hence the "paradox." The light curve plotted shows the periodic eclipses, which tell us not only the orbital period but also the relative sizes and brightnesses of the two stars; spectra reveal Doppler shifts that tell us how fast the stars are orbiting about each other; with this information we can determine the masses. Astronomers can therefore learn a great deal about Algol (and similar systems) even though the stars are too close for optical telescopes to separate the two images.*

larger (but dimmer) red star. The brightness of the binary source drops by a factor of 3 and the apparent color changes. The orbital parameters are well enough known that the masses of both stars can be determined, leading to the famous "Algol paradox." It turns out that the blue star, which is still burning hydrogen in its core in its main sequence phase, is more massive than the red star, which is already on its way to becoming a red giant. But stellar evolution theory says that the more massive star should become a red giant first, if we assume (as must surely be the case) that both stars formed at the same time.

The resolution of the paradox is that the less massive star *used to be* the more massive star and did indeed reach the red giant phase first. As it swelled, however, it overflowed its own zone of gravitational influence (its teardrop-shaped *Roche lobe*) and spilled across the *Lagrange point* into its companion's zone of influence. The companion absorbed so much mass in this way that it became the more massive star.

If we follow this process to its logical conclusion in a sufficiently massive binary system, we can understand how an X-ray binary can form. The details were first worked out by the Dutch astrophysicist Edward van den Heuvel in the early 1970s, and they have been refined by many workers since. We start out with an Algol-type situation in which mass transfer has left a star heavier than 5 solar masses that has evolved beyond the main sequence and an even more massive star still on the main sequence. Although the evolved star has lost most of its mass, it proceeds steadily on its march toward gravitational Armageddon (since its rate of evolution is determined by the *core* of the star, which does not participate in the mass transfer), and within a few hundred thousand years it blows up as a supernova. Here is where a crucial aspect of the Algol-type configuration comes in. If the *more* massive star in a binary explodes, leaving behind a neutron star or black hole, the result is almost always to destroy the binary: the compact star and its former companion go flying off in different directions. But if the exploding star is the *less* massive member of the system, the

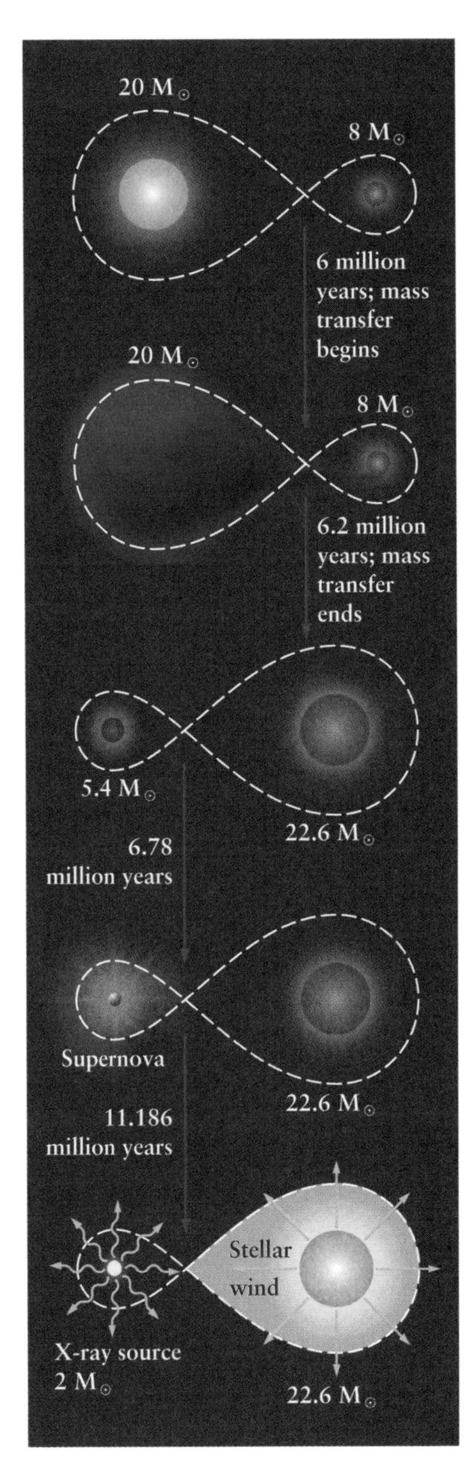


*Evolution of a close binary system into an X-ray binary. In this example, the two stars are initially 20 and 8 solar masses. The heavier star is the first to exhaust its hydrogen fuel. Its swells up into a giant and loses mass, most of which is dumped on its companion. Its remaining core, however, continues to evolve and explodes as a supernova. Because the exploding star is by this stage less massive than its companion, the sudden mass loss is not enough to disrupt the binary. The compact remnant stays in orbit around its companion—a (now much more massive) bright blue star with a strong stellar wind. Accretion of gas from this wind gives rises to a strong compact X-ray source. Cumulative times are shown.*

binary remains intact. The latter condition is automatically satisfied in an Algol-type binary. So we have a binary consisting of a neutron star (or black hole) and a very massive main sequence star. Now it is the latter that transfers mass onto its now-compact companion. This mass comes at first from a strong stellar wind; within a few million years the massive star evolves into a red giant and overflows its Roche lobe.

A second phase of mass transfer, onto the neutron star or black hole, thus follows the first phase, which set up the Algol-type configuration. This second phase of mass transfer is more dramatic than the first, for a very simple reason: the gravitational field near the surface of the neutron star (or the horizon of the black hole) is much stronger than the gravitational field near the surface of any ordinary star. When matter is accreted by a gravitating body, energy is released in what we can conceptualize as a two-stage process. (In reality, the two stages occur simultaneously.) First, gravity accelerates the gas and imparts kinetic energy to it. Since energy can neither be created nor destroyed, but can only change from one form to another (that is, energy is *conserved*), this kinetic energy must come from somewhere. Its source is the gravitational interaction between the gas and the compact star. We say that matter in the presence of a gravitational field possesses gravitational *potential* energy, which is converted to energy of motion through gravitational acceleration. In the second stage of the energy-release process, the kinetic energy is *dissipated,* being ultimately converted to heat or radiation or both. The exact way in which this energy conversion occurs can be complicated, and may involve collisions among turbulent cells of gas, interactions with a magnetic field, and other processes. However, the "bottom line"—the net amount of energy available to be radiated—is determined by the conservation of energy and is insensitive to all the messy details: it just depends on the amount of gravitational potential energy released as the material drops deeper and deeper into the gravitational field.

We can quantify the amount of energy released per gram of matter falling into a gravitational field by specifying the depth of the *gravitational potential well.* This quantity is often expressed in units of $mc^2$, the energy equivalent of the mass being dropped into the field. As a useful reference, thermonuclear burning of some amount of material would release at most 0.8 percent of $mc^2$. Expressed in units of $mc^2$, the gravitational field of the Sun is puny. The gravitational potential at the surface of the Sun is only a few times $10^{-6}$, meaning that you would get several thousand times more energy out of a gram of matter by placing it in the core of the Sun and burning it than by dropping it onto the surface. For neutron stars and black holes, the potential is about 0.1, or more than 10 times the peak efficiency of nuclear burning.

As sources of accretion power, white dwarfs are intermediate between ordinary stars and neutron stars or black holes. The gravitational potential of a white dwarf is about 100 times higher than that of the Sun, but is still considerably lower than the maximum energy released by nuclear burning. Nevertheless, binaries containing accreting white dwarfs produce spectacular effects, through both gravitational energy release and nuclear reactions. A white dwarf by itself will not support nuclear reactions, but such reactions can sometimes take place in a layer of "fresh," hydrogen-rich nuclear "fuel" deposited on the surface by accretion. Since white dwarfs are so faint to begin with, the release of gravitational potential energy alone can create an interesting observational show, characterized by intense and highly variable flares of ultraviolet radiation. Accreting white dwarfs in binaries are generally known as "cataclysmic variables." They include "dwarf novae," and "recurrent novae," as well as other variants, depending on the nature of the accretion and on whether the white dwarf has a strong magnetic field.

Generally, the gas settling onto a white dwarf is not hot enough to support steady thermonuclear burning, but as more material accumulates the gas near the bottom of the layer is compressed by the weight of the overlying gas. If the density becomes large enough, it can trigger *pycnonuclear* reactions (from the Greek *pyknos,* for "dense"), which, in contrast to *thermo*nuclear reactions, do not require a high temperature. The high density alone causes the atomic nuclei to develop high random speeds, according to the Pauli exclusion principle, just like the electrons holding up the interior of a white dwarf. These nuclei slam into each other with high enough speed to undergo nuclear reactions. This is a form of "cold fusion" that works, but only at densities thousands of times higher than those attained in the notorious laboratory experiments of a few years ago. Once these pycnonuclear reactions get under way, they rapidly heat the gas to temperatures high enough for thermonuclear reactions to take over, and the energy generation rate climbs steeply. The nuclear energy released easily overwhelms the gravitational binding energy of the accreted layer, and the layer is blown off the star in an explosive event called a *nova.* Unlike a supernova, which permanently alters or destroys the exploding star, a nova is only a surface event, and can repeat every time a new layer of gas is accreted.

What happens if a white dwarf accretes so much matter that it slightly exceeds the Chandrasekhar limit of 1.4 solar masses? It will start to implode, and if no nuclear reactions take place it will certainly collapse all the way to a neutron star. But if there is substantial nuclear fuel left to burn inside the white dwarf, then the collapse itself may trigger the onset of runaway nuclear burning. Provided that nuclear burning begins before the collapse has proceeded too far, the energy released by the nuclear reactions will be sufficient not only to reverse the collapse but also to completely disperse the star. The energy released in this thermonuclear explosion is very similar to the kinetic energy carried off by the remnants of the gravity-powered core-collapse supernova we discussed in the last chapter. In fact, the thermonuclear explosion is believed to lead to a "Type I" supernova, in contrast to the core-collapse kind, which is labeled Type II. But the total energy released is far greater in a core-collapse supernova. The reason the two types of supernova appear to be comparable is that 99 percent of the energy released in a core-collapse supernova is carried away by nearly undetectable neutrinos. Only 1 percent is released as kinetic energy, whereas a Type I supernova puts most of its energy into the expansion of its ejecta. So the gravity-powered explosion wins energetically, even if the competition appears to be a draw!

Neutron stars and black holes are the real gravitational powerhouses; their gravitational potentials of roughly 0.1 mean that infalling objects will release an amount of energy equivalent to 10 percent of their mass. There is no way that steady nuclear burning can compete with this rate of energy release, due entirely to gravitational infall. Occasionally the accumulated nuclear fuel on the surfaces of certain accreting neutron stars flares up, producing *X-ray bursts*. Unlike a nova explosion on a white dwarf, these bursts are not able to blow the surface layer off the neutron star, but they do dominate the star's luminosity for a brief time. Over the long haul, however, it is accretion power that reveals both neutron stars and black holes in close binary systems, so it is essential that we understand the accretion process as thoroughly as possible.

*Gas captured by a neutron star or black hole would generally not fall radially inward: because of its angular momentum, it forms a swirling disk or vortex. Arthur Rackham's illustration for Edgar Allan Poe's* A Descent into the Maelström *dramatizes the effect of a whirlpool.*

## Accretion Disks

The accretion of matter from a binary companion is a much more complex process than accretion from a uniform gaseous background. Since the stars in the binary are orbiting each other, any gas flowing from one star to the other does so with a lot of sideways motion or, more precisely, angular momentum. It cannot simply fall "straight in." Like water going down a drain (which generally has too much angular momentum to go straight in), the gas forms a kind of vortex centered around the accreting object.

There are some important differences between a vortex in a bathtub and the gas swirling around a black hole or neutron star. Gas is always much more compressible than water, and it becomes even more so as it cools down. The accreting gas in most X-ray binaries cools quite effectively, so instead of a three-dimensional vortex the flow gets flattened

*The gas closer to the center is orbiting faster than the gas farther out. Frictional effects tend to transfer energy (and angular momentum) outward, allowing the gas to drift slowly inward. The nature of the friction is uncertain: it could arise from turbulent eddy motions; magnetic fields may also play a role. The friction heats the disk, and the inner parts radiate more intensely than the outer parts. This radiant energy comes ultimately from the energy released as gas swirls deeper into the gravitational potential well of the central object.*

down to a spinning platter-shaped configuration, called an *accretion disk*. (There are some conditions under which an accretion flow may not flatten into a thin disk, especially flows into supermassive black holes in active galactic nuclei, which we will discuss in Chapter 5. In X-ray binaries this is probably more the exception than the rule.) The platter in this case does not rotate rigidly, however, in contrast to, say, a spinning phonograph record or compact disc. Instead, the inner parts are swirling around more rapidly than the outer parts. It is this *differential rotation* that allows the star or black hole to absorb the gas, since the rapidly rotating inner parts drag against the slower outer parts, lose some angular momentum, and spiral slowly inward. Basically the same effect allows water to go down the bathtub drain rather than remain in the vortex indefinitely.

For inflow to occur, there must be some kind of *friction* between the more rapidly spinning parts of the disk and the slower parts. In the bathtub, we know what causes this friction: the viscosity of the water, arising from interactions among water molecules. But under astrophysical conditions, in an accretion disk or almost anywhere else, this kind of "molecular" viscosity is much too small to be important. Uncertainty about the nature and magnitude of the "viscosity" has given accretion disk theorists a chronic headache from the earliest days of the subject. One possibility is that random motions and collisions of turbulent cells of gas (like the turbulent "air pockets" that jostle an airplane) give rise to a kind of viscosity. Another is that irregular, rapidly changing magnetic fields have a similar effect. Fortunately, many of the observable predictions of accretion disk theories, such as the spectrum of the radiation and characteristic timescales of variability, do not depend on the exact prescription for the viscosity. A great deal of progress has been made by expressing the strength of the viscosity in terms of parameters "to be determined."

The strong magnetic field of a neutron star (and some white dwarfs) introduces further complications into the basic disk theory. If the field is strong enough, it will disrupt the disk far from the stellar surface, deflecting the accretion flow toward the magnetic polar caps. If the neutron star is spinning, the result can be the pulsations mentioned earlier, in sources like Her X-1 and Cen X-3. But if the neutron star is spinning fast enough, the magnetic field can inhibit accretion altogether, by flinging the gas away from the star.

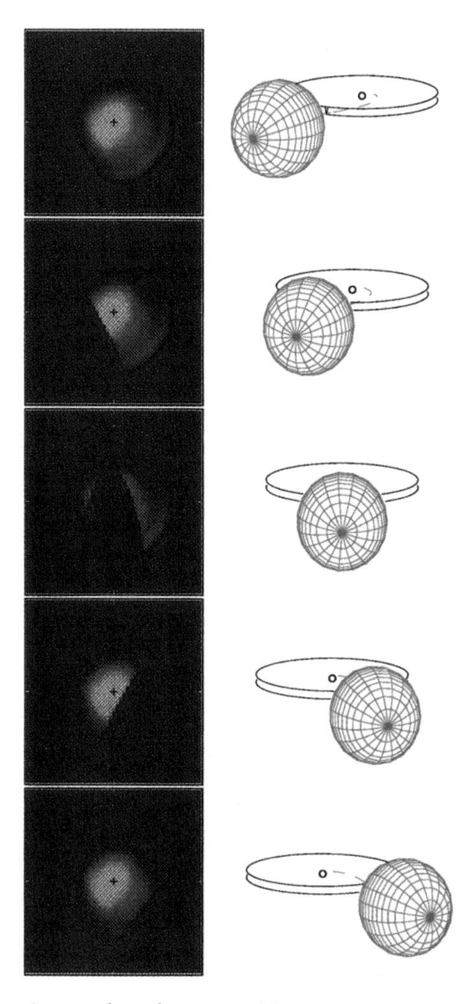

*Some close binaries, like UU Aquarii, are so oriented that the normal star passes in front of the accretion disk once per orbit. By observing the changes in brightness and color as the eclipse proceeds, the appearance of the disk can be reconstructed, even though it is too small to resolve with any telescope. Keith Horne has pioneered this technique.*

By observing cataclysmic variables, astronomers have been able to verify some key theoretical predictions for accretion disks around white dwarfs. A by-product of frictional drag is the heating of the gas in the disk, much as a "rope burn" may result from the friction of a rope sliding through your hands. According to theory, some heating should occur at all radii in the disk, and not just where the gas hits the object at the center. The entire disk should glow, with a temperature that decreases with increasing radius. The accretion disks around white dwarfs are far too small to image directly, but Keith Horne has produced images of accretion disks that show the expected thermal structure using an ingenious technique akin to medical tomography (which relies on a series of two-dimensional "slices" to reconstruct a three-dimensional image of the internal organs).

All told, about half the gravitational potential energy released by accretion will be radiated away throughout the disk. What happens to the other half depends on the nature of the compact object at the center. You may have been struck by how often the words "neutron star" and "black hole" have appeared together in the preceding pages. In many ways it is hard to distinguish an accreting black hole from an accreting neutron star just by observing its X-ray emission. But there is one important difference in the way they absorb matter that is crucial for the detection of stellar-mass black holes in X-ray binaries. In contrast to a white dwarf or neutron star, a black hole has no "hard" surface. By the very definition of the black hole horizon, no energy is radiated as gas crosses into the hole. In fact, angular momentum can no longer keep gas swirling around a black hole when the gas is within three times the radius of the horizon (for a nonrotating hole; the corresponding distance is somewhat smaller if the hole is rotating). Any gas that spirals down to the radius of the "innermost stable orbit" can drop into the hole without further need of friction. Thus, an accreting black hole swallows whatever fraction of the gravitational potential energy is not radiated by the disk. If gravitational energy were not radiated from all radii in an accretion disk, but just emerged from the center, then accreting black holes would be undetectable. In a white dwarf or neutron star, the energy left over at the center makes a "splash" as the gas hits

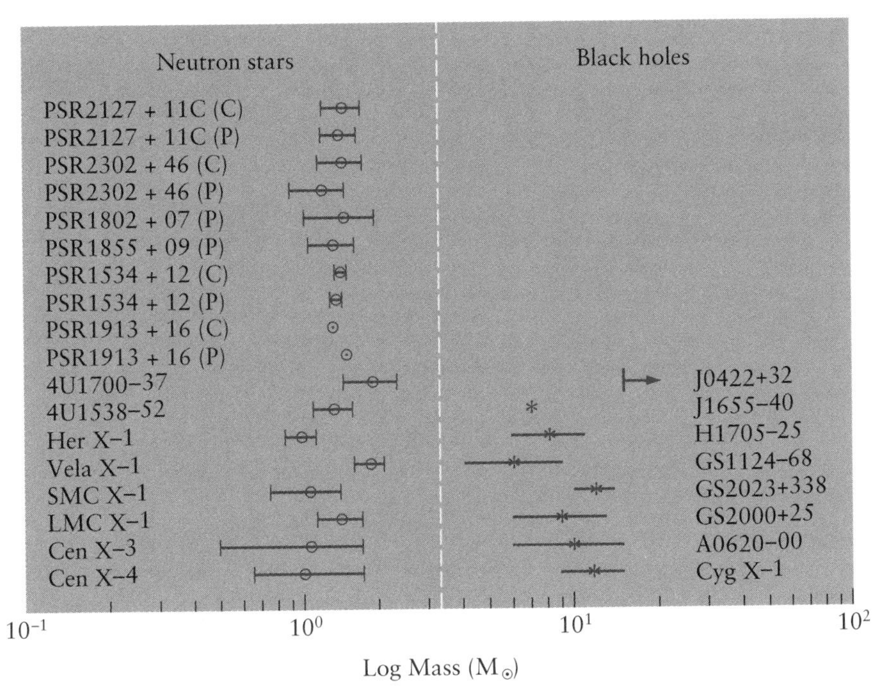


*This diagram shows the masses of compact objects which are believed to be either neutron stars or black holes. The best-determined are those of the binary pulsar PSR 1913 + 16 (P) and its companion: these are pinned down as 1.441 and 1.387 solar masses, respectively. The masses of several other pulsars (P) and their compact companions (C), and of some pulsing X-ray sources, are not quite so precisely determined (horizontal error bars), but all lie in the range consistent with their being spinning neutron stars (to the left of the dashed vertical line). In contrast, the black hole candidates have much higher masses, with error bars that do not extend down into the neutron star range.*

the stellar surface. We would be able to observe this "boundary layer" even if the accretion disk itself were invisible. But in the case of a black hole, the disk is all that we can see. It is an enduring hope of astronomers that we will learn how to distinguish the radiation from the boundary layer from that of the disk on purely observational grounds. But so far the only surefire signature of a black hole in an X-ray binary is its mass.

## The Mystery of Gamma-Ray Bursts

We do not want to leave the impression that the story of energetic phenomena from neutron stars and stellar black hole remnants is nearly all written save for a few details. One need not look far to find huge gaps in our understanding. In Chapter 6, we shall meet some peculiar X-ray binaries that occasionally produce enormously fast (and we mean fast—as much as 90 percent of the speed of light!) jets of gas, for no apparent reason. One object, SS 433, even seems to behave as a cosmic lawn sprinkler, rotating the direction of its jet around a broad cone once every 164 days. The mechanisms of these objects are understood poorly, if at all. But nearly all "high-energy" astrophysicists would agree that the award for "mystery of the decade" must go to the phenomenon of *gamma-ray bursts.*

The story of gamma-ray bursts dates back to the 1960s. At that time the American *Vela* satellites, whose prime purpose was to detect clandestine tests of nuclear weapons in space, started to record flashes of gamma rays. Gamma rays are the form of electromagnetic radiation with the shortest wavelengths and, correspondingly, the highest energies—higher than X-rays. It took several years before the Los Alamos scientists became convinced that these gamma-ray bursts were natural rather than sinister in their origins, and weren't instrumental artifacts; not until 1973 was the discovery announced in *The Astrophysical Journal.* Typical bursts lasted for several seconds, though there was often flickering on much shorter timescales.

Not surprisingly, this discovery quickly stimulated a variety of theoretical speculations. Less than two years later, at the 1974 Texas Symposium on Relativistic Astrophysics, the American physicist Malvin Ruderman reviewed a long catalogue of exotic ideas that had already appeared in the literature. He noted a tendency, still apparent in later work, for theorists to "twist rather strenuously to convince ourselves and others that observations of new phenomena fit into our chosen specialties." An updated list of gamma-ray–burst theories, published in 1993, identified no fewer than 140 distinguishable models!

During the late 1970s and 1980s, satellites accumulated further data on bursts. The Pioneer Venus Orbiter (PVO) detected about 200 strong bursts over a thirteen-year period. The intensity of the radiation from a typical burst—the amount of energy per square centimeter per second collected at the Earth—exceeds (albeit for only a few seconds) that of any known celestial object beyond the Solar System. There is no standardized spectrum or time profile of a burst, although "classical" gamma-ray bursts can be clearly distinguished from X-ray bursts, which are known to come from nuclear explosions on neutron stars accreting from binary companions, and from a rare class of objects known as "soft gamma-ray repeaters."

Between 1973 and 1991, most astrophysicists came to favor the view that gamma-ray bursts originated at neutron stars in our own Galaxy. The bursts were envisaged, for example, as random magnetic flares on the surfaces of old neutron stars whose pulsar activity had died out. These flares could be triggered by "starquakes" in the stars' rigid crust. Or perhaps neutron stars are surrounded by planetary systems, and bursts signal the destruction of an asteroid or comet converting 10 percent of its mass-energy into radiation as it crashes into the star. However, not all theorists "bought" the old-pulsar hypotheses. The problem was that the distances to gamma-ray bursts were completely unknown, and positional measurements were too crude to tell whether the bursts coincided with known objects such as galaxies or star clusters. Some suggested that the bursts might originate as close by as the comet cloud on the outskirts of our own Solar System. Others supposed them to be at the edge of the Universe, perhaps the discharges of "superconducting strings" or other exotica.

Even without directional information, one can learn something about how the bursts are distributed in space by comparing the relative numbers of strong bursts and weak ones. If the bursts originate from uniformly distributed sources, then it is easy to calculate how the number of detected bursts would go up as the sensitivity of the detector improves. However, observations have shown that the sources are clearly *not* uniformly distributed in space. Weak bursts are rarer, relative to intense bursts, than would be expected if the bursts came from a uniformly distributed population. It seems that the sources of bursts thin out beyond a certain distance. Unfortunately, this test alone does not tell us what that distance is. Could it be that the bursts do come from Galactic neutron stars, and that we see the population declining with increasing distance from the Milky Way?

If we could see gamma-ray bursts from neutron stars all the way out to the edge of the Milky Way, then their distribution on the sky ought to reflect the asymmetric shape of the Milky Way as viewed from our vantage point—the fainter bursts, at least, should be concentrated in a band, just like the stars in the Milky Way. This was nearly everybody's expectation when, in 1991, NASA launched the *Compton Gamma Ray Observatory,* which carried an instrument for monitoring bursts. To nearly universal surprise, the bursts seemed to come from random directions over the entire sky. More than 2000 have now been detected: there is no tendency for them to be concentrated toward the center of our Galaxy, nor toward the plane of the Milky Way.

This observation of *isotropy,* coupled with the relative brightness distribution, made it exceedingly difficult to explain the gamma-ray bursts as originating from *any* kind of object located in our Galaxy. Even a population of objects located in a halo around the Solar System would show detectable asymmetry, because the Earth is not located at the center of the Solar System. The same would be true for a hypothetical population of objects located in an extended halo around the Milky Way. Even if this halo were perfectly spherical, we would detect an asymmetry due to the Sun's 25,000-light-year offset from the Galactic center.

The only known system that shows the required level of isotropy is the Universe itself—the distant galaxies are distributed around us with remarkable uniformity. This does not mean that we are located at some special "center" of the Universe. In fact, one of the strongest tenets of astronomers is that there is *nothing special* about our location. (This is an extension of Copernicus's assertion that the Earth does not lie at the center of the Solar System, and is therefore often referred to as the Copernican principle.) Theoretical models for the expanding Universe, based on general relativity, have the property that the distribution of distant galaxies should appear isotropic when viewed from *any* location. Thus, the isotropy of gamma-ray bursts had, by 1996, led most astrophysicists to suspect, contrary to the pre-1991 majority view, that the bursts were a remote extragalactic phenomenon. The uniform distribution is then no problem, and if their distances are large enough—approaching the observable "Hubble radius" of the Universe—the redshift due to the expansion of the Universe would tend to diminish the number of fainter (and more distant) bursts detected (see the box on pages 69–70). Our inability to detect the fainter bursts would make it appear that we were seeing out to the "edge" of the distribution of burst sources.

*Edwin Hubble, shown here at the prime focus of the 200-inch telescope on Mount Palomar, discovered that the light from each distant galaxy arrives with a redshift that is proportional to the galaxy's distance from us—the first evidence that we live in an expanding Universe.*

But the case for "cosmological distances" was still circumstantial. No redshift could be measured from the gamma-rays themselves. The only hope was to detect the bursts in a different wavelength band, such as optical or radio, for which accurate positional or distance measurements are feasible. Unfortunately, the bursts detected by the *Compton Observatory* were only pinned down to within a circle several degrees across—a patch of sky bigger than the field of view of most optical telescopes. A network of interplanetary probes, including spacecraft like the PVO, were able to do somewhat better for especially bright bursts—about an arc minute—by using a technique known as triangulation. This is still not good enough to fix whether (for instance) a burst originates in a particular galaxy, but it would suffice to tell optical astronomers where to point their telescopes in order to search for a visible "afterglow." However, triangulation was a laborious process; by the time a position had been calculated the burst had long faded.

The breakthrough came in 1997, after the launch of an Italian/Dutch satellite known as *Beppo-SAX* (the prefix was added to the original acronym in honor of the eminent Italian physicist Giuseppe [Beppo} Occhialini). This instrument was less sensitive than the *Compton Observatory,* and only detected about one burst per month. But *Beppo-SAX* had a crucial advantage: it combined arc-minute positional accuracy with rapid response time. Positions of gamma-ray bursts, unlike those obtained by the triangulation method, could be made available within a few hours—a specially important advantage if the afterglow is shortlived. The first afterglow was discovered on 28 February, 1997 and faded over the next few months. A second one, detected on May 8, nailed down the distance scale. Astronomers were able to take an optical spectrum with the Keck 10m telescope in Hawaii. This spectrum displayed strong absorption features caused by gas with a redshift of 0.835. The burst was, therefore, either in

## The Expanding Universe

The expansion of the Universe has been known since the 1930s, when Edwin Hubble realized that distant galaxies are receding from us at speeds proportional to their distances. His discovery proved so momentous that it has received the designation Hubble's law.

Although we see galaxies moving away from us no matter which direction we look, this doesn't imply that we are in a special central position. To see why, consider the lattice in M. C. Escher's *Cubic Space Division,* below. Imagine that the rods in this lattice all lengthened at the same rate; then the vertices would all move away from one another, and an observer on any one of them would see the others receding. Moreover, every time a neighboring vertex receded by 1 unit of length, successively more distant vertices would recede by 2, 3, 4, . . . units of length. In other words, the vertices recede in accordance with Hubble's law. This is a good "broad brush" analogy

*M. C. Escher's infinite lattice has no center. If the rods all lengthened at the same rate, an observer at any vertex would see the others receding according to Hubble's law.*

*Because of light's finite speed, we do not actually see a uniform lattice: distant parts of the Universe are seen as they were in the remote past. When telescopes probe regions at high redshifts (near the observational "horizon"—equivalent to the regions near the boundary of Escher's* Angels and Devils), *they reveal a phase of cosmic history when everything was packed more closely together.*

for the expanding Universe, except that the galaxies are not laid out in a regular lattice. Actual galaxies are grouped into clusters and superclusters, and consequently they move locally under the gravitational influence of other cluster members. The galaxies within a cluster are not receding from one another: Hubble's law shows up clearly only on scales larger than clusters.

The light from very remote objects set out when the Universe was at an earlier stage in its expansion and galaxies were more closely packed together. Because of light's finite speed, what we actually observe is better represented by another Escher picture, above: the scale of the pattern (i.e., the separations of the galaxies) is compressed as we look farther away. The wavelengths of light reaching us from an earlier era have stretched by a factor equal to that by which the Universe has expanded since the light was emitted. Since light with a longer wavelength is "redder," the radiation we receive from distant galaxies is said to be "redshifted." The amount of the redshift can be associated with the speed of recession, and we can then use the calculated recession speed to derive the distance to a remote galaxy, using Hubble's law.

Astronomers normally quote redshifts in terms of the number $z$, defined such that the received wavelength is $(1 + z)$ times the emitted wavelength. For example, the redshift of the quasar 3C 273 is 0.15, meaning that the Lyman alpha line of hydrogen, which was emitted with a wavelength of 1216 Å, reaches our telescope with a wavelength of 1398 Å, or a factor of 1.15 times longer. For the (relatively nearby) galaxies that Hubble himself observed, $z$ was very much less than 1. However, some especially luminous galaxies—the quasars and radio galaxies that we will meet later—are now routinely observed with redshifts of up to 5. The number $(1 + z)$ tells us how much the Universe has expanded while the light has been on its journey. The light from an object with $z = 3$, for instance, set out when the Universe was compressed by a factor $(1 + 3) = 4$, and was denser by a factor $4^3 = 64$.

If the Universe expanded at a uniform speed, then when it was four times more compressed it would have been four times younger. However, the mutual gravitational pull of everything in the Universe decelerates the expansion. Since the early stages of expansion were faster, the Universe at $z = 3$ was less than $\frac{1}{4}$ its present age. In the simple and philosophically attractive cosmological model in which the Universe contains just enough matter to bring its expansion gradually to a halt (but not enough to cause it to recollapse), the scale goes as the two-thirds power of time, so that when the Universe was $\frac{1}{4}$ its present scale it would have been only $\frac{1}{8}$ its present age.

To determine actual ages, rather than just relative ages, we need to know the so-called Hubble constant. When an object's redshift is small (and $z$ much less than one), its distance is just $z$ divided by the Hubble constant. To determine the Hubble constant, one needs to make accurate measurements of the distances to a sample of galaxies, using measurements *other than their redshifts*. Obtaining these measurements is a difficult and still uncertain procedure, but we are confident that we now know the Hubble constant to within a factor of two. Thus, we can state that 3C 273, with its redshift of 0.15, lies about 2 billion light-years from us.

The relationship between distance and redshift is more complicated for high-redshift objects. The light now reaching us from high-$z$ objects set out when the Universe was substantially more compressed and not necessarily expanding at the same rate as it is now. Without also knowing the cosmic deceleration (which depends on the total amount of dark matter in the Universe), we cannot assign an unambiguous distance to a quasar or galaxy with high redshift. (We therefore avoid quoting such distances in this book.) Cosmologists often, however, talk loosely of the "Hubble radius." This is the distance we would calculate for an object with $z = 1$ if the simple proportionality between redshift and distance applied out to such a large redshift. This distance, divided by the speed of light, gives an equally rough "age" for the Universe—about 10 to 20 billion years.

a galaxy with that redshift, or was even further away, so that the light from its afterglow passed through this galaxy on its way to us.

Very little doubt remains that gamma-ray bursts come from great distances—after more than 25 years of debate, this basic issue has been settled. But what are they? Few phenomena pose such severe challenges to theorists. At such immense distances the typical luminosity during a burst would have to approach $10^{44}$ watts, about $10^{18}$ times the luminosity of the Sun. How can a gamma-ray source be switched on in only a few milliseconds, emit power for a few seconds at 10 thousand times the rate of the most powerful steady emitters of energy—quasars—and then fade away? The most likely possibility involves a pair of neutron stars (or maybe a black hole and a neutron star) in a close binary orbit. The two stars spiral together, slowly at first and then faster and faster, until they finally merge. The final coalescence, perhaps lasting only a millisecond, would certainly trigger a large enough release of energy. Such events are rare—they would occur only once every 100 thousand years in a typical galaxy. But there are at least a billion galaxies within the Hubble radius, so the rate at which bursts are detected (about one a day) poses no real problem.

Coalescence of a compact binary releases more energy than the collapse of a stellar core that triggers a supernova. But the main difference between a gamma-ray burst and a supernova—and the reason why the effects are far more spectacular—is that the released energy escapes within a few seconds. In a supernova, the "trigger" in the stellar core is surrounded by the rest of the star: the energy percolates out slowly (taking several hours) and is degraded into much "softer" radiation. The actual mechanism for producing the gamma-rays is controversial. The emission is probably boosted and speeded up by the effects of relativity; it may even be directed along jets rather than being emitted in all directions. Theorists still cannot explain the wide variety of burst profiles.

When a supernova explodes, the expanding remnant sweeps up external matter in a hot shell. The debris from a gamma-ray burst would likewise explode into the surrounding medium. When two neutron stars merge, much less mass is thrown off than in a supernova. The ejecta therefore move much faster—indeed they create a blast wave that initially moves at nearly the speed of light, but slows down as it runs into external material. The resultant afterglow is in itself a remarkable phenomenon, shining 100 times more brightly than a supernova for several months.

Thus the mystery of gamma-ray bursts may signal yet another context in which neutron stars and black hole stellar remnants reveal their presence with displays of cosmic fireworks. The bursts may represent the coalescence of at least a subset of the kinds of X-ray binaries we have been studying closer to home; otherwise they are something even more exotic.

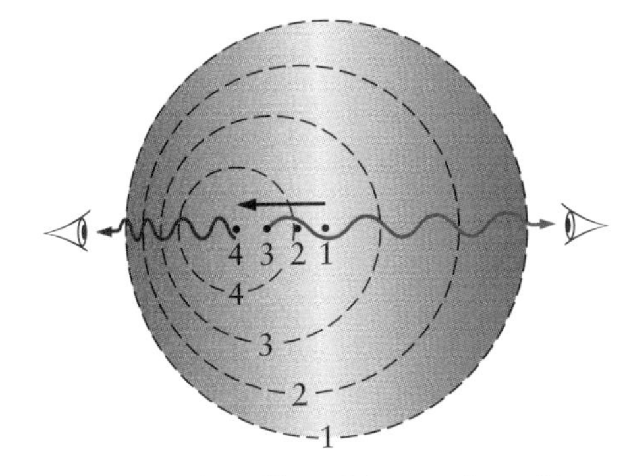


*The Doppler effect. If a moving object (like the source depicted here, moving from position 1 to position 4) emits a succession of waves or pulses of light, the waves (represented here by circles numbered 1 to 4) will be more closely bunched together (and will arrive at higher frequency) toward the direction of motion; they will be more spread out (and have lower frequency) in the opposite direction. A source of light appears bluer (and brighter) if it is approaching and redder (and fainter) if it is receding.*

## Stellar-Mass Black Holes: How Strong a Case?

If the credit for discovering neutron stars must be shared between the radio and X-ray astronomers, it is the X-ray astronomers who have led in the search for stellar-mass black holes in our Galactic backyard. It had already been recognized by several theorists in the late 1960s that black holes in close binary systems should be accreting material from their companions and should manifest themselves as X-ray sources. Moreover, such X-rays, though not varying periodically, could display irregular flickering on timescales as short as a millisecond (this being the characteristic period of the innermost orbits around a hole of a few solar masses, where most of the energy is released).

But flickering alone would not be a guarantee that the compact object was a black hole rather than a neutron star. As Zel'dovich pointed out, the clincher would be if the mass of the compact object exceeded the highest possible mass of a neutron star. Such mass estimates can be obtained by measuring the orbital period and speed of the normal star (using the Doppler effect). In this way Cygnus X-1, an X-ray source in a binary system, was inferred to be at least 6 times heavier than the Sun thereby becoming the first (and still perhaps the firmest) example of a black hole candidate. The X-rays displayed no periodicity, but flickered in a manner consistent with an irregular swirling of material toward a compact body with no hard surface. Unfortunately, the minimum time resolution of *Uhuru*'s detectors was no better than 0.1 second, which is much longer than the predicted minimum timescale for flickering. (This limitation was an economy measure reluctantly forced on Giacconi and his colleagues.) Observations at a higher time resolution were subsequently made by a Naval Research Laboratory rocket flight, revealing fluctuations on timescales as short as milliseconds.

Cygnus X-1 is still a prime candidate for a black hole, along with five or six other strong binary X-ray sources. The masses of these other compact objects are also too high for them to be neutron stars, and their X-ray spectra and characteristic variability are similar to those of Cygnus X-1. The black hole binaries seem to be of two distinct types. In some, as in Cygnus X-1 itself, the companion star is more massive than the hole. But there are others (with, presumably, a different evolutionary history) in which the companion is less massive. Over the last fifteen years, researchers playing "devil's advocate" have proposed many alternative models to the black hole hypothesis, but these models are generally too contrived to be convincing. As Edwin Salpeter puts it, "A black hole in Cygnus X-1 is the most conservative hypothesis."

Half a dozen convincing black hole candidates among X-ray binaries is far from a disappointingly small harvest. There are on the order of 100 million neutron stars in our Galaxy, yet maybe only one in a million formed (and managed to remain) in a close binary and is currently accreting mass from its companion. A conservative guess would be that at least 1 percent of all massive stars are heavy enough to collapse to black holes. Thus, we might expect at least 1 percent of X-ray binaries to contain black hole stellar remnants, with the vast majority of these remnants moving solo through interstellar space, unseen.

Nevertheless, it is still impossible to construct a completely watertight case, since ad hoc and contrived models that do not involve a black hole can always be devised. One's assessment of the odds obviously depends on whether one regards black holes as inherently absurd or as plausible end points for stellar evolution. The X-ray data are fully consistent with the presence of black holes having the properties that are predicted by general relativity.

However, the observations have not yet offered any signature that allows us to test whether the geometry of spacetime around these objects (which determines the flow pattern of infalling gas, etc.) is precisely that of a Schwarzschild or Kerr black hole, as Einstein's theory would predict. In principle, the spacetime geometry could be probed if the observational data were of high enough quality, and if the relevant gas dynamics and radiation processes could be modeled with sufficient accuracy, but such testing is not yet realistic in practice. We know only that in some relatively nearby binary systems, there is an object that has roughly the gravitational potential of a black hole and that is too massive to be a neutron star.

The black holes we have discussed so far—and the ones for whose existence the astrophysical arguments are most straightforward—are those with masses of a few suns and radii of approximately 10 kilometers. These represent the final evolutionary state of massive stars. But one might speculate about the possibility that larger black holes exist. In the next chapter we discuss the nuclei of galaxies, the likely setting of such holes. We shall see that conditions in galactic nuclei may be propitious for the accumulation of a very large amount of mass, which could perhaps collapse to form a black hole. Is there direct evidence that such holes exist? Much of the evidence is indirect, but persuasive. It suggests that the most powerful sources of energy in the Universe, quasars and "active" galactic nuclei, involve accretion onto massive black holes—scaled-up versions of what may be happening in Cygnus X-1. This indirect evidence is just now being bolstered by direct measurements of gravitational fields that can be associated plausibly only with supermassive black holes.

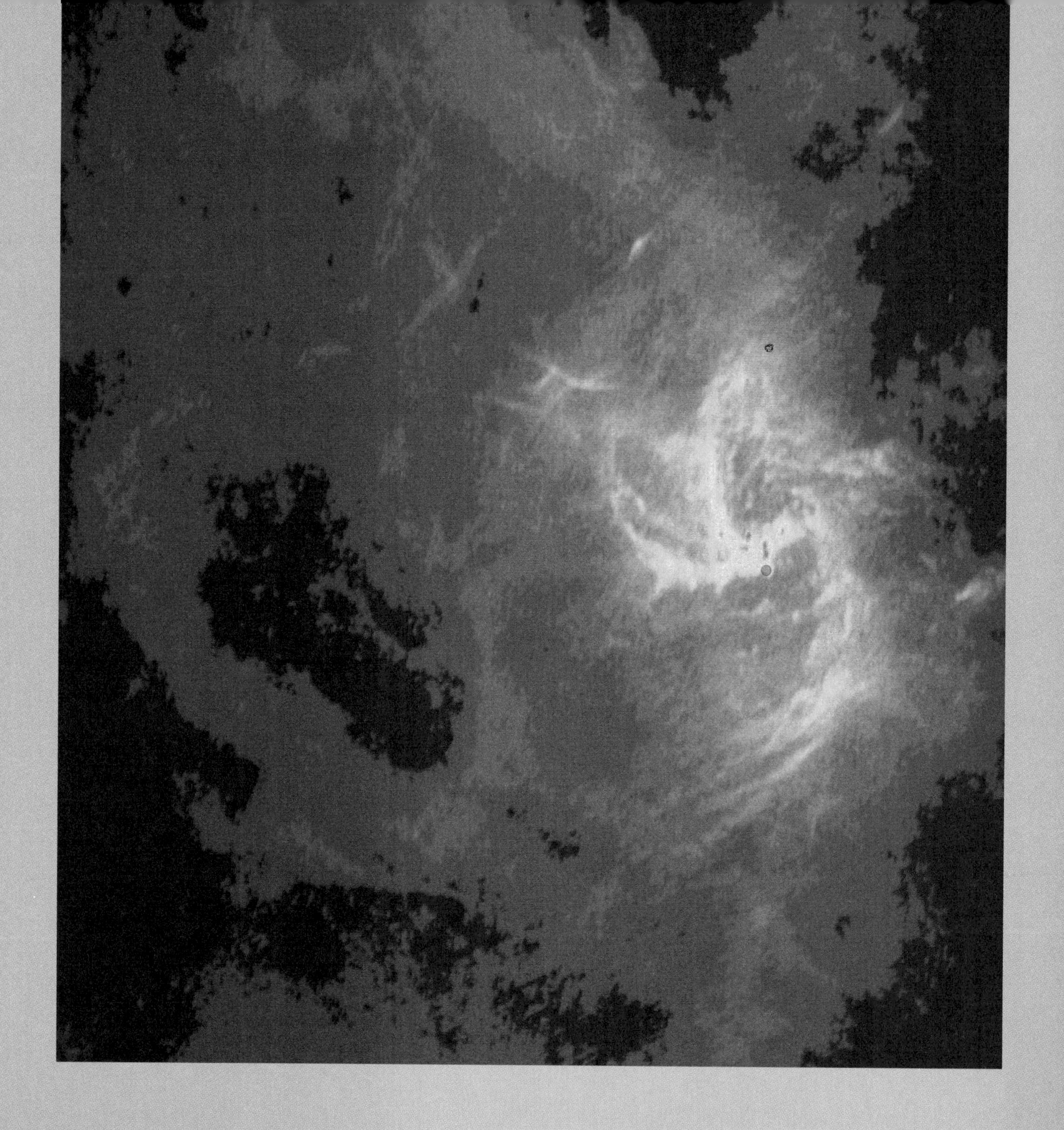

Chapter 4

# Galaxies and Their Nuclei

*The center of the Milky Way Galaxy, as viewed by the Very Large Array radio telescope. The intense point source of radio waves at the center of the spiral pattern, known as "Sagittarius A*," is suspected of being a million-solar-mass hole. The region shown is about 33 light-years on a side.*

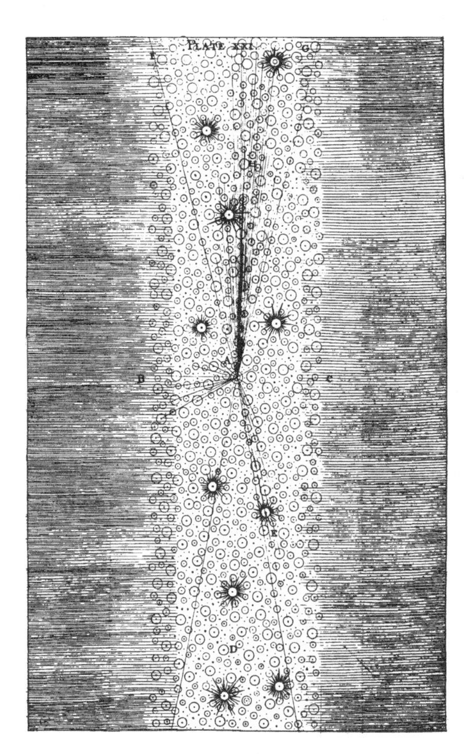

*Thomas Wright's Milky Way. From inside the slab of stars an observer would see starlight as a band on the sky.*

Over four hundred years have passed since Copernicus dethroned the Earth from the privileged central position accorded it by Ptolemy's cosmology and described the general layout of the Solar System as it is accepted today. But the eventual abandonment of Copernicus's heliocentric picture—in which the Sun is regarded as the central body in the Universe, not just the focus of the Solar System—and the full realization of the scale of the cosmos came about much more gradually and was completed only recently. In 1750, Thomas Wright of Durham, a self-educated teacher of astronomy, argued that the Milky Way, which appears to the naked eye as a diffuse band of light on the sky, could be explained as a thin slab or disk of stars in which the Solar System is embedded; William Herschel put this idea on a more precise footing in the 1780s. Wright's prescient ideas were popularized (in somewhat distorted form) through a pamphlet written by the philosopher Immanuel Kant, who added his own, equally remarkable conjecture that the Milky Way, our Galaxy, is just one fairly typical galaxy similar to the hundreds of millions of others that can be registered by a large telescope, and that galaxies are the basic units making up the large-scale Universe. Despite this startlingly modern view, the "spiral nebulae" photographed by late-nineteenth- and early-twentieth-century telescopes were not widely regarded as distant "island universes" until Edwin Hubble succeeded in establishing their enormous distances in the 1920s.

Our foray into the study of galaxies aims squarely at their centers—their *nuclei*—where we will find the conditions ripe for the formation of very massive black holes. Before focusing on the pyrotechnics engendered by these holes, we need to understand their environments and why the holes exist at all. To reach that point, we first describe how galaxies are put together and how they evolve with time.

## Normal Galaxies

Galaxies—giant assemblages of stars, gas, and dust—are supported against gravitational collapse through the motions of their stars. Beyond this general statement of gravitational equilibrium, no theory analogous to the well-developed theory of stellar structure exists to explain the gross properties of galaxies. Indeed, we do not know why such things as galaxies should exist at all, let alone constitute the most conspicuous large-scale features of the cosmos. Only through diligent observation have we learned that galaxies possess fairly standardized properties, from which we can infer their structures and something about their fates. One can classify them according to their appearance even if one does not understand how they came to be.

The most familiar types of galaxy are the disklike, or spiral, systems. On a photograph, such a galaxy seems to be composed essentially of stars and gas concentrated in a circular disk. A typical spiral galaxy contains about $10^{11}$ stars, and about 10 percent as much material in the form of gas, spread through a region about 50,000 light-years in radius. The stars and gas execute approximately circular orbits about the center of the disk, with centrifugal force balancing gravity. A typical star would take about 200 million years to orbit the center of the galaxy, at a speed of 200 kilometers per second; such a star would have completed 50 orbits since the time the typical galaxy formed. If our own Milky Way were viewed by a distant observer, it would appear disklike, the Sun lying about two-thirds of the way out toward the edge of the visible disk. The Andromeda Galaxy, our nearest major neighbor in space, is about 2 million light-years away. Like other spiral galaxies, it contains stars of all ages, as well as gas. The Andromeda Galaxy is similar in size to our own, but spiral galaxies in general have a spread of at least a factor of 10 in mass.

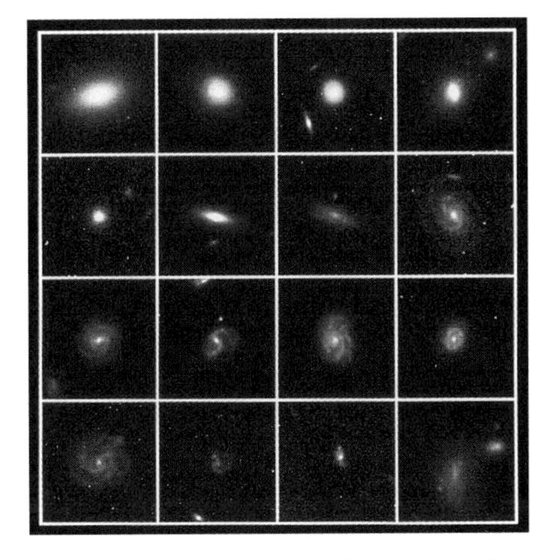

*Most large galaxies can be classified according to appearance, as either ellipticals or spirals. These deep exposures taken by the Hubble Space Telescope show that the mix of galaxy types was already well established when the Universe was only half its current age.*

In any galaxy classification scheme the main distinct class apart from the disk galaxies must be the "ellipticals." These are amorphous swarms containing between $10^8$ and $10^{12}$ stars, each star tracing out a complicated path under the influence of the overall gravitational field. The smoothed-out appearance of an elliptical galaxy indicates that the orbits have become thoroughly randomized. An equilibrium has been established such that the tendency of gravitation to pull the system together is balanced by random motions, which by themselves would cause the galaxy to fly apart. Even though a typical photograph does not resolve the individual stars, over virtually the whole galaxy these stars are so widely spaced relative to their actual physical sizes that there is no significant chance of collisions, or even of close encounters, during the entire 10-billion-year lifetime of such a system (the same is also true of disk systems). Only in the densest parts of galactic nuclei is there some risk of stellar collisions; we shall see later that such encounters may contribute to the formation of black holes. Elliptical galaxies, generally less photogenic than spirals, come in a variety of apparent shapes ranging from nearly circular to highly flattened. The variety of shapes that we observe can be attributed partly to the different orientations of these galaxies relative to our line of sight, but also partly to genuine differences in shape.

Astronomers have developed elaborate taxonomies for classifying galaxies according to their shapes. Gérard de Vaucouleurs—the world authority on galactic morphology—formulated a scheme with over a hundred categories, and even this classification failed to include all "peculiar" systems! Most galaxies can be crudely categorized as either "disk" or

*The Sombrero Galaxy, some 40 million light-years away, is classified as a spiral because of its disk of stars and gas. It sports a prominent central bulge of stars, which, if seen in isolation, would resemble an elliptical galaxy. The shadowlike feature outlining the disk is obscuration caused by a layer of dusty gas.*

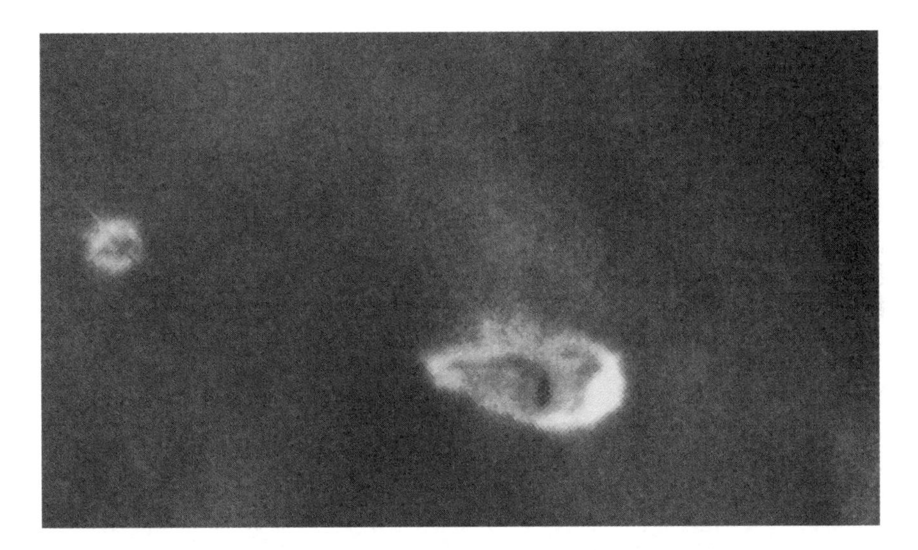

*The blue linear feature, centered in its comet-shaped trail, is a "proto-planetary disk," or "proplyd" for short. This structure, located in the Orion Nebula , is believed to be an edge-on disk of gas in which a star and its retinue of planets are forming before our eyes.*

"elliptical," but even a disk galaxy often contains an elliptical-like "bulge" in its center, while some ellipticals show vestigial disks. A slightly more sophisticated classification is based on the relative prominence of disk and elliptical components. The Sombrero Galaxy displays an average combination of disk and bulge components; most other disk galaxies are basically similar except that the bulge is either more or less prominent relative to the disk.

Disk galaxies contain cool gas and young stars, but elliptical galaxies contain neither. These seemingly incidental facts yield powerful clues to the nature of these systems. Stars are fueled by nuclear reactions in their interiors, and spend most of their time on the so-called main sequence, drawing their energy from the conversion of hydrogen into helium. A star like the Sun can remain in this phase for about 10 billion years. On the other hand, more massive stars (which can be identified observationally by their bluer colors, indicating a hotter surface layer) burn up their fuel at a much greater rate. They consequently have shorter lifetimes and eventually explode in violent supernova explosions, as we discussed in Chapter 2. In places such as the Orion gas cloud, we can see the actual process whereby gas is being converted into stars; and we can also see large numbers of blue stars that must have formed very recently in the history of the galaxy. When these stars die, they will eject much of their material back into the interstellar gas, from which new generations of stars can subsequently condense. The cool gas and young stars are different stages in a recycling

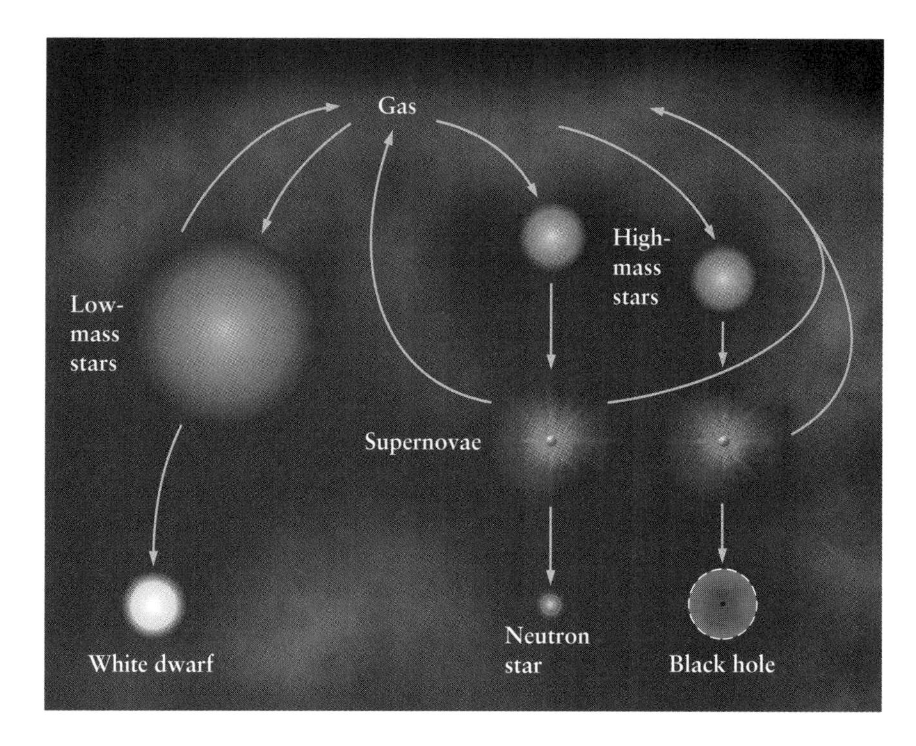


*Gas incorporated into stars is recycled when the stars explode as supernovae or emit winds during the red giant phase of their evolution. Gas released during the red giant phase can be enriched in elements such as carbon, oxygen, and nitrogen; that released during a supernova explosion contains, in addition, a variety of heavier chemical elements. The matter remaining behind in white dwarfs, neutron stars, and black holes is taken out of circulation.*

process, and it should therefore not surprise us that the kind of galaxy that lacks gas clouds also lacks young stars.

Galaxies are not only dynamical units but chemical units as well. The atoms we are made of come from all over our Milky Way Galaxy, but few come from other galaxies. All the heavy elements present in the Solar System—every atom of carbon, nitrogen, oxygen, and iron—must have been produced in a generation of stars that formed, evolved, and died before the Sun formed.

In their classic book on galactic dynamics, James Binney and Scott Tremaine assert that galaxies are to astronomy what ecosystems are to biology. The process of star formation, evolution, death, and rebirth going on in our Galaxy is, in a sense, an ecological cycle in which diffuse gas and stars share a symbiotic relationship. The ecological analogy reflects other features of galaxies: their complexity, ongoing evolution, and relative isolation. The cycling of matter between stars and gas is not perfect, however. Part of the gas that turns into stars becomes permanently trapped, in the sense that it becomes incorporated into long-lived (low-mass) stars, white dwarfs, neutron stars, or black holes.

One can now understand why elliptical galaxies have no young stars, whereas those galaxies where star formation is still going on should be disk-shaped. Let us suppose that all galaxies started their lives as turbulent gas clouds contracting under their own gravitation. The collapse of such a cloud is highly dissipative in the sense that colliding globules of gas produce shock waves that radiate away some of their energy of motion. Having lost energy that could have made them bounce apart from each other, the gas globules are able to merge—in other words, they "stick together." If the cloud has some weak rotation to begin with (as will certainly be the case), it will settle into a rotating disk—but only if the collapse proceeds without all the gas first condensing into stars. The only stars that will join in the disk configuration are those that form after the gas has settled into a disk. Those that form before will have random orbits characteristic of an elliptical configuration.

The foregoing scheme suggests that the thoroughness with which gas is turned into stars is crucial in determining the kind of galaxy that results. In elliptical galaxies, almost all the stars formed billions of years ago. In disk galaxies, on the other hand, gas is being converted into stars at the present time, either because the conversion rate is low or because gas from outside the galaxy is still raining down, pulled by the galaxy's gravity. Disk galaxies can therefore be viewed as those with slower metabolisms; they have not yet approached the final state in which essentially all the gas is tied up in low-mass stars or dead remnants. "Irregular galaxies," such as the nearby Magellanic Clouds, are even more extreme cases of arrested development. These are systems in which maybe less than half the gas has so far been incorporated into stars.

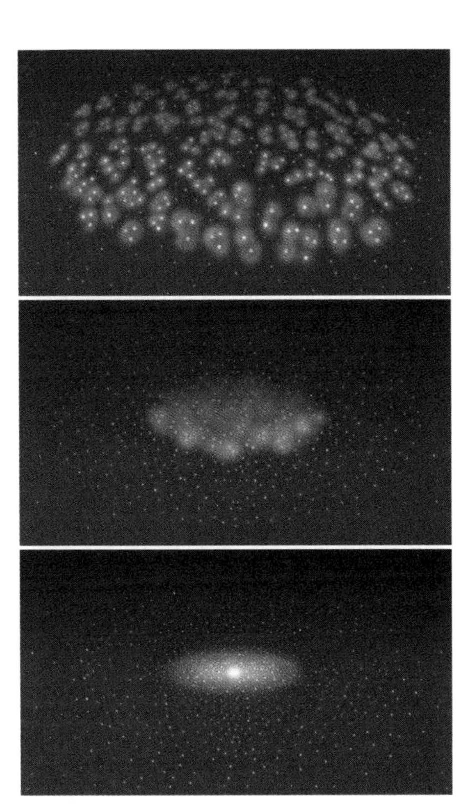

*A schematic illustration of galaxy formation. Stars that form early in the collapse of the gas cloud are destined to travel on random orbits. If the cloud is rotating, gas that does not quickly condense into stars will settle into a disk. Stars that form later, within this disk, will share in a highly organized rotation. The final galaxy will resemble an elliptical or spiral, depending on the relative amount of star formation that occurred at early and late stages—in other words, on the rate at which gas is converted into stars.*

The spiral arms that are such a conspicuous feature of some disk galaxies delineate regions where star formation is proceeding with unusual rapidity. The most spectacular arms seem to correspond to some kind of persistent wave pattern in the disk, but there is still no completely satisfactory explanation of what excites and maintains such a wave.

The classification scheme that we have outlined is somewhat confused by the existence of a class of disk galaxies (known as S0 galaxies) that do not appear to contain gas. "S0s" are found preferentially in clusters of galaxies; some may be disk systems in which the gas has been swept out by interactions with an external medium.

Although most galaxies are disklike or elliptical, some show a disturbed appearance, from gravitational interaction with a companion. The difference in mutual gravitational attraction between the near and far sides of an interacting pair raises tides, just as the Moon and Sun cause tides on the Earth. The tidal forces draw out from each galaxy dramatic "bridges" and "tails" of stars and gas. The disturbance in the gas

*The Large Magellanic Cloud, an irregular galaxy in orbit around the Milky Way Galaxy, at a distance of about 170,000 light-years. Sites of active star formation glow pink with the light of ionized hydrogen.*

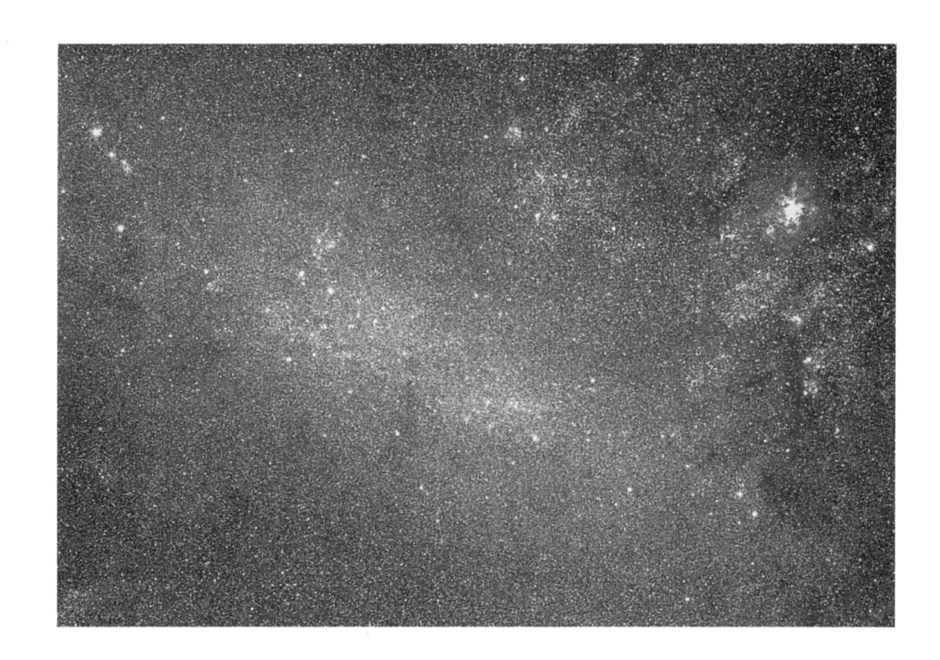

component can create strong shock waves, triggering star formation.

If we could handle a galaxy in the laboratory, we would probe and perturb it in various ways and see how it responded. In these pairs of interacting galaxies, nature has performed just such an "experiment" for us. Such strongly interacting systems are relatively rare in the recent (and nearby) Universe, but should have been more frequent at earlier times, when galaxies were much closer together on average. Indeed, results from the refurbished Hubble Space Telescope are showing that violent galactic interactions were more common in the early Universe. If the encounters

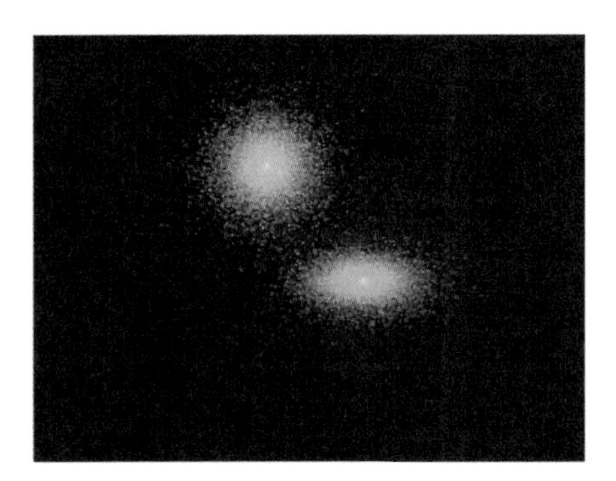

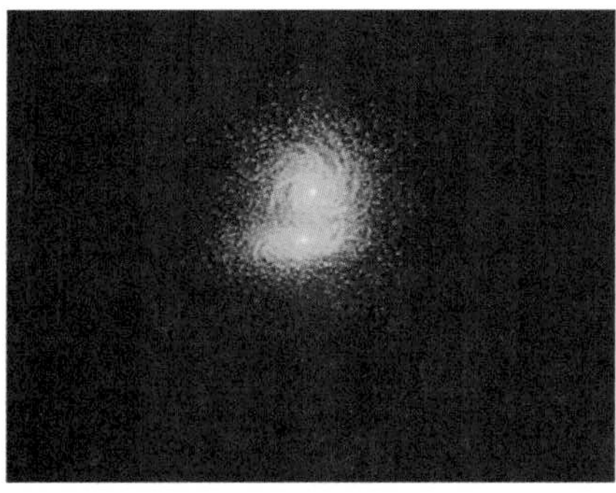

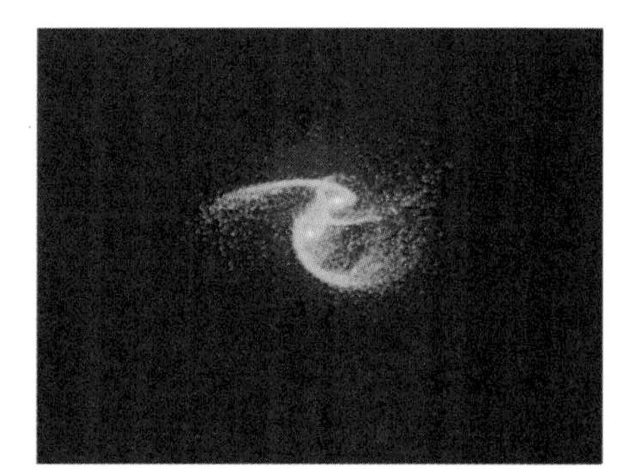

are sufficiently close, the interacting systems are expected to merge. A merger of two spiral galaxies would disrupt the disks, creating a more amorphous elliptical-like structure. Some elliptical galaxies could have formed in this way.

Single stars, the individual organisms in the galactic ecosystem, can be traced from their birth in gas clouds through their life cycle, and we have come to understand why stars exist with the general properties we see. The question of why galaxies exist is less straightforward than the equivalent question for stars. Galaxies formed at an earlier and more remote cosmic epoch. We don't know how much can be explained in terms of ordinary processes accessible to study now, and how much has its causes in the exotic conditions that existed in the early Universe. Why, for example, are most galaxies roughly the same size? Is there any physics that singles out clouds of galactic dimensions, just as, since the work of Eddington and Chandrasekhar, we have understood the natural scale of stars? To some extent, the scales of galaxies must be determined by the early history of the Universe—galaxies couldn't exist unless the "initial" conditions and dynamics of the expanding Universe allowed gravitationally bound gas clouds to condense out of the background. However, there is obviously something that determines where in the mass hierarchy individual galaxies end and clusters of galaxies begin—why, for instance, does the Coma Cluster consist of distinct galaxies, instead of being a huge amorphous agglomeration of $10^{14}$ stars? Yet it would be unfair to our colleagues to leave the impression that no significant progress has been made toward answering these questions. Simple calculations suggest, for example, that a collapsing turbulent gas cloud would be unable to cool and fragment into stars if its mass exceeded $10^{12}$ solar masses, or if its radius were much larger than 300,000 light-years. Such arguments may explain why there is a fairly well defined maximum mass and size for galaxies.

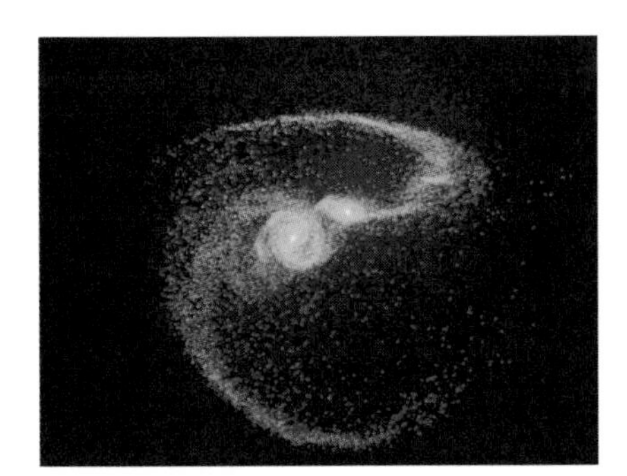

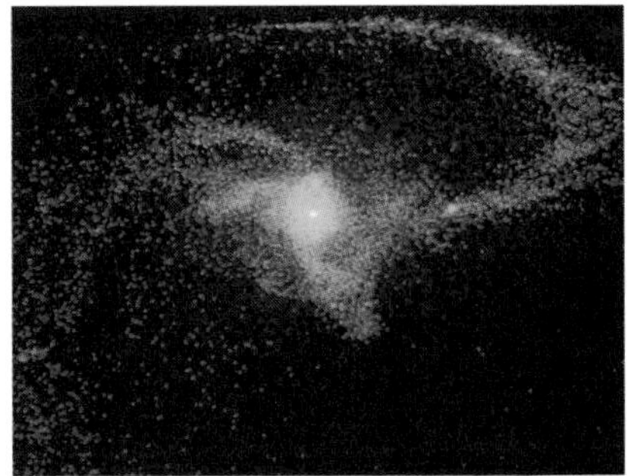

*The collision and eventual merger of two spiral galaxies, as simulated by Joshua Barnes and Lars Hernquist. The rapidly changing gravitational field draws out long streamers of stars in both galaxies and mixes the halo stars (red) and disk stars (blue). The chaotic structure shown in the last frame will settle down to a structure that resembles an elliptical galaxy.*

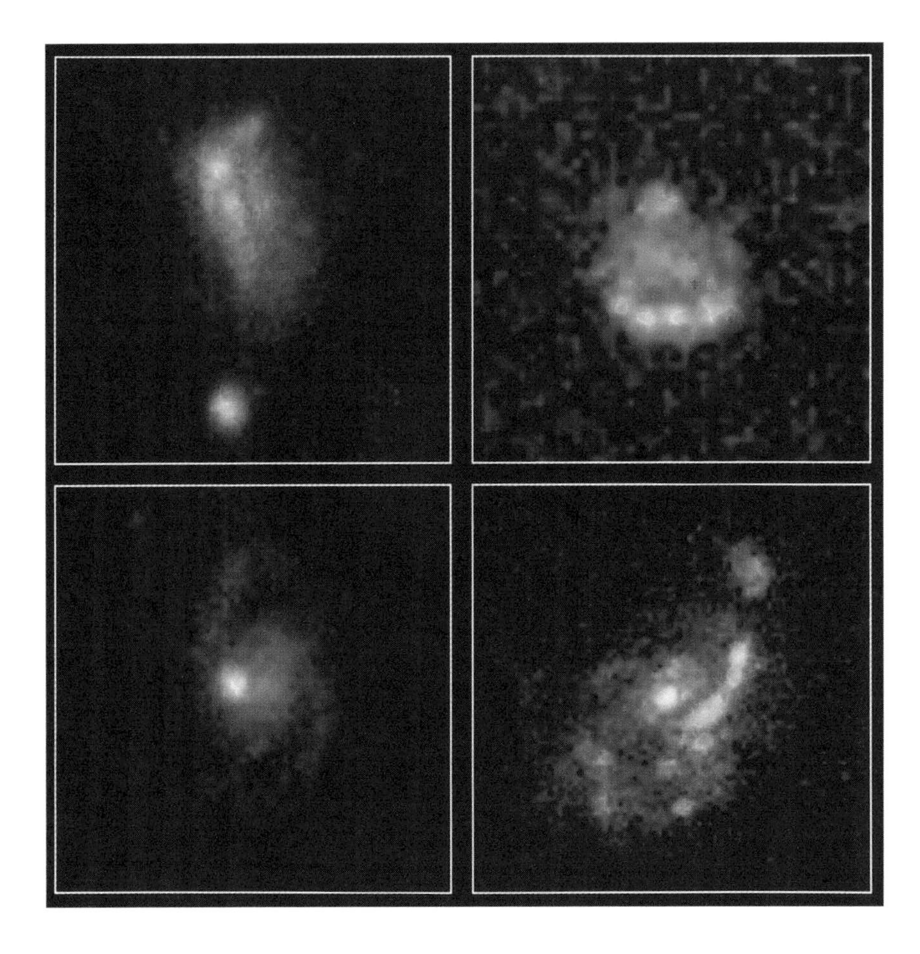

*Galaxies with irregular or asymmetric structures were once much more common than they are today. It may take some time for galaxies to establish an equilibrium structure; collisions and mergers would also have been more frequent in the past. When light left these odd-looking galaxies, the Universe was only about a third of its current age.*

Eddington claimed that a physicist on a cloud-bound planet could have predicted the properties of the gravitationally bound fusion reactors that we call stars. We have not yet found an analogous set of arguments that can be used to "predict" the properties of galaxies with anything approaching the same level of confidence. To achieve a full explanation of galaxies will require setting them in a cosmological context—the "seeds" that developed into galaxies must have been implanted in the early Universe. Also, there is the embarrassing fact that most of their mass, maybe as much as 90 percent, is unaccounted for—it isn't in the stars and gas that we see but instead takes some unknown "dark" form. To a large extent this dark matter determines the gravitational fields of galaxies, and therefore it affects the accumulation of matter all the way in to the central regions.

## Dark Matter

Perhaps nothing confounds our understanding of galaxies so much as the realization that the gas and stars we observe constitute only a small fraction of the total mass of all large structures in the Universe, from galaxies on up. The observed stars and gas may be little more than tracers for the material that dominates the gravitational fields of these structures. The evidence for "dark matter" dates back more than sixty years, to the time when Fritz Zwicky realized that the aggregate mass of all the galaxies we could see in the Coma Cluster was inadequate, by a wide margin, to prevent the system from flying apart. The case for the existence of dark matter is now quite compelling; what dark matter consists of, however, remains a mystery.

Our own Galaxy and other spiral systems like the Andromeda Galaxy apparently have halos of dark matter, extending well beyond the outer-

*The Coma Cluster. In this system there must be several times more mass in some "dark" form than in the stars and gas we detect—otherwise the galaxies would fly apart and the cluster would disperse.*

## The 21-Centimeter Line of Hydrogen

Explaining the hydrogen spectrum—the Lyman lines, the Balmer lines, etc., discussed in the box on pages 37–39—was one of the triumphs of the early quantum theory in the 1920s. But some complications were appreciated later. One subtle effect, which turns out to be especially important for astronomy, depends on the fact that each electron and proton has an intrinsic "spin." Just as a charged spinning ball, or a loop carrying an electric current, creates a magnetic field, a charged particle with spin behaves like a tiny magnet.

Two magnets placed side by side repel each other when they are aligned and attract when they point in opposite directions. An electron "in orbit" around a proton is similarly pushed slightly farther out if the magnetic fields are aligned than if they are misaligned. Energy can therefore be released if the electron "flips" so that it becomes magnetically misaligned.

The effect is tiny: only a small quantum of energy is released when an electron's magnetic field flips from being aligned to being misaligned with the proton's. These so-called hyperfine transitions are therefore signaled by radiation of very low frequency—in other words, of very long wavelength. This radiation can be detected by radio astronomers who tune their receivers to a wavelength of 21.1 centimeters.

Radio telescopes are sensitive enough to pick up the "21-centimeter line" from hydrogen in interstellar clouds—indeed, this proves to be the best technique for probing diffuse hydrogen in other galaxies as well as our own. Another feature of the line is that it is very sharp: a radio receiver has to be tuned exactly right to pick it up. The wavelength will of course be stretched by the Doppler effect if the gas is moving away from us; the extent of the Doppler shift tells us the speed of the gas. Since radio telescopes can routinely measure the wavelength to one part in 100,000, speeds can be measured with a precision of 1/100,000 the speed of light—to within 3 kilometers per second. The 21-centimeter line can therefore be used to map out how gas is moving within galaxies. Astronomers can infer the strength of the gravitational field that causes these motions; this technique provides one of the main lines of evidence for dark matter.

most stars. One line of evidence for these dark matter halos comes from mapping the speeds at which the disk rotates at various distances from the galactic center. Clouds of neutral hydrogen, themselves just a small fraction of the total mass, serve as tracers of the orbital motion. In spiral galaxies that are observed almost edge-on (the Andromeda Galaxy is the closest example), radio astronomers can infer the speed of the gas from the Doppler shift of the 21-centimeter spectral line, explained in the box on the opposite page. Moreover, this gas can be observed at radii far outside the edge of the conspicuous stellar disk. If the outermost hydrogen clouds were feeling just the gravitational pull of what we can see, their speeds should fall off roughly as the square root of distance outside the visible limits of the galaxy: the outer gas would move more slowly, just as Neptune and Pluto orbit the Sun more slowly than the Earth does. Instead, the velocity remains almost constant, implying that the outer gas is "feeling" the gravitational pull from a larger mass than the inner gas. It is as though we found that Pluto was moving as fast as the Earth, even though it is much farther from the Sun; if that were the case, we would have to postulate an invisible shell of matter, outside the Earth's orbit but inside Pluto's. Measurements like those shown above for Andromeda have been made for many disk galaxies. In some galaxies the inferred mass keeps increasing out to distances larger than 250,000 light-years, implying that "dark" matter contributes at least five times the mass of the "luminous" disk.

*The Andromeda Galaxy contains a spinning disk viewed almost edge-on. By observing the 21-centimeter line of hydrogen, radio astronomers can detect cool gas in the disk out to distances far beyond the apparent boundary of the galaxy. The orbital speed of the gas, which can be inferred from the Doppler effect, is plotted on the graph. The fact that the orbital speed remains high even at the largest distances indicates that the outermost gas must be "feeling" the gravitational pull of extra matter outside the visible part of the galaxy. Similar evidence exists for many other galaxies, and suggests that very extended "dark halos" contain several times more mass than all the visible stars and gas in galaxies.*

Further evidence for dark matter within our own Galaxy comes from the measured velocities and inferred orbits of dwarf satellite galaxies and outlying globular clusters found in our Galaxy's halo. The total mass of the Local Group of galaxies (of which our Galaxy and Andromeda are the dominant members) can be derived from the so-called timing argument, whereby one calculates how much mass would be needed in order to bring about the present situation in which Andromeda and our Galaxy, now 2 million light-years apart, are observed to be "falling" toward each other. These estimates suggest that the two halos may be so massive that they extend out until they effectively merge.

Similar conclusions hold for elliptical galaxies, though here the evidence is different. These galaxies contain gas so hot that its pressure prevents it from condensing into the center (or into a disk). This gas is detected by its X-ray emission. Moreover, the X-rays tell us how the temperature of the gas varies with radius. Even far out, the gas is so hot that it wouldn't be bound to the galaxy at all unless it were being held down by a stronger gravitational force than the stars alone provide. So elliptical galaxies must have extended dark halos, just as disk galaxies do.

*The gravitational field of the galaxy cluster Abell 2218 distorts the images of more distant galaxies, stretching them into concentric arcs. By measuring the pattern of distortion, it is possible to map out the distribution of matter—dark as well as luminous—in Abell 2218. The results show that this cluster is heavily dominated by dark matter. The longest arcs are magnified more than 10 times. The cluster is therefore acting like a "telescope," allowing astronomers to study galaxies at high redshifts (up to at least 2.5) that would otherwise be invisibly faint.*

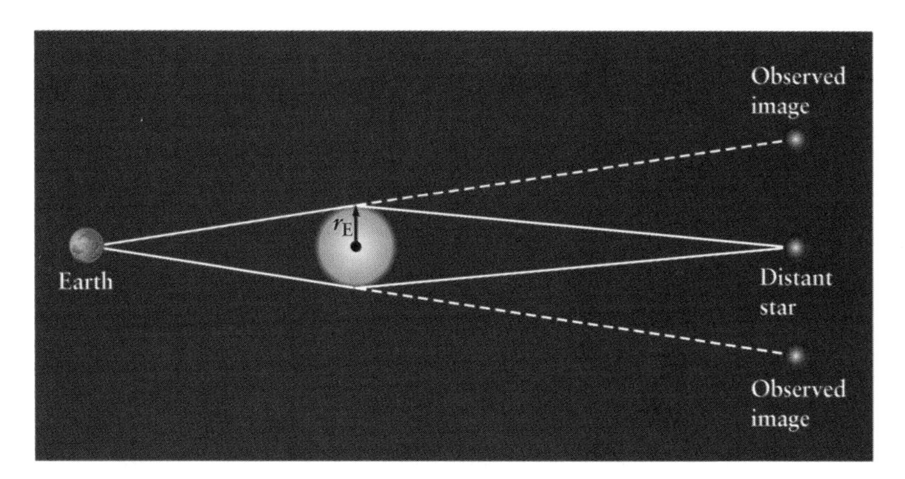


*Gravitational lensing occurs when a compact object lies along the line of sight to a star. If the object is exactly in front of the star, the latter's brightness is enormously magnified; observing with very sharp resolution, we see the star distorted into a ring whose radius is the "Einstein radius,"* $r_E$. *If the alignment isn't quite perfect, two highly magnified images, separated by* $2r_E$, *are seen. Significant (though less extreme) magnification occurs whenever the offset is less than* $r_E$.

Clusters of galaxies appear to have even larger concentrations of dark matter, compared to the luminous matter forming the galaxies. For these clusters to be in equilibrium, there must be a balance between the effect of the relative motions of the galaxies (which tends to disperse the cluster) and the effect of gravity (which, if the galaxies weren't moving, would cause them to fall together at the cluster center). The speeds of the galaxies (or at least the component of their motion that is directed along our line of sight) can be measured from the Doppler effect. Once we know the motions, we can calculate what the cluster's total mass must be. This is a simple application of the "virial theorem," discussed in the first chapter, which asserts that the gravitational binding energy—which is related to the mass—must equal the internal kinetic energy if a system is in equilibrium. This method of calculating a cluster's mass is now complemented by X-ray studies—analogous to those performed for elliptical galaxies—of the hot gas that fills the space between galaxies in clusters.

An interesting new method should enable astronomers to trace how the mass is distributed throughout a cluster. The method makes use of the deflection of light rays by the cluster's gravitational field. That the path of a light ray is deflected by a gravitational field is one of the principal predictions of general relativity, and the first one to be confirmed observationally. This effect distorts the images of objects lying behind the cluster in the background, very much as a glass distorts an image—hence the phenomenon is often referred to as "gravitational lensing." Some remarkable arc-like objects are observed, which are highly stretched and magnified images of faint galaxies lying far behind the cluster. More detailed studies reveal that many background galaxies viewed through rich clusters are somewhat distorted, though less dramatically than the arcs. Such data in principle

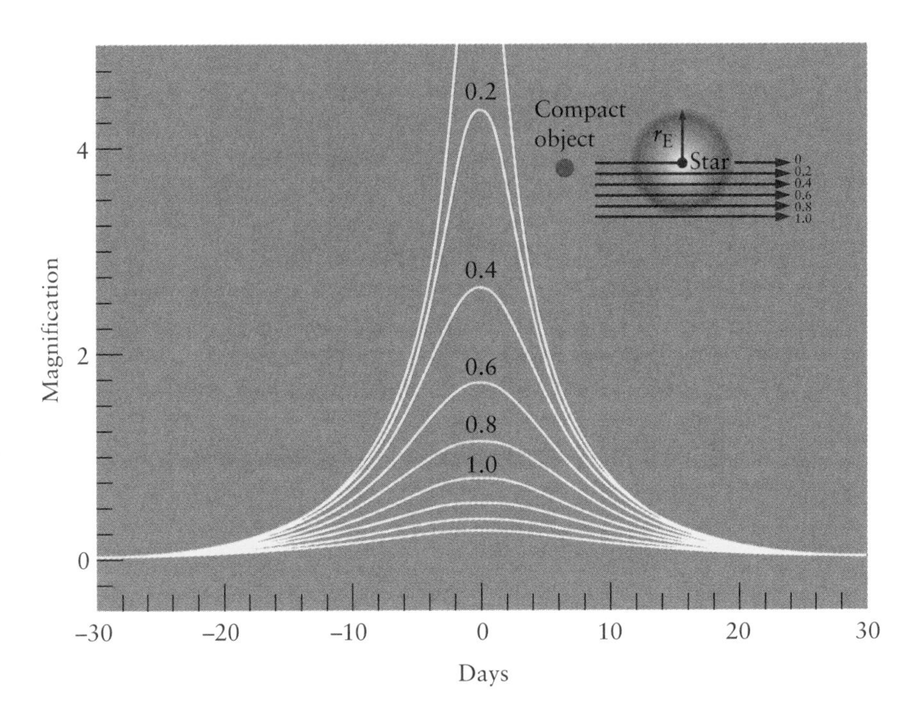


*This graph shows how the total magnification of a star rises and falls when a compact object moves steadily across the line of sight to a background star. Curves are shown for cases when the foreground object passes exactly in front of the background star, and also when the closest alignment is 0.2, 0.4, 0.6, 0.8, and 1.0 times $r_E$ (top right). The timescales correspond to an event in which a star in the Large Magellanic Cloud is lensed by a compact object of 0.01 solar masses in our Galactic halo, which moves with the typical velocity expected of halo objects. For different "lens" masses, the times would scale as the square root of mass.*

allow astronomers to reconstruct the two-dimensional projection of the cluster's mass density on the sky. This new line of evidence would have delighted Zwicky: not only a pioneer investigator of dark matter, he was also one of the first to speculate about gravitational lensing by galaxies.

All these techniques are more or less in agreement, and imply a discrepancy of a factor of 5 to 10 between the total mass of galaxy clusters and the sum of the masses of the stars and gas in the individual galaxies. That is, there seems to be at least 5 to 10 times more dark matter in the Universe than luminous matter. Determining the total amount of dark matter is of fundamental importance for cosmologists, who wish to know whether there is sufficient matter in the Universe that its gravitational pull will eventually bring a stop to the receding motion of the galaxies. The quantity of dark matter inferred so far amounts to 10 to 20 percent of the "critical density" above which the expansion of the Universe will eventually be halted and reversed. Is there still more dark matter between the clusters, which as yet goes undetected? Is there enough to make the Universe grind to a halt, or even recollapse?

Of comparable importance is the question, What is the dark matter made of? All we can state with confidence is that the dark matter must be mainly in some unknown form—it can be neither ordinary stars nor the gas that emits X-rays. One shouldn't be too surprised that most of the matter in the Universe is dark: there is no reason why everything should shine, and it isn't difficult to think of possible candidates for the makeup of the dark matter. Indeed, the problem is rather to decide among a long list of possibilities. Obviously, direct detection of the dark matter itself would be the "cleanest" and most decisive discriminant: low-mass stars and planetlike objects in the Galactic halo, too faint to be seen directly, would reveal themselves by deflecting and magnifying the light from more distant stars; exotic particles that pervade the Galaxy (and therefore are continually passing through every laboratory) might be detected by sensitive experiments.

The first and most "conservative" guess is that the dark matter in our Galaxy is in the form of stars too low in mass to ignite nuclear fuel in their centers. The masses of these so-called brown dwarfs or "Jupiters" would have to be below 7 percent of a solar mass in order for these objects to be faint enough to have escaped detection by conventional astronomical techniques. Even if they emit no radiation (apart, presumably, from very faint infrared emission), the objects would still disclose themselves by occasionally magnifying the light from a background star, a by-product of the gravitational lensing effect. The strength of the magnification would depend on how far the dark foreground star is from Earth, compared to the background source, and the precision with which the dark object is "lined up" in front of the source. Because the mass of a brown dwarf is so small, the alignment has to be near perfect in order to obtain an observable effect. Brown dwarfs in the halo of our Galaxy would create magnified images of the more distant stars in the Magellanic Clouds with an apparent angular size of a few microarcseconds ($10^{-6}$ arcsecond), far too small to be resolved with existing telescopes. (A telescope with a resolution of a microarcsecond would enable us to read a newspaper on the Moon.) One can nevertheless identify instances of lensing by watching for the characteristically symmetric rise and fall in magnification that is visible when a halo object moves almost across the line of sight to a background star.

Unfortunately, even if low-mass stars contributed all of our halo's dark matter, the probability of there being significant lensing, at any instant, along a typical line of sight would be only 3 chances in 10 million. (This result does not depend on the typical mass of the brown dwarfs.) To stand a chance of detecting this effect, one must monitor the lines of sight to millions of different stars. Two independent groups have succeeded in doing this within the last few years, and there are now nearly a dozen candidate lensing events. (Other groups have used this technique to find lensing by stars in the direction of the bulge at the center of our Galaxy.)

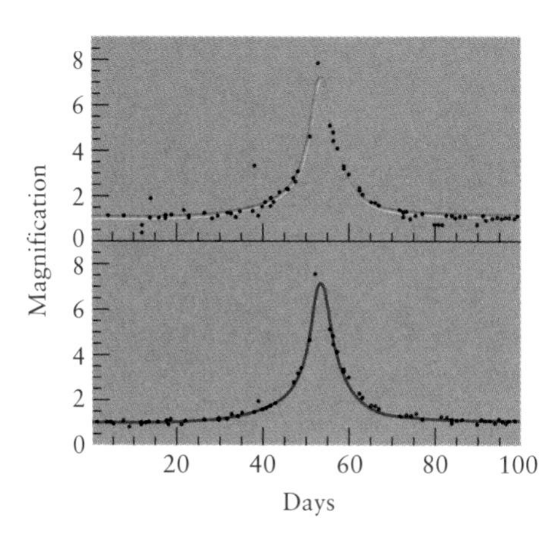


*Data for a candidate "microlensing" event in the Magellanic Clouds, observed by Charles Alcock and collaborators. The maximum magnification is about a factor of 7. The rise and fall is symmetrical, and follows the behavior expected. Further evidence that the brightening of the star resulted from lensing is that the red and blue curves track each other closely; gravitational lensing should be achromatic.*

Although there is a great deal of room for improving the statistics, the results so far suggest that the contribution of brown dwarfs to our Galaxy's halo is *not* sufficient to account for most of the dark matter.

The remnants of a hypothetical population of very heavy stars ("very massive objects" or VMOs), which might have formed early in galactic history, are another widely discussed candidate for the dark matter. These so-called VMO remnants could be black holes with masses in the range of 1000 to 1 million solar masses. The prospects for detecting such objects by their lensing signatures are less hopeful than for brown dwarfs because the individual events would be much slower and less frequent in the VMO case. A survey would probably have to wait decades before detecting even a single event.

One feature that both brown dwarfs and VMO remnants share is that they are forms of "baryonic" dark matter—that is, matter composed of ordinary subnuclear particles such as protons and neutrons (or, in the case of black holes, resulting from the collapse of such matter). In the standard theory of the expanding Universe (the "hot big bang"), most of the helium, all the deuterium, and some of the lithium observed today were created by thermonuclear reactions during the first few minutes following the initial explosion. So far, no other theory has been proposed that comes close in its ability to produce all of these elements in roughly the correct quantities. In order for the reactions to yield the observed abundances, however, the baryon density must not exceed 3 to 10 percent of the total density of matter needed to "close" the Universe (that is, to cause it to stop expanding). Given this cap on the baryon density, it is not clear whether baryonic material could account for all dark matter inferred to exist in galaxies and clusters; it most certainly could not account for the amount of dark matter that seems to pervade larger structures known as "superclusters."

This argument has led to serious suggestions that most of the dark matter in the Universe may be of a "nonbaryonic" form, probably consisting of fundamental particles not ordinarily found in matter. The most obvious candidate is neutrinos. Neutrinos are particles much less massive than electrons, which apparently come in three types. According to the standard big bang theory, there should be, on average, 100 million neutrinos of each type per cubic meter of space. Because neutrinos outnumber baryons by such a large factor (by more than $10^8$, on average), they would not need to have a very high mass in order to have an important cumulative effect on cosmic dynamics: they would close the Universe if their mass were only one ten-thousandth that of an electron.

Twenty years ago, Ramanath Cowsik, John McClelland, George Marx, and Alex Szalay conjectured that neutrinos could provide the dark mass in galactic halos and clusters. Unfortunately, the masses of neutrinos are still unknown. They may be completely massless, like photons, in which case they would be out of the running as candidates for dark matter. For this reason, the suggestion that they could provide dark mass was not followed up very extensively at the time it was originally made. But by the 1980s physicists had become more open-minded about nonzero neutrino masses. (Moreover, there was an experimental claim, now generally discounted, that the electron neutrino mass had been measured at $\frac{1}{14,000}$ of an electron mass.) Theorists were then stimulated to explore scenarios for galaxy formation in which neutrinos played key roles.

New evidence on neutrino masses comes from analysis of the neutrino burst from Supernova 1987A in the Large Magellanic Cloud. In traversing the distance of 170,000 light-years from the supernova, a neutrino of finite mass would be delayed an amount of time proportional to the square of its mass and the inverse square of its energy. The sharpness of the measured burst implies that the electron neutrino (one of the three types) weighs less than $\frac{1}{25,000}$ of an electron mass (i.e., less than $4 \times 10^{-36}$ kilograms). This result doesn't rule out the possibility that neutrinos are massless, nor does it preclude one of the other types of neutrinos from contributing significantly to the dark matter. Firmer clues to neutrino masses may soon come from terrestrial neutrino beam experiments, or by studying neutrinos from the Sun.

Neutrinos have an "edge" over other nonbaryonic candidates in that they are at least known to exist. But theoretical physicists have a long shopping list of hypothetical particles that might have survived in the requisite numbers from the early Universe. One widely discussed possibility goes by the whimsical name of WIMPs, which stands for "Weakly Interacting Massive Particles." (Not to be outdone, the proponents of dark stars and VMO remnants call their targets MACHOs, for "Massive Compact Halo Objects.") WIMPs could be comparable to protons in mass, or they could be even more massive. Several groups are attempting to detect WIMPs directly, by measuring the recoil that would occur when one of these particles hits an atomic nucleus in a laboratory apparatus. However, even if WIMPs exist in sufficient numbers to provide all the dark matter in our Galaxy's halo, these collisions must be extremely rare. Moreover, the detection apparatus is bedeviled by extraneous collision events produced by the internal radioactivity of the equipment and by cosmic rays (energetic particles of normal matter from space, possibly produced in supernova explosions). Consequently, the experiments must be carried out deep underground. Since the Solar System sweeps through the halo on its orbit around the Galactic center, most WIMPS hitting the Earth would come from the direction of this motion. Moreover, the predicted collision rate would have a seasonal variation, because as the Earth moves round the Sun its velocity relative to the Galactic halo (and thus to the population of WIMPs) changes. Either the directional information or the seasonal variation would suffice to distinguish WIMP-induced events from other collision events.

Ingenious schemes for detecting a halo population of exotic particles seem among the most worthwhile and exciting high-risk experiments in physics or astrophysics. A null result would surprise nobody; on the other hand, such experiments could reveal the existence of fundamental new forms of matter as well as determine what 90 percent of the Universe consists of. The late Professor Roderick Redman of Cambridge, a "no-nonsense" observer with little taste for speculation, once claimed that any competent astrophysicist could reconcile any theory with any set of facts. An even more cynical colleague extended this claim, asserting that the astrophysicist often need not even be competent. Dark matter theorists have in the past all too often exemplified Redman's theorem, and its extension as well. But searches may soon yield definitive results; in any case, various constraints are now restricting the tenable options. Moreover, it is not "wishful thinking" to expect that there may be more than one important kind of dark matter—for instance, nonbaryonic dark matter could control the dynamics of large clusters and superclusters, even if individual galactic halos contained a lot of brown dwarfs or VMOs.

It would be especially interesting if astronomical techniques were to reveal some fundamental particle that has been predicted by theorists, but not yet encountered in the laboratory. If such particles turned out to account for the dark matter, however, we would have to view the galaxies, the stars, and even ourselves in a downgraded perspective. Copernicus dethroned the Earth from any central position. Early in this century, Harlow Shapley and Edwin Hubble demoted the Sun from any privileged location in space. But now even "baryon chauvinism" might have to be abandoned: the protons, neutrons, and electrons of which we and the entire astronomical world are made could be a kind of afterthought in a cosmos where exotic elementary particles control the overall dynamics. Great galaxies could be just a puddle of sediment in a cloud of invisible matter 10 times more massive and extensive.

We should be pleased that we have been able to deduce the existence of dark matter, but sobered by our ignorance of its nature. The entities that collectively constitute most of the mass of the Universe may have individual masses anywhere from $10^{-36}$ kilogram up to $10^{36}$ kilograms—an uncertainty of more than 70 powers of 10! Astrophysics may not always be an exact science, but seldom is the uncertainty as gross as this.

The range of options could be narrowed if there were less uncertainty about what particles might be created in the earliest stages of the expansion of the Universe. Computer simulations of galaxy formation can also yield indirect (but potentially very important) evidence on the nature of the dark matter. Since 90 percent of the gravitating material in a galaxy is "dark," the way in which galaxies form and cluster plainly depends on what the dark matter is. Under specific assumptions about conditions shortly after the big bang, one can compute how clustering develops with time, and then compare the various outcomes with observations. We shall return to these simulations in Chapter 9, when we consider whether rare supermassive black holes could form as a by-product of the early galaxy formation process. For now, we journey inward toward the centers of galaxies, where more observable, if no less exotic, processes hold sway.

## Galactic Nuclei

If you examine a photograph of a galaxy, it is easy to see that the concentration of stars increases toward the center, the so-called nucleus. The nucleus of the Milky Way is obscured at visible wavelengths by a thick veil of dust and gas, but observations at infrared wavelengths have revealed as many as a million stars within the innermost single light-year. Compare this to the concentration of stars near us, which is about 0.006 suns per cubic light-year. On average, stars are 300 times closer together in the Galactic center than they are in the vicinity of the Sun.

The nuclei of galaxies would appear to be rather dangerous places, in which stars are not only packed together but are also moving around randomly at speeds of up to several hundred kilometers per second due to their mutual gravitational attractions. The only other places where we know such high stellar densities to exist are in the centers of globular clusters, but there the random velocities are 10 to 100 times slower. (Some astronomers have speculated that galactic nuclei formed from the coagulation of globular clusters.) Violent collisions between stars are inevitable in galactic nuclei, and indeed are predicted to occur about once every 10,000 years in the nucleus of the Milky Way.

The fate of such a dense central concentration of stars is not easy to predict, and it forms the focus of a debate that started in the 1960s and still has not been resolved. Part of the problem is that the outcome of even a single stellar encounter is not well understood theoretically. When considering the structure of a galaxy as a whole, it is acceptable to treat individual stars as "point masses" of infinitesimal size. But such an approximation will not do when considering encounters in which two stars pass within a few stellar radii of one another. Then the tidal forces deform the stars' internal structures, heating them slightly and placing a drag on their motions. If this drag slows their relative motions enough, the stars may fall into orbit about one another. In this way, two stars, initially moving independently of one another, can become locked in a "tidal capture binary."

*An intense concentration of starlight marks the nucleus of M100, a spiral galaxy in the Virgo Cluster.*

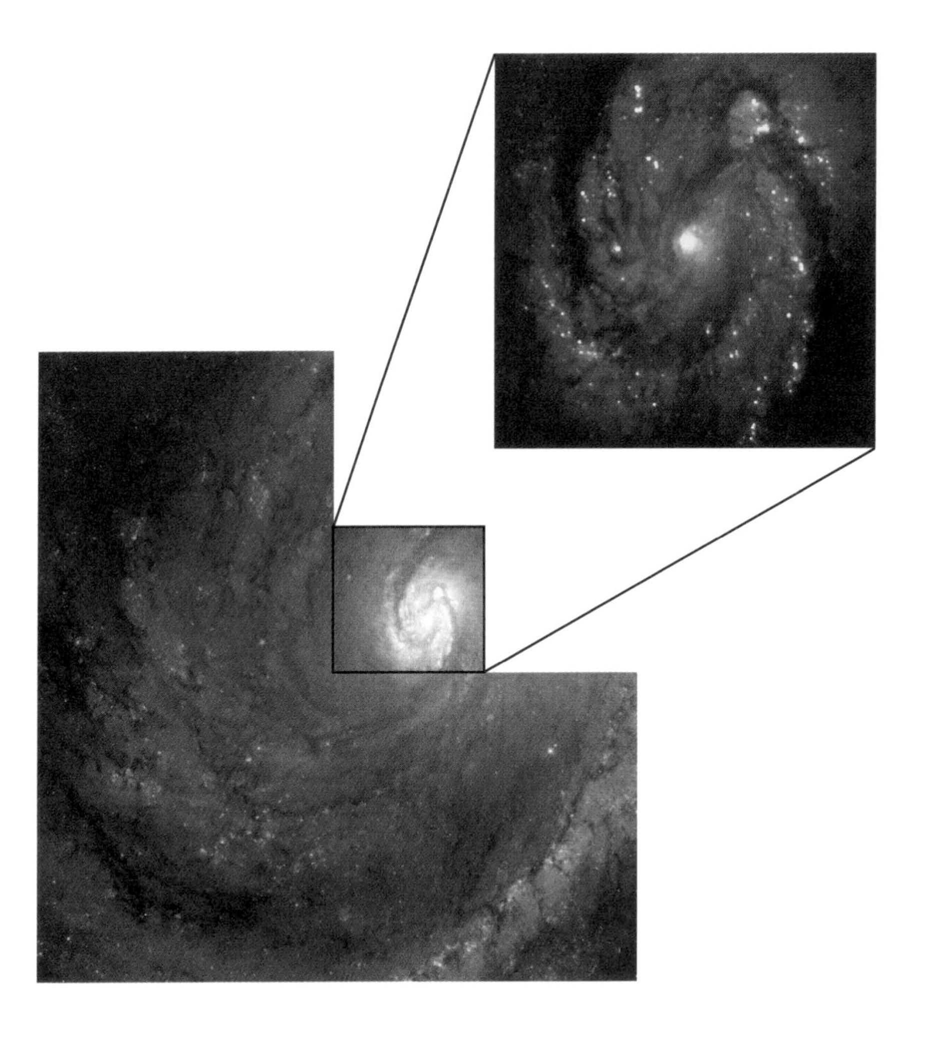

(The tidal capture mechanism probably accounts for most of the close binaries identified in globular clusters.) If the tides are strong enough, the stars may eventually spiral together and coalesce into a single, more massive star.

The fate of a coalesced star depends on exactly how the merger takes place, and especially on the evolutionary state of the participants in the merger. Stellar evolution theory tells us how to predict the subsequent evolution of a star from some point in time, provided that we know the star's chemical *structure* at that point. This means that we need to know not only the total ratio of hydrogen to helium in the star, but also how that ratio changes with depth. If we start with a pure hydrogen star at the beginning of its life, we can follow the chemical changes step by step as nuclear reactions alter the composition. But what is the initial compositional structure of a star formed by coalescence? We can guess the overall composition by adding up the contents of each participant in the encounter, although mass may be lost during the coalescence process, making even this estimate uncertain. Variations of the composition with depth, however, will be affected by complicated and hard to predict mixing processes. Our best hope is to study the coalescence process "experimentally," using computer simulations. These are in their infancy now, but their sophistication is increasing rapidly.

Coalescence is the likely outcome for two stars encountering each other at velocities that are small compared to the escape speed from the surface of either star. The escape speed from the Sun, for example, is 618 kilometers per second. At much higher relative velocities, the stars literally smack into one another, releasing high-speed plumes of gas into space. Whereas low-speed collisions build up high-mass stars, high-speed encounters demolish the existing stars. In the latter case, as in the former, it is impossible to make detailed predictions of the outcome (especially the fraction of mass liberated) without computer simulations. Collisions are mainly disruptive when they take place in clusters of fast-moving stars having typical random velocities of more than 1000 kilometers per second. In the central star cluster of the Milky Way, the typical random speed is only about 200 kilometers per second, and so encounters between these stars should be in the coalescent regime. Indeed, our Galactic center displays an unusually large abundance of massive blue stars, which quite possibly have been produced by coalescent encounters within the last million years or so.

Stellar encounters will certainly affect the evolution of a galactic nucleus, but we are still not sure exactly how. Some of the possibilities are illustrated in the flow chart on page 98. By itself, widespread stellar coalescence would tend to make the average star in the cluster slower and more massive. In response, gravity would pull the star cluster more tightly together, and the stars would end up moving *faster,* as predicted by the virial theorem. As the random speeds of stars in the cluster increased, we could imagine a steady progression from a nucleus in which most encounters lead to coalescence to a nucleus in which most encounters lead to disruptive collisions.

During the phase dominated by coalescence, other effects would also be important. Adding mass to a star leads to a short life and violent death, so the rate of supernova explosions would increase as the number of massive stars did. By flinging matter out of the nucleus at high speed, numerous supernova explosions would oppose and possibly reverse the contraction of the cluster—the loss of mass would decrease the strength of

*In the central part of a galaxy (the innermost few light-years) the concentration of stars can be so high that collisions are frequent. Gravity causes gas to accumulate right in the center. This diagram depicts the various evolutionary tracks that might occur, giving rise to violent activity in galactic nuclei. The almost inevitable end point (literally the "bottom line") is the formation of a black hole with a mass of millions or even billions of suns. Such black holes are implicated in the most energetic phenomena in galactic nuclei, the subject of our next two chapters; they are also now being found (see Chapter 7) in the centers of nearby "quiet" galaxies.*

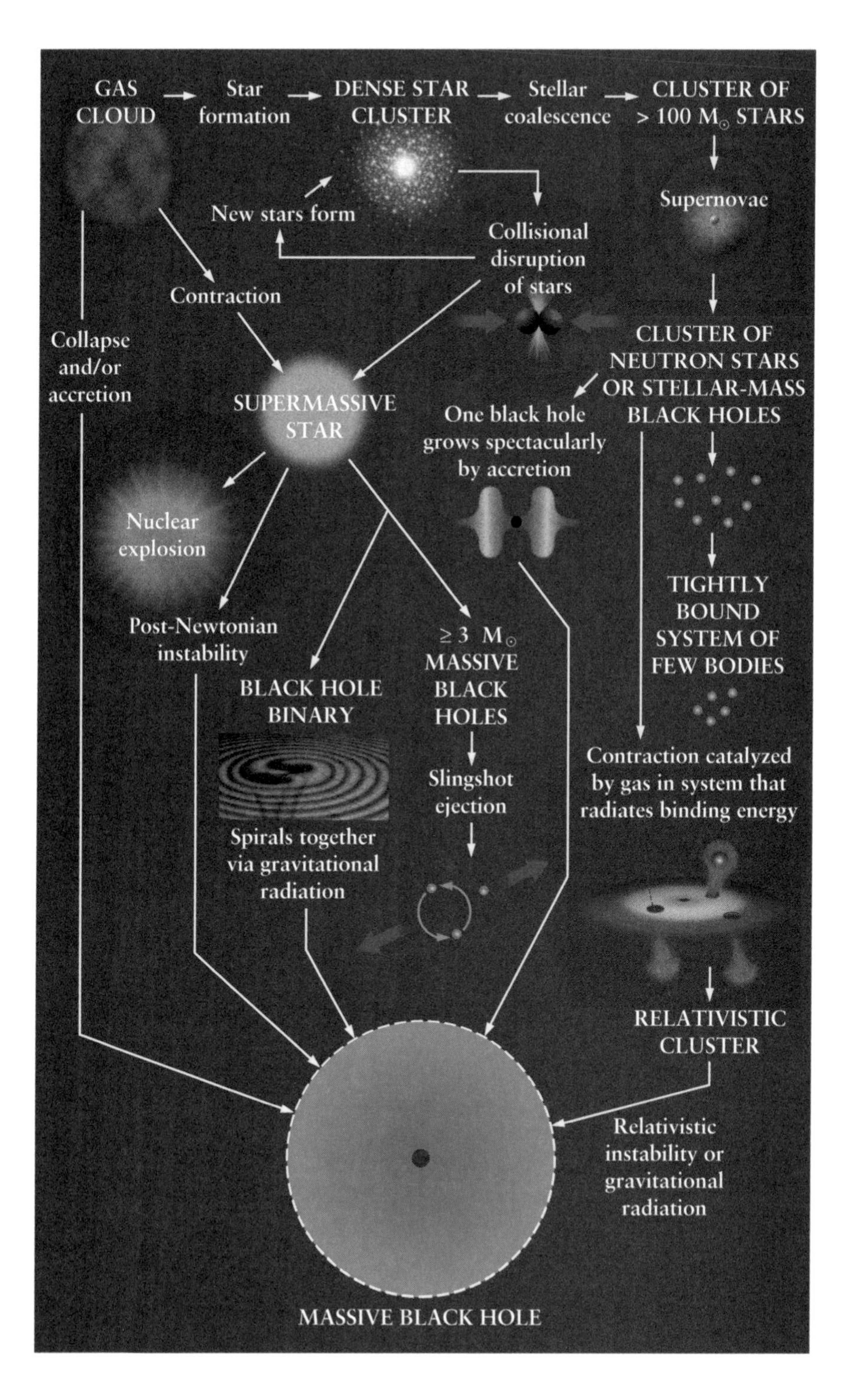

gravity. The supernovae would leave behind an admixture of neutron stars and stellar-mass black holes, which would suffer fewer direct collisions because of their compact size but would still interact gravitationally. Weak gravitational interactions between stars that never physically touch will tend to drive the nucleus toward a "core–halo" structure: most of the stars would be ejected to larger distances and relatively few would remain in the contracting central regions (similar to the evolution of a globular cluster that we discussed in Chapter 1). The fate of the cluster depends on the outcome of the three-way competition among the processes of direct collisions, weak gravitational encounters, and explosive mass loss. At one extreme, the coalescence process might lead to the creation of a few enormous stars that would sink to the center of the cluster, merge, and ultimately collapse to a black hole. At another, a small fraction of the initial cluster might contract to form a tightly bound stellar core in which most collisions are violently disruptive.

Once a galactic nucleus enters the disruptive-collision phase, its evolution becomes even more uncertain. In contrast to the ejecta from supernova explosions, the matter liberated by disruptive stellar collisions is unable to escape the gravitational pull of the cluster. Most of the stray gas would collect at the cluster's center. In one of the earliest treatments of the evolution of a dense star cluster, in 1966, the Princeton astrophysicists Lyman Spitzer and William Saslaw suggested that the liberated matter could form a new generation of stars, more condensed than the original cluster and with higher random velocities. These in turn would eventually suffer even more violent disruptive collisions, and the process could repeat. Whether stars could form efficiently in such a compact and turbulent system is questionable, but the basic idea of a dense cloud of gas collecting in the center of a collision-dominated galactic nucleus is very plausible. Once such a cloud begins to contract under its own gravity, collapse to a black hole is almost inevitable.

To make matters still more complicated, one must not treat a galactic nucleus as a closed system. Supernova explosions or powerful stellar winds could expel much of the gaseous debris from the central region, preventing it from accumulating. This is especially likely in a galactic nucleus of modest mass and random velocity, such as the center of the Milky Way (which contains at most a few million solar masses). But in more massive or more condensed galactic nuclei, which can contain upward of a billion solar masses, the accumulation of matter seems unavoidable. Some of the gas far from the nucleus—whether released by stellar evolution in the galaxy, accreted from intergalactic space, or left over from the epoch of galaxy formation—is bound to drift toward the center as it loses angular momentum. Once in the nucleus, it has nowhere to go and must accumulate. What is its fate? Will it form stars, adding to the crowding in the central

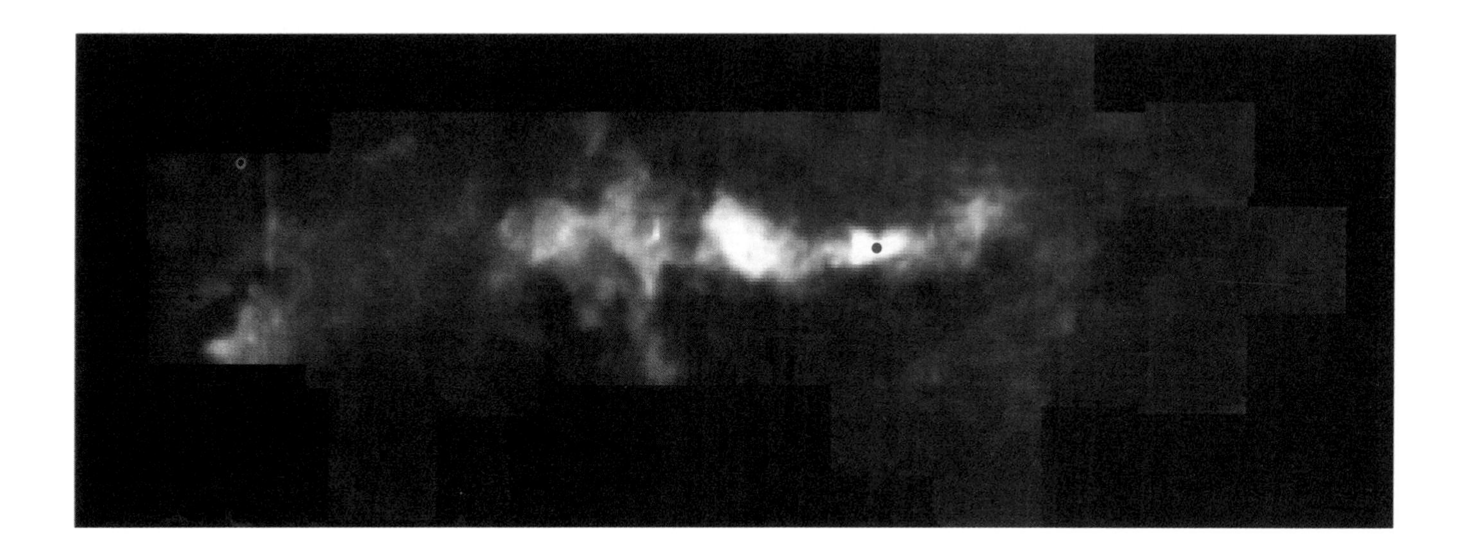

*This image shows the intensity of radio waves emitted by carbon monoxide molecules in the direction toward the Galactic center. The cold clouds of gas that populate the disk of the Galaxy are rich in molecules, and carbon monoxide is one of the easiest to detect. The horizontal band corresponds to the band of the Milky Way in the direction of the constellation Sagittarius; the actual location of the Galactic center is marked by a red dot. Different colors on this image correspond to different velocities of the molecular gas, as deduced from the Doppler shift.*

star cluster? Or will it collect to form an amorphous gas cloud, dragging on the orbits of passing stars and hastening their coalescence? Could it contract under its own gravity to form a single "superstar," so unstable according to theory that it would collapse to a black hole all at once? Our answers to any of these questions remain highly speculative, but we should not be surprised to find very massive black holes, of perhaps millions or even billions of solar masses, in the nuclei of many galaxies.

In the face of so many complex, competing phenomena, the only safe course is the empirical one. What do we *observe* to be going on in the only galactic nucleus that we can study in some detail, that of the Milky Way, which lies only 25,000 light-years away? Although optical and ultraviolet light from the Galactic center is almost totally obscured by gas and dust, we can readily map out the region at radio, infrared, X-ray, and gamma-ray wavelengths. What we see can only be described as a mess. A dense star cluster is embedded in an extremely complex array of gaseous filaments, disks, and blobs. Large filaments with linear striations, mapped in the radio synchrotron emission of energetic electrons spiraling in a magnetic field, remind us of the prominences that erupt from the surface of the Sun. At the center of it all, there is a pointlike source of radio emission with the unromantic name "Sagittarius A-star" (SgrA*). For years a debate has raged over whether this object might be a black hole of as many as a million solar masses. Proponents cite its compact size (it must be small enough to fit inside the orbit of Jupiter) and unusual radiation spectrum, which resembles the kind of spectrum that might be produced by gas swirling into a black hole at a very low rate. Another point in favor of a massive black hole is the slow motion of SgrA* on the sky. If it were only

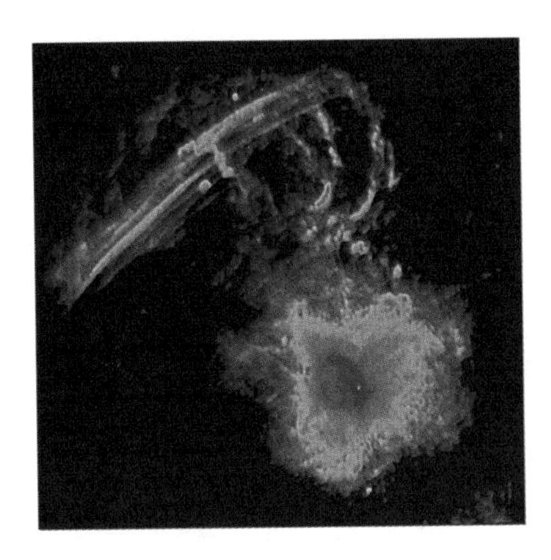

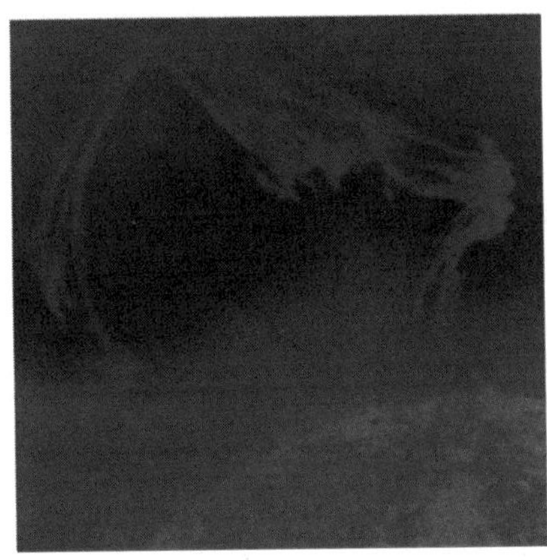

*Radio observations of the Galactic center (left) show striated filaments of synchrotron radiation. The appearance, similar to that of a solar prominence, suggests that we are witnessing huge eruptions of the magnetic field.*

a few times as massive as a normal star, it would most likely move at a sizable fraction of the other stars' velocity; but in fact it is virtually stationary. Detractors point out that there could well be enough ordinary stars to account for all the mass present in the domain of the central star cluster, so a black hole is not *needed* to explain the stellar motions. They also worry that the energy output of SgrA* is implausibly *low* for a very massive black hole sitting in such a dense environment of stars and gas.

Arguments for and against a black hole in the Galactic center may now be approaching a conclusion. A few years ago a pessimist might have found in this controversy a warning against expecting black holes to signal their presence in any obvious way. But optimists now rule in the search for black holes. As we explain in the following chapters, it seems that many black holes in galactic nuclei fairly scream out their presence with astonishing displays of pyrotechnics. Through these dramatic signals, we can confidently claim to have discovered truly supermassive black holes in the nuclei of other galaxies. Some of them approach a billion solar masses or more, dwarfing any hypothetical black hole in our own Galactic center. They produce the most luminous phenomena known anywhere in the Universe—the quasars—and can also be detected through their gravitational influence on the motions of stars and gas in the nuclei of galaxies.

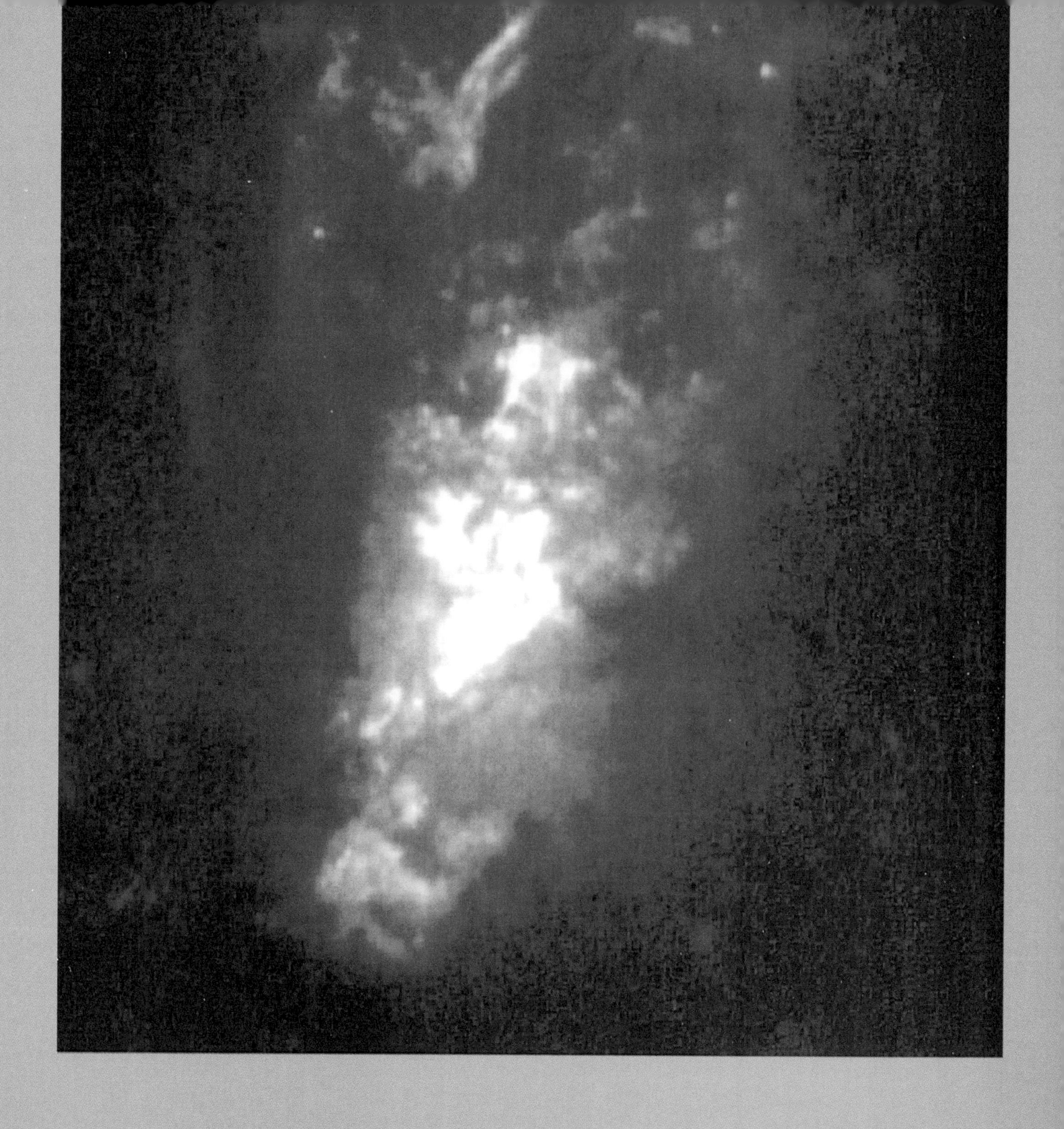

*Chapter* 5

# Quasars and Kin

*Like a searchlight beam piercing a misty sky, the active nucleus of the nearby Seyfert galaxy NGC 1068 lights up surrounding wisps of dust and gas. Hidden from our direct view (at lower left) inside a doughnut-shaped ring of opaque gas clouds, the intense ultraviolet radiation from the nucleus can escape only through the doughnut's opening. This Hubble Space Telescope image shows the cone of interstellar gas illuminated and ionized by the escaping rays. If we could look along the axis of the doughnut, we would be able to see the unattenuated radiation from the "central engine."*

Not long after "spiral nebulae" and their elliptical cousins were recognized as "island universes" of stars and gas separate from the Milky Way, astronomers began to realize that something strange was going on in the centers of many galaxies. Often, there lurked intense concentrations of blue light. This light had characteristics quite unlike the radiation associated with aggregates of stars and gas, the normal components of galaxies. Its spectrum contained too much blue and ultraviolet radiation to come from ordinary stars, even if they were very hot. Sometimes these central sources of energy were as bright as the entire surrounding galaxy, and later it was discovered that their brightness often varied. Galaxies containing these central sources came to be known as *active galaxies,* and the central sources themselves are called *active galactic nuclei,* or *AGN*s for short. The most extreme examples are called *quasars.* Observations made in the radio band revealed even more bizarre behavior—the narrow, fast jets of gas to which we devote the next chapter.

The discovery and interpretation of active galaxies and quasars was a convoluted process, with many false starts and wrong turns. Perhaps more than any other phenomenon discussed in this book, the discovery of the nature of AGNs involved serendipity in its purest form. AGNs even trump pulsars on this score, since astronomers were already looking for neutron stars in 1968 and Franco Pacini had even suggested that spinning magnetized neutron stars might radiate—although he hadn't guessed that they would pulse! *No one* had predicted or even speculated that galaxies could be "active." It required accidental discoveries by both radio and optical astronomers before it became clear that observers had found one of the most energetic phenomena in the Universe—and, almost certainly, the "smoking gun" of massive black holes in the nuclei of galaxies.

## Signs of Activity

Evidence that some form of violent activity was occurring in the centers of galaxies had trickled in for nearly half a century before quasars were finally discovered in 1963. Heber D. Curtis, then Director of Lick Observatory near San Jose, California, had discovered in 1917 that the nearby galaxy M87 had a jetlike feature emanating from its nucleus. He remarked on its peculiarity but did not follow up his observation. Other nearby galaxies, such as M82, seemed to be undergoing some kind of violent disruption, although it later turned out that M82 belongs to a subclass of galaxies, called "starbursts," whose activity results not from a central black hole but rather from hyperactive star formation.

No firmer clue that something unusual was taking place in some galaxies appeared until World War II. Then Karl Seyfert, at the Mount Wilson Observatory in California, noted that a sizable subset of spiral galaxies—as many as a few percent—have intense blue pointlike nuclei. Spectra of what we now call "Seyfert galaxies" show strong emission lines of the type produced by clouds of ionized gas. Instead of being sharp like emission lines observed in the laboratory, however, these lines are smeared over a surprisingly large range in wavelength. This smearing is believed to be due to the Doppler effect and indicates that the gas emitting the lines is extremely turbulent, with random velocities as large as a few percent of the speed of light—10 to 100 times faster than the ordinary gas motions found in a galaxy. The Armenian astronomer B. E. Markarian, while making a survey similar to Seyfert's, later found that some elliptical galaxies also have intense blue nuclei.

But it was not until the radio astronomers weighed in that the astronomical community was forced to confront the possibility that some galaxies might be more than just aggregates of ordinary stars and gas. In 1954 Walter Baade and Rudolf Minkowski, of the Mount Wilson and Palomar Observatories, were trying to identify a source of radio waves called Cygnus A, the second "brightest" object in the radio sky. At that time astronomers knew only that something was emitting a lot of radio waves, but they didn't know what, or even how distant, the object was. Baade and Minkowski's breakthrough was to associate Cygnus A's position on the sky with a very faint galaxy having a redshift of 0.05, thus

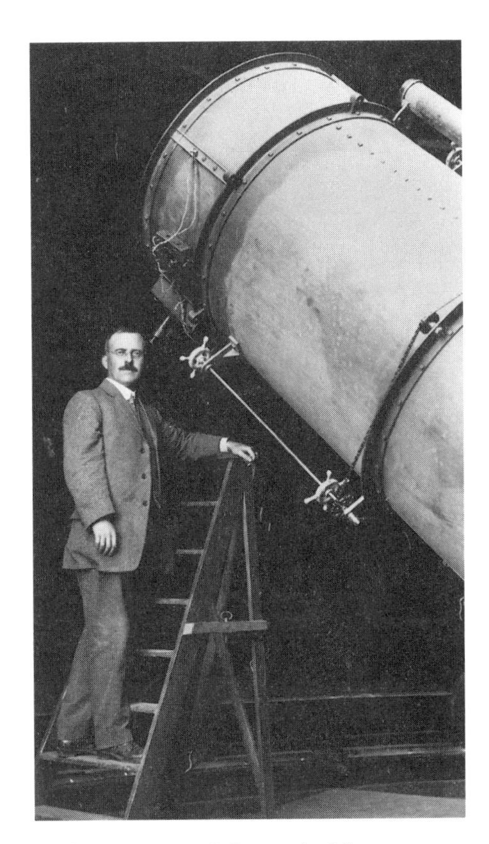

*Heber Curtis, of the Lick Observatory in California, discovered as early as 1917 that there was a "jet" emerging from the center of the giant elliptical galaxy M87 in the Virgo Cluster.*

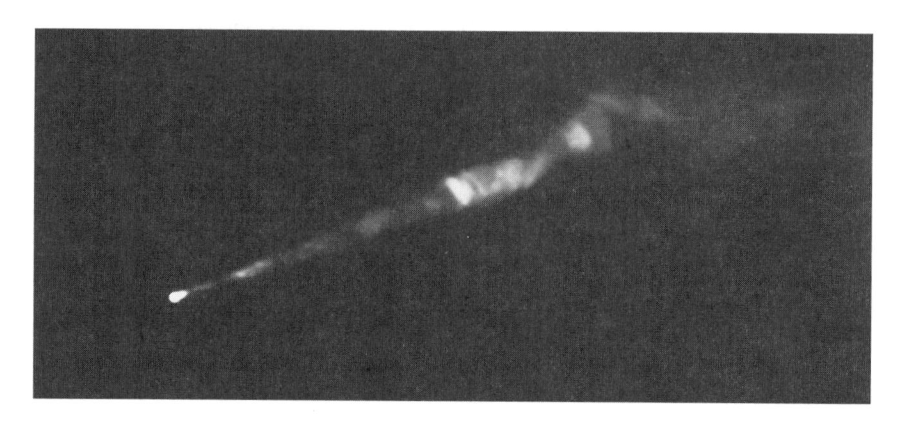

*The jet that emerges from the nucleus of the elliptical galaxy M87 is about 10,000 light-years long. This image shows the complex pattern of radio emission from the jet. The bright "knots" could be shock waves, and the side-to-side "wiggles" might represent instabilities or magnetic structures. Images at visible and ultraviolet wavelengths reveal strikingly similar features.*

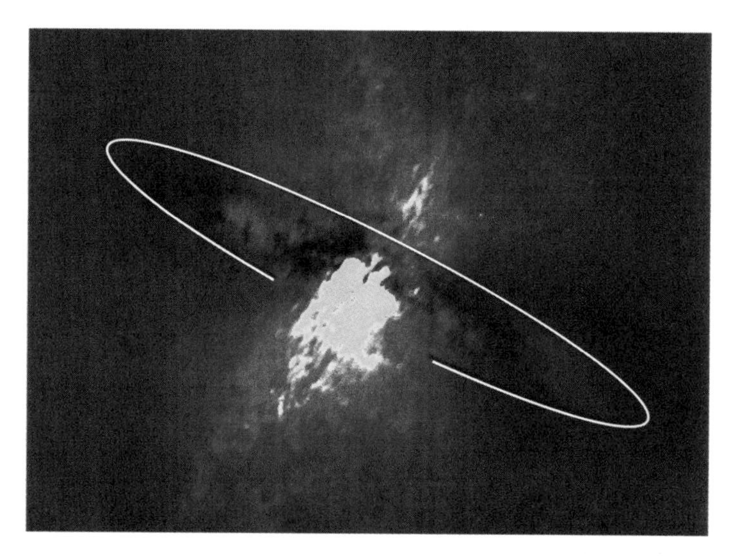

*The nearby "starburst" galaxy M82, imaged through a filter that blocks all but the light of the Hα spectral line. Starlight and emission from cool gas and dust is invisible in this view; an ellipse indicates the outline of the nearly edge-on galactic disk. New stars are forming at a furious rate near the center of M82. Frequent supernova explosions create a hot wind that escapes through two conical channels along the rotation axis of the disk. The Hα emission is produced by 10,000-degree gas on the fringes of the wind; the interior, hotter still, is observable only in X-rays.*

*If an array of randomly moving gas clouds emits light in a spectral line, the receding clouds will appear reddened relative to the approaching clouds. The light from the whole system, when spread out into a spectrum, yields a line that is much broader than that from an individual cloud. The overall width is determined by the spread in cloud velocities: the shape of the line indicates the relative number of clouds with different velocities.*

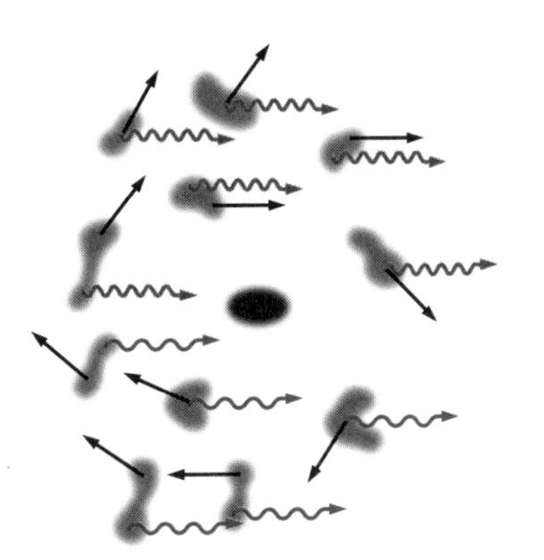

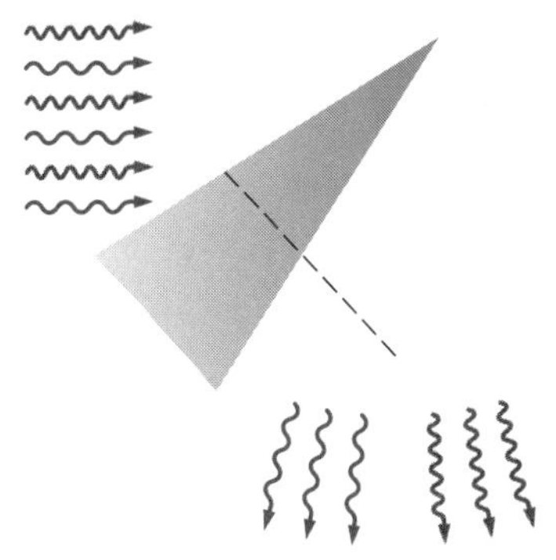

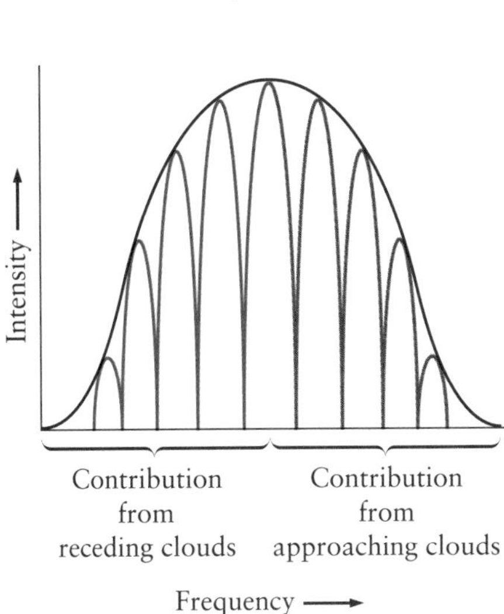


placing it about 300 times farther away than the Andromeda Galaxy. This distant galaxy was evidently the source of the radio emission. If a galaxy that produced such faint optical signals could produce such loud radio signals, then the hope was raised that some peculiar galaxies might be detectable by radio techniques even if they were so far away that the combined light from $10^{11}$ stars failed to register optically.

The faint optical image of Cygnus A showed a seemingly disturbed structure made of two distinct parts, which Baade and Minkowski interpreted as the signature of two colliding galaxies. Shortly thereafter, Roger Jennison and M. K. Das Gupta, of the Nuffield Radio Astronomy Laboratories at Jodrell Bank near Manchester, England, showed that the radio emission from Cygnus A was not coming from the galaxy at all, but from two giant patches placed symmetrically about the galaxy on the sky, a quarter million light-years to either side! These radio patches are thought to have a three-dimensional structure like a dumbbell and hence are often called "lobes." This double radio structure is now known to be characteristic of the strongest sources of radio waves, and we shall have much more to say about its interpretation in the next chapter. Jennison and Das Gupta's discovery cast doubt on the collision hypothesis (and the double *optical* structure proved to be an illusion created by an obscuring lane of dust), but left the mystery of the energy source for the radio emission more urgent than ever.

It was already recognized in the 1950s that the radio emission from cosmic sources was synchrotron radiation emitted by electrons gyrating in magnetic fields at nearly the speed of light, just as in the Crab Nebula. To produce the observed amount of synchrotron emission, the lobes would have to contain energy in the forms of both fast-moving electrons and magnetic fields. But how much energy? In celebrated papers published in 1956 and 1959, Geoffrey Burbidge showed that the total energy content of the radio lobes of an extended source such as Cygnus A—the combination of magnetic fields and fast particles—must exceed the amount that would be released by the *complete annihilation* (via $E = mc^2$) of a million solar masses of material.

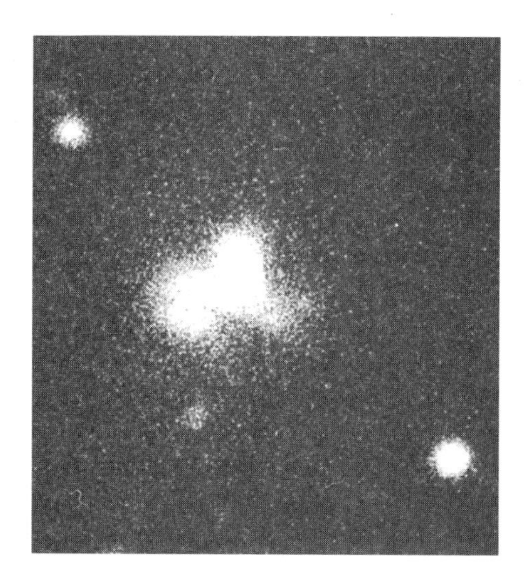

*Walter Baade and Rudolf Minkowski identified the radio source Cygnus A—the second brightest source in the sky (not counting the Sun)—with this faint galaxy 300 times farther away than the Andromeda Galaxy. The dust lane running across the galaxy gave it a double appearance, which misled Baade and Minkowski into proposing that two galaxies were colliding. The radio output is actually energized by a concentrated power source in the galactic nucleus.*

Burbidge's result was the first indication that galactic nuclei can release energy on scales that vastly exceed even that of a supernova explosion, and that somehow this energy is channeled into the form of particles moving very close to the speed of light—in other words, a "relativistic" plasma—and magnetic fields. Where did this energy come from? The great Armenian astrophysicist Victor Ambartsumian (who, as director of Byurakan Observatory in Soviet Armenia, had sponsored Markarian's survey) speculated that these "exploding galaxies" were the places where matter was being created, as required by the then-popular "steady state" theory of cosmology propounded by Hermann Bondi, Fred Hoyle, and Thomas Gold. (According to this theory, the gross structure of the Universe was unchanging despite the cosmic expansion, because matter was continuously being created. Despite its philosophical appeal, this theory has been generally discarded because of compelling evidence that the expansion of the observable Universe began at a fixed time in the past, and that distant galaxies at high redshifts, whose light set out when the Universe was younger, look systematically different from their nearby counterparts.) Although Ambartsumian's conjecture was forgotten when the steady state theory fell into disrepute, other speculations abounded as well—for example, it was suggested that galactic nuclei were giant matter–antimatter annihilation reactors. No one yet suspected black holes.

The major contribution of optical astronomy to this story came in 1963. Cyril Hazard, M. B. Mackie, and John Shimmins, using the recently completed Parkes Radio Telescope in Australia, managed to pin down the position of a particular radio source, 3C 273, to a precision of a few arcseconds, by noting the time at which it was occulted by the Moon. They had to make special adjustments to the telescope to observe something as far north—and consequently as low in the sky—as 3C 273. ("3C" denotes the third Cambridge catalogue of radio sources—the source had been discovered by means of a radio telescope in Cambridge, England.) Fixing the position of a radio source was a real coup at that time, since radio telescopes had much poorer angular resolutions than they do today. The ability of any kind of telescope to pinpoint a location on the sky is fundamentally limited by the ratio of the aperture of the telescope to the wavelength of the radiation being observed. Radio waves have wavelengths of a few centimeters, so to obtain a resolution of 1 arcsecond ($5 \times 10^{-6}$ radian) would require a telescope several kilometers across! In photographs taken in visible light, the only object coincident with the radio position of 3C 273 on the sky was a 13th-magnitude starlike ob-

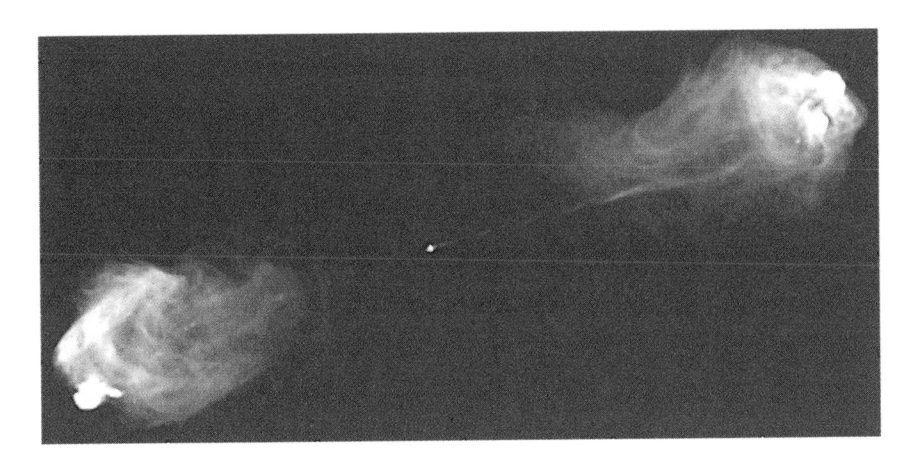

*The VLA reveals the double-lobed radio structure of Cygnus A. The energy produced in the nucleus is carried to the lobes by jets, the traces of which are visible as thin lines. "Hot spots" at the end of each lobe are thought to be shock fronts where the jets run into the intergalactic gas. The source extends about half a million light-years end to end; the galaxy photographed in visible light is less than a tenth this size.*

ject. Although an object with this magnitude is about a thousand times fainter than the faintest stars visible to the unaided eye, it is not especially faint by the standards of modern astronomy. Ground-based optical telescopes equipped with electronic sensors routinely observe stars and galaxies that are more than 10,000 times fainter than 3C 273.

Once astronomers had located an object that seemed to be the radio source 3C 273, Maarten Schmidt used the 200-inch Hale telescope on Mount Palomar to take a spectrum of it. At first the spectrum seemed like gibberish—none of the spectral lines seemed to be at the wavelengths produced by known atomic elements. However, Schmidt eventually noticed that the *relative* wavelengths of the lines fit the pattern formed by glowing hydrogen, but redshifted by 15 percent. At first he was baffled, since an object at that redshift would be nearly 2 billion light-years away, if it were receding from us according to Hubble's law of cosmic expansion. At that distance, even an entire galaxy would have been much fainter than 3C 273. Yet Schmidt finally convinced himself that 3C 273 was actually at a cosmological redshift of 0.15, and was therefore by far the most luminous object ever detected. Once the intellectual block had been removed, it was quickly realized that the hitherto mysterious spectra of other objects of the same class also displayed high redshifts. A check of photographic archives, kept at Harvard College Observatory since the late nineteenth century, revealed that 3C 273 fluctuated in brightness on a timescale as short as a month. Thus, there seemed to be a class of objects that looked like ordinary stars on photographic plates, but showed spectral emission lines with high redshifts. These objects, dubbed quasars or QSOs (for quasi-stellar objects), could outshine an entire galaxy by more than a factor of 100!

*Maarten Schmidt was the first to recognize that quasars are ultra-luminous distant objects. After Cyril Hazard and his colleagues in Australia had pinned down the position of the radio source 3C 273, Schmidt took a spectrum of the starlike object in that location. The unusual features of the spectrum could be understood only if there was a 15 percent redshift, implying an enormous distance. Schmidt has, over the last thirty years, continued to discover and study distant quasars. He has been a leader in analyzing how the quasar population depends on redshift (or cosmic epoch), as described in Chapter 7.*

Part of the reason that quasars were not recognized sooner was that they superficially resemble ordinary stars. They surrendered their extraordinary secret only upon careful study with spectrographs, performed after their radio emission had been detected by chance. Quasars have just the properties envisaged by John Herschel in 1820, in a draft of his address to the newly founded Royal Astronomical Society: "Yet it is possible that some bodies, of a nature altogether new, and whose discovery may tend in future to disclose the most important secrets in the system of the Universe, may be concealed under the appearance of very minute single stars no way distinguishable from others of a less interesting character, but by the test of careful and often repeated observations."

Quasars are believed by most astrophysicists to be just very luminous versions of the same blue nuclei that Seyfert observed in the centers of nearby spirals. The reason that they appear to us as isolated stars is that they so outshine the surrounding galaxy as to render it invisible. That most of the early quasars discovered had strong radio emission is not surprising, since optical astronomers used the radio emission as a tipoff to decide which objects to study. However, we now know that quasars with strong radio emission are the exception rather than the rule. In nine out of ten cases, the intense pointlike source with broad (that is, Doppler-smeared) emission lines is unaccompanied by strong radio emission.

Whereas quasars and Seyfert galaxies are especially luminous at optical, ultraviolet, and X-ray wavelengths, other galaxies show anomalous activity of a very different sort. *Radio galaxies,* of which Cygnus A and M87 (sometimes called Virgo A by radio astronomers, since it is the brightest radio source in the constellation Virgo) are archetypes, somehow manage to channel the bulk of their energy outputs into the radio band, producing large diffuse patches of radio emission on the sky. Astronomers love to invent classification schemes for new phenomena, and the many different varieties of "active galaxy" have provided a fertile outlet. The major classes of Seyferts, quasars, and radio galaxies are all subdivided in a variety of ways, often according to overlapping observational criteria. Unfortunately, observers were often more perceptive in distinguishing the different classes than they were imaginative in naming them. As a result, we routinely use the nondescriptive terminology of Type 1 and Type 2 for Seyfert galaxies, "Fanaroff-Riley types" 1 and 2 for radio galaxies, and other such nomenclature. (This seems to be a general problem in astronomy: supernovae, for example, are classified as Type 1, Type 1a, Type 1a-peculiar, Type 1b, Type 2, etc.; stellar populations are Pop I and Pop II, etc.) Additional subcategories were invented to represent transitional objects. For example, highly variable, polarized AGNs with *no detectable emission lines* are placed in a special category called BL Lac objects, so named because the prototype had been misclassified as a variable star, BL Lacertae.

All the varieties of quasars, Seyferts, and radio galaxies are considered to be members of the observationally heterogeneous class of objects known collectively as active galactic nuclei. But how much do these objects really have in common? Although spiral and elliptical galaxies have very different appearances, stellar populations, and internal dynamics, they share fundamental structural similarities as self-gravitating systems of stars and gas. Therefore, lumping them together as *galaxies* amounts to a powerful generalization that has enabled astrophysicists to uncover principles that unify the types of structure found in the Universe. Does the term "active galactic nucleus" similarly reflect a unity among these seemingly diverse objects, at a deep level?

One obvious link is that all forms of activity in these galaxies originate in the nucleus. (The connection is obvious for quasars and Seyferts, because of their bright compact nuclei; we shall see in Chapter 6 how the connection has been established for radio galaxies.) But one can go further. Generic similarities among the various forms of activity suggest that they all represent different manifestations of the same kind of phenomenon. One might group these generic properties into four categories, as follows:

*AGNs can emit energy at an enormous rate.* The power outputs of the most luminous quasars and the most powerful radio galaxies can exceed the total luminosity of *all* the stars in a large galaxy by more than a factor of 100. Episodes of activity can apparently last for many millions of years. AGNs are the most powerful sustained phenomena known in the Universe.

*AGNs are extremely compact.* We have already noted that the quasar 3C 273 changes in brightness in the course of a month. Rapid variability—on timescales as short as a few hours in some cases—is a generic feature of AGNs. This variability puts an upper limit on the size of an AGN. If an astronomical body routinely varies in brightness by a factor of 2 over a certain time interval, one can generally infer that it must be smaller than the distance light can travel during the same interval. No signal can travel faster than the speed of light to tell different parts of the object to vary in step.

Consider the following analogy. The final game of the Intergalactic World Cup is being held on Earth, with the Milky Way team playing the Andromeda Galaxy. Four planets—Earth, Mars, Jupiter, and Saturn—all happen to be lined up on one side of the Sun and lie in the plane of the sky as viewed from Andromeda. To take advantage of this unusual arrangement, an executive from the ABC (Andromeda Broadcasting Corporation) Network proposes the following stunt: the four planets will be wired so that when a switch is flipped on Earth, they light up to spell "GOAL." Each planet is to display one letter—"G" on Earth, "O" on Mars, "A" on

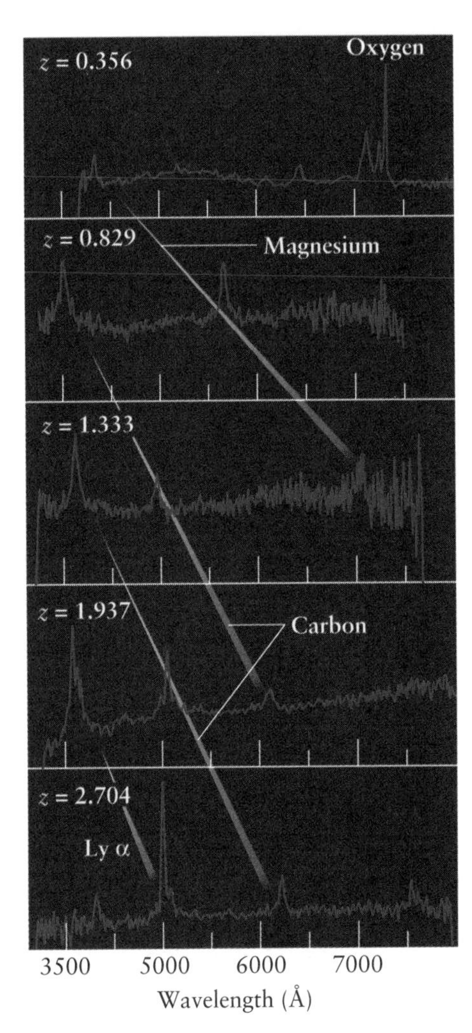


*Quasar spectra of different redshifts, z, show a variety of prominent lines. The redshift stretches the wavelengths by a factor (1 + z) between emission and detection. In quasars with redshifts of 2 or more, the strongest line observed at visible wavelengths is the Lyman α line of hydrogen, which is emitted in the far ultraviolet . Since far-ultraviolet radiation cannot penetrate the Earth's atmosphere, this line is never observed from the ground except when it is redshifted.*

*The interplanetary "GOAL" sign at the Intergalactic World Cup. Since it takes 71 minutes for signals from Earth to reach Saturn, traveling at the speed of light, the Andromedans do not see all the letters of the sign light up simultaneously. Reversing the argument, we can estimate the maximum possible size of a variable radiation source—such as a quasar—by assuming that the typical variation timescale is the light travel time across the source.*

*What the Andromedans saw . . .*

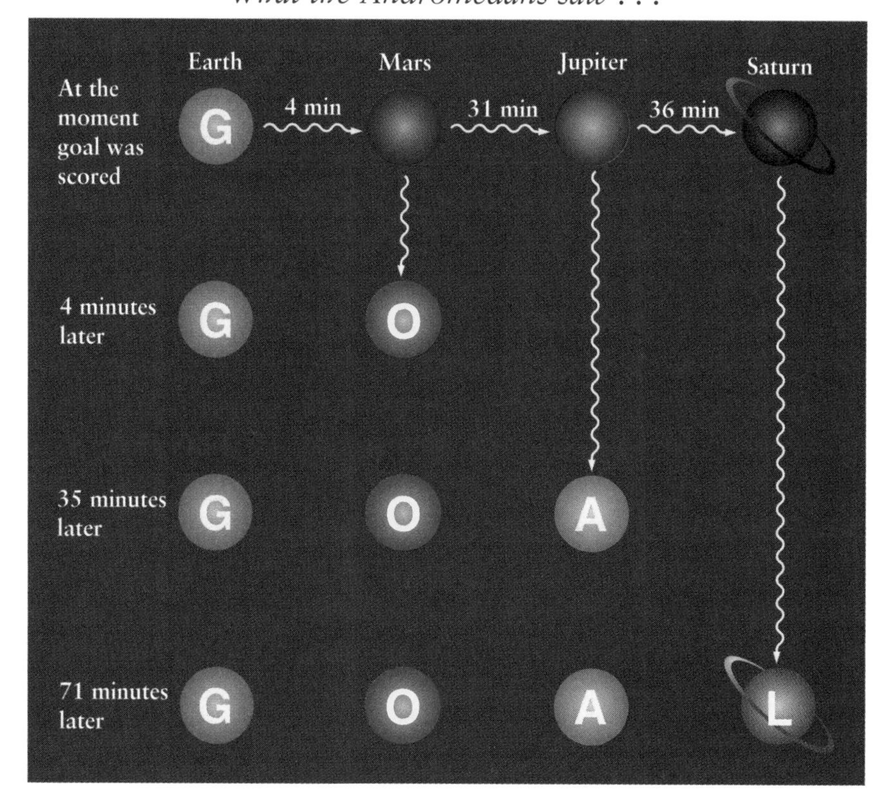


Jupiter, and "L" on Saturn. The idea is that Andromedan sports fans with powerful telescopes can then see the Solar System light up as soon as a goal is announced over the ABC Network. Unfortunately, the plan doesn't work very well. True, as soon as they hear about the goal via a direct radio link from the Earth, they see the Earth light up with a "G." But 4 minutes pass before the "O" on Mars switches on, another 31 for the "A" on Jupiter, and an additional 36 for the "L" on Saturn. Altogether, it has taken one hour 11 minutes for the sign to turn on, by which time the game is over. The reason, of course, is that it takes light 71 minutes to travel from the Earth to Saturn. There is no way, even in theory, that Saturn could know about the goal less than 71 minutes after it occurred. Likewise, if a quasar were as big as the distance from the Earth to Saturn, it could not change in brightness significantly in less than about 71 minutes. In fact, quasars seldom vary this rapidly, but the observed rate of variability does imply that they cannot be too much larger than the Solar System—a maximum size that is tiny by galactic standards.

*AGN radiation is unlike that normally produced by stars or gas.* Although AGNs radiate most strongly in one part of the spectrum or another, they always produce substantial radiation over a very wide range of wavelengths. Some AGNs emit comparable amounts of power in the radio, infrared, optical, ultraviolet, X-ray, and gamma-ray portions of the spectrum, a factor of more than $10^{10}$ in wavelength! Stars and gas, by contrast, tend to emit the bulk of their radiation over a relatively narrow band of wavelengths. That characteristic band of wavelengths is inversely correlated with the temperature of the radiating material, according to the so-called blackbody laws describing radiation from solids, liquids, and dense gases. Thus a very hot body will radiate mostly in a band of very short wavelength, in the X-ray or gamma-ray portion of the spectrum. Radiation characterized by a well-defined peak in the spectrum is called "thermal," because it resembles that produced by a radiating solid of fixed temperature, such as the filament in a light bulb or the heating element in an electric stove. AGNs have "nonthermal" spectra, because the power is so evenly distributed in wavelength that it is impossible to associate a temperature with it.

*AGNs contain gas moving at extremely high speeds, sometimes approaching the speed of light.* We can deduce the velocities of gas clouds in quasars and other types of AGN by measuring the Doppler shifts of spectral emission lines. The velocities measured by this method typically range up to 10 percent of the speed of light. As we discuss in Chapter 6, the velocities of jets produced by radio galaxies are usually estimated by more indirect methods; nevertheless, the evidence that they contain gas flowing away from the nucleus at very nearly the speed of light is compelling.

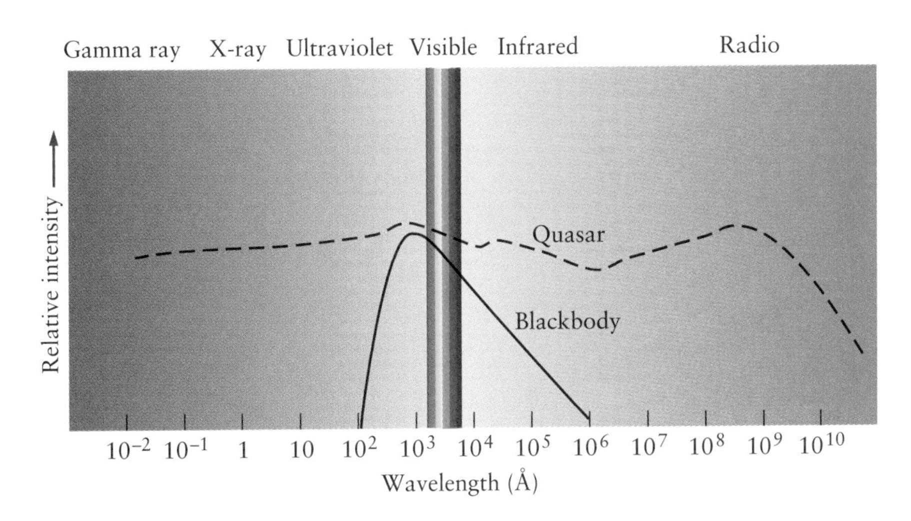


*A "black body" absorbs all the energy that falls on it and radiates at a well-defined temperature. The surfaces of stars, for instance, emit approximately "blackbody" radiation whose spectrum peaks at shorter wavelengths (bluer colors) for hotter bodies. A similar plot for a quasar, however, shows that energy is distributed over a much broader range of wavelengths. This is because several regions with different temperatures contribute to the total spectrum, and also because some of the emission comes from processes (such as synchrotron radiation) that do not have a well-defined temperature. The solid curve indicates what shape a quasar spectrum would have if the radiated power came from immediately around the hole and had a spectrum like that of a star.*

## The Black Hole Paradigm

What can we deduce from these shared characteristics of AGNs? We shall argue that all these features can be explained most simply as manifestations of massive black holes at the centers of galaxies. The crux of the argument is contained in the first two characteristics, which together are so extraordinary that they bear restating: AGNs are objects that can produce a galaxy's worth of power (or more) in a region the size of the Solar System. Since by definition the gravitational field near a black hole can accelerate gas to near the speed of light, it is not surprising that the fourth characteristic can also be accommodated readily within the black hole paradigm. However, the details of how high-speed gas streams are produced seem to be rather complicated; we shall explore some aspects of their creation in Chapter 6. Nonthermal radiation spectra (the third characteristic) are also consistent with the black hole paradigm, but here the details are more debatable; we reserve further discussion on this topic until the end of this chapter.

The reasoning leading to the black hole paradigm is surprisingly simple. Since AGNs vary rapidly in brightness, we can place *upper limits* on their sizes ($R$). At the same time, we can place *lower limits* on the amount of mass contained in the region where the luminosity is produced ($M$). This means that we can derive a *lower limit* on the ratio of mass to size, $M/R$—let's call this the "mass compactness." The mass compactness turns out to be a direct measure of the strength of the gravitational field in an object, and for a black hole it attains a unique maximum value that depends only on constants of nature—$c^2/G$, where $c$ is the speed of light and $G$ is Newton's universal constant of gravitation. Thus, by comparing $M/R$ (estimated from observations) to $c^2/G$ (predicted by theory), we can tell how "close" an object is to being a black hole. Needless to say, general relativity predicts that no object can have a mass compactness that exceeds $c^2/G$!

We estimate size as follows. If the luminosity changes significantly (say, by a factor of 2) during a typical time interval $t$, then it is probably safe to assume that the source cannot be much bigger than the distance light can travel during the same amount of time—that is, $R$ is less than $c \times t$, where once again $c$ is the speed of light. (There is an exception to this rule, when the source is moving toward the observer at nearly the speed of light. This exception is crucially important if one wants to estimate the properties of jets, as we shall see in Chapter 6, but is not believed to be very important for optical emission from quasars and Seyfert galaxies.)

Mass estimates can be made in several ways. One method utilizes Einstein's law of mass-energy equivalence, $E = mc^2$. If we can find out the total energy radiated by an AGN over its lifetime, we can use this relation to calculate the amount of mass converted into energy. No plausible process can convert matter into energy with perfect efficiency; the best one can hope for is an efficiency ranging from 1 percent (for nuclear reactions) to about 40 percent (for matter spiraling into a spinning black hole). Matter–antimatter annihilation, which does yield 100 percent conversion, can be ruled out as the main power supply by other arguments. Thus, at least half the "fuel" that enters the "central furnace" of an AGN retains its form as matter. What happens to this unspent fuel? Given the inexorable pull of gravity, the better part of it remains behind, accumulating in the center of the galactic nucleus. Thus, given the lifetime energy output of an AGN, we can estimate a *minimum* mass for its furnace.

Astronomers have devised at least two techniques for estimating the lifetime energy output of an AGN. If the active nucleus contains a powerful radio source, we can use Burbidge's method to calculate the minimum energy that would be required to produce the observed synchrotron radiation (see page 43). This method works because the radio lobes store a large fraction of the *cumulative* energy output of the AGN. Alternatively, if we can guess the *age* of a source, we can estimate the lifetime output by multiplying the current luminosity by the age. Nobody has found a way to measure the age of an AGN with any reliability, but we can use statistical arguments to place a lower limit on the average lifetime of an AGN. That limit is deduced from the fact that about 1 percent of all galaxies are observed to be active at any one time. If the typical duration of activity were shorter than about 100 million years, then there wouldn't be enough galaxies to account for all the active ones we see (since 100 million years is about 1 percent of the age of the Universe). With an estimate of the age of an AGN in hand, we can perform the sequence of steps needed to calculate a minimum mass: age times luminosity gives total lifetime energy output; assuming an energy conversion efficiency of (say) 10 percent gives the total mass "processed" through the nucleus, of which (say) at least half remains behind in the AGN.

Using a different argument, we can obtain a lower limit on the mass directly from the luminosity of the AGN. Radiation exerts a force, called radiation pressure, that is negligible in the laboratory but must be reckoned with in an object as powerful as a quasar. If the ratio of luminosity to mass in the AGN is too great, the force exerted by the radiation is great enough to blow the matter out of the gravitational field. The limiting luminosity for a given mass is called the "Eddington limit" (after the British theorist who appeared in Chapter 2 as Chandrasekhar's nemesis). Since AGNs seem to persist for a considerable time without dispersal, we can guess that their luminosities do not exceed the Eddington limits for their respective masses, and this assumption leads to a lower limit on the mass.

How do the numbers work out? Fortunately, the masses estimated by different methods seem to be in agreement with one another. A typical luminous quasar, or powerful radio galaxy, must have a central mass on the order of 100 million to a few billion suns. Less luminous Seyfert galaxies can get by with somewhat lower masses. Having obtained a size and a mass, one can now compute the mass compactness $M/R$. But the variability of AGNs is seldom fast enough, and thus the estimated value of $R$ small enough, to clinch the black hole identification without further argumentation. The best we can do by direct measurement is to conclude that $M/R$ must exceed $0.001c^2/G$ in typical AGNs. This is 100 to 1000 times more compact than a normal star and much too compact to be a star cluster. We should remember that this value for $M/R$ represents the minimum possible compactness—we have no reason to expect that the luminosity of an AGN fluctuates at the maximum rate consistent with its size and the speed of light. But this lower limit on $M/R$, by itself, does not offer ironclad proof in favor of black holes.

Yet we believe that massive black holes lie at the centers of active galactic nuclei. Why? Partly by a process of elimination. Just as we saw in Chapter 3 that compact stellar remnants more massive than about 3 solar masses can hardly be anything but black holes, so have astrophysicists failed to imagine any realistic alternative to massive black holes in AGNs. From time to time alternatives have been proposed, but none has survived close scrutiny. As we saw in Chapter 4, a variety of runaway processes are expected to occur in galactic nuclei composed of ordinary stars and gas. Could these processes result in highly compact and luminous nuclei of some form other than black holes? We are not yet able to predict exactly where such runaway processes will lead. Not only would the stars interact with one another gravitationally, but also stellar collisions and interactions with gas would be important. Computer models are starting to grapple with these complexities, but we still do not know whether a dense cluster of ordinary stars would dissolve into a single supermassive star or into a cluster of neutron stars and stellar-mass black holes that would then evolve further via gravitational interactions. Some of the collisions that hasten the collapse of an ordinary star cluster can be avoided once the stars themselves are compact. But even these superdense star clusters would be subject to instabilities (predicted by general relativity) that would lead to their prompt collapse to a single massive black hole. The prospects for avoiding the formation of a black hole do not look promising.

Could a "supermassive star" be the long-lived "central engine" of an AGN? Probably not. Any star much more massive than 100 solar masses would be supported against gravity by the force of radiation pressure and would have a luminosity very nearly equal to the Eddington limit corresponding to its mass. But this equilibrium is precarious. Such stars are barely stable according to the Newtonian theory of gravity, and the corrections required by general relativity, even when small, are destabilizing. Thus, it appears that "supermassive starhood," if it occurs at all, is unlikely to represent more than a brief phase on the way to a massive black hole. The stages leading up to the formation of such a hole become increasingly short-lived as the galactic nucleus becomes more centrally condensed. Moreover, a black hole offers a more efficient power source for an AGN than any conceivable progenitor.

All the above arguments suggest that we should associate at least the most powerful and long-lived objects—quasars and strong radio sources—with black holes, and then consider, as a secondary issue, whether some other categories of AGN (such as starburst nuclei) could be precursor stages in any of the evolutionary pathways toward a black hole shown in the figure on page 98. But one should keep in mind that not *all* pathways of cluster collapse have been fully explored. For example, not much is known about supermassive stars with rapid rotation. Perhaps such an exotic object, endowed with a magnetic field, could become a sort of supermassive pulsar. But most astrophysicists regard such possibilities as far-fetched, and are optimistic that quasars can be explained by a theory in which the prime mover is a black hole as massive as 100 million suns or more. Just as X-ray binaries are fueled by the transfer of gas from the companion star onto the compact remnant, the massive black holes in AGNs would be fueled through the capture of gas from the surrounding galaxy, or even by the swallowing of entire stars. This captured debris would swirl downward into the intense gravitational field, reaching nearly the speed of light before it is swallowed. The gravitational energy thus liberated would provide the power for both the luminosity and the outflowing gaseous jets that characterize AGNs.

## Is "New Physics" Necessary?

Imagine an energy source equivalent to 10 trillion Suns, rather than just one, crammed within the Earth's orbit! The stupendous furnace is rather unstable and turbulent, with pronounced power surges and lulls often occurring within the space of a few hours. This description gives a fair representation of the conditions that must exist in a typical quasar. Now, does your intuition tell you that quasars can be explained in terms of well-accepted theories of physics?

We, and most astrophysicists today, would answer "yes"—all of these phenomena can be understood in terms of the interactions between a black hole and surrounding matter. But our intuition is steeped in 25 years of

*For more than 25 years, Fred Hoyle was perhaps the most versatile and original figure in the world of astrophysics. He pioneered the study of how the chemical elements are synthesized inside stars, co-invented the "steady state" cosmology, and contributed many ideas on stellar and galactic structure. In the early 1960s, he was one of the first to emphasize the importance of gravitational energy and supermassive objects in galactic nuclei.*

black hole lore. If you had answered "no" back in 1963, you would have been in the company of many eminent astrophysicists. Starting when quasars were discovered, and lasting at least into the mid-1970s, there was a lively debate over whether the laws of physics as we know them are adequate to explain objects like quasars, or whether there may be new basic laws that manifest themselves only on a cosmic scale. Note that black holes should be considered "established physics" in this context, since by 1970 their properties, as predicted by general relativity, were well delineated. The plausibility of their existence was also becoming well established, although at the time many physicists and astronomers were still uncomfortable with the concept that black holes were an almost inevitable feature of the Universe.

Much of the debate over the need for "new physics" centered on whether quasars really were ultraluminous objects at huge distances. Could it be that astronomers were wrong to apply Hubble's law relating redshift to distance, which worked so well for ordinary galaxies, to such extraordinary objects as quasars? If the redshifts arose some other way, then quasars could be much closer, and their inferred power outputs would not be so enormous. One group of prominent researchers, among them Fred Hoyle, Geoffrey Burbidge, and Halton "Chip" Arp, proposed that quasars were projectiles shot out of individual galaxies. These astrophysicists produced data claiming to show statistical correlations linking high-redshift quasars to low-redshift galaxies that appeared nearby on the sky; they claimed, moreover, to find "bridges" of gas and stars leading from galaxy to quasar. The motion of quasars outward bound from the Milky Way would produce the observed redshifts. But where are the quasars coming *toward* us from neighboring galaxies? These should have blueshifts, yet no blueshifted quasars have been found.

Others postulated quite new physical effects to explain the redshifts. Advocates of these ideas have from time to time, in support of their case, pointed to various statistically odd patterns that they have observed, arguing, for example, that quasars are preferentially found at certain redshifts. If these statistical arguments had survived scrutiny, they would have posed serious challenges to the interpretation that quasars really did lie at huge, "cosmological" redshifts. However, even when these effects were first claimed, it was hard to assess their significance. One can all too easily perceive patterns in random data. It is one thing to search for an improbable pattern that has been specified in advance, but quite another to select one from among many possible patterns after the fact. The methodological dangers here are well known, if not always fully appreciated. When there is said to be only one chance in (say) 200 of getting a particular effect by pure coincidence, we should not blindly accept this result as statistically significant without applying a "discount" to allow for all the other similarly improbable effects that might equally well have been found but were not. The crucial test is whether a hypothesis applies not just to the objects in which the alleged effect was first noticed, but to some new samples as well. Astronomers have by now measured the redshifts of more than 10,000 quasars. Most of the previously claimed peculiarities in the distribution of quasar redshifts or positions have been weakened, not strengthened, as more data have accumulated. As studies of quasars continue, additional surprising effects will inevitably be discovered. But unless these effects can be incorporated into a well-defined theory with predictive power, they add no cumulative weight to a particular unorthodox viewpoint.

The debate over the nature of quasars has been fascinating, not only for its intrinsic scientific importance, but also because of the way it has highlighted the contrasting attitudes of different scientific personalities. Many who have supported the view that quasars can be explained within the envelope of standard physics would have been genuinely disturbed if anomalous redshifts really existed, because it would have meant that we were farther from delineating the Universe's large-scale structure. On the other hand, those who espoused radical or iconoclastic views would have been elated had astronomical observations upset the apple cart and revealed some fundamentally new physics, since it would have meant that astrophysics was more than merely an exercise in applying laboratory physics under extreme conditions. Philosophers of science who subscribe to Kuhn's theory of "paradigm shifts" would be surprised at the many astronomers who are eager rather than reluctant to join a revolutionary bandwagon.

*Geoffrey Burbidge recognized, as early as 1956, the colossal energy requirements of extragalactic radio sources, and played a leading part in the debates, during the 1960s and thereafter, about the nature of quasars. He has maintained a collaboration with Hoyle that has lasted forty years, exploring nucleosynthesis, galactic nuclei, and cosmology.*

By now, the vast majority of astronomers accept the huge distances of quasars as indicated by their redshifts. An important turning point was the discovery that some quasars clustered on the sky with ordinary galaxies and shared the same redshifts. Since the redshift–distance relation of galaxies is well established, this result converted most, but not all, of the skeptics. Even more compelling evidence for the remoteness of quasars came later when astronomers realized that some lay far behind galaxies that themselves had substantial redshifts. Sometimes the galaxy imprints a signature with its own redshift on the quasar light passing through it; in other cases, the galaxy's gravity deflects light from the quasar, producing multiple "gravitationally lensed" images.

Many aspects of quasars remain problematical, and much remains to be explained in detail. But the same can be said of many better-known and better-studied phenomena in astrophysics (the eleven-year cycle of sunspots, for instance), and even in terrestrial and laboratory physics. It would have been more surprising had the quasar puzzle yielded immediately to the limited efforts of the relatively few scientists working on the subject. But if we look back at the insightful early work of Edwin Salpeter and of Yakov Zel'dovich and Igor Novikov, who as early as 1964 were arguing that supermassive collapsed objects were implicated in the quasar phenomenon, progress indeed seems slow. This is partly because quasars were discovered so early. If they had been discovered in the early 1970s when pulsars and compact X-ray sources had familiarized us with the efficiency of gravitation as a power source, a consensus would have developed more quickly, and there would have been less inclination to invoke "new physics."

This is not to say that we advocate excessive complacency. It is not inherently implausible that some basically new law of nature may operate in the nuclei of distant galaxies. After all, a physicist whose laboratory floated freely in space probably would never have discovered gravity, because this force is very weak unless a large mass such as the Earth is nearby. Maybe there are other effects, insignificant even on the scale of the Solar System, that are nonetheless crucial on the millionfold-larger mass and size scales of galactic nuclei.

The question of whether quasars are "near or far" reminds us of another astronomical controversy. This one, which took place two hundred years ago, concerned the reality of binary stars. Many instances were known of pairs of stars that lay close together on the sky, and in 1767 John Michell (the same man who conjectured about black holes) showed statistically that there were too many such pairs for them to be merely chance alignments of foreground and background stars. He therefore argued that these stars must be physically associated "either by gravity . . . or by some other law or appointment of the creator." One of the leading astronomers of the time, William Herschel (whose work on the structure of the Milky Way was mentioned in Chapter 4), contested this conclusion on the grounds that the members of the pairs often differed greatly in apparent brightness. Since it was then "well known" (though, as we now realize, completely false!) that all stars had the same intrinsic luminosity, Herschel thought that brightness on the sky had to be a measure of distance. Being of different brightnesses, the members of the alleged binary pairs must, he claimed, be at very different distances from us. It took 36 years for Herschel to change his mind.

We readily see a parallel between this debate and the controversy between those who believe that the redshifts of quasars are a true measure of their distances and those who adduce statistical evidence to the contrary. We doubt very much that the upholders of orthodoxy will be forced to recant this time. But we shall never be sure that there are absolutely no anomalous redshifts. We should recall that another momentous debate early this century ended in a draw when some of the "nebulae" proved to be distant galaxies while others were nearby clouds of glowing gas.

## Modeling Quasars

It is one thing to accept that supermassive black holes interacting with local matter cause the phenomena of active galactic nuclei, and quite another to understand how these interactions lead to AGN phenomena *in detail.* We are still far from a detailed model for AGNs, but certain pieces of the puzzle seem to be falling into place.

Massive black holes can generate a high luminosity in two quite distinct ways: straightforwardly by accreting mass, like the X-ray binaries discussed earlier; or via electromagnetic processes that tap the spin energy of the black hole. In the latter case the black hole behaves like a flywheel, and the surrounding matter behaves like a "brake" that extracts energy by slowing down the spin. The flywheel mechanism may be key to the production of jets, as we shall see in Chapter 6, but there is no reason that both mechanisms couldn't operate simultaneously. The properties of an AGN must depend, among other things, on the relative contributions of these two mechanisms, which in turn depend on the accretion rate, the magnetic fields, and the spin of the hole. The observed properties of AGNs must also depend on other parameters, such as the mass of the black hole and the orientation and properties of the host galaxy. Ideally, one would like a unified model that explains the multifarious types of AGN in the same way that theories explain the correlations among various properties of stars.

A fairly close analogy can be drawn between accretion in X-ray binaries and in AGNs. Mass transferred from the binary companion supplies the fuel in an X-ray binary; in the case of an AGN, it is uncertain whether the accreted matter comes from normal interstellar gas in the surrounding galaxy, gas captured in a close encounter or merger with another galaxy, or (particularly in the less luminous cases) gas released in the disruption of individual stars passing too close to the hole. In any of these cases the gas is likely to have enough angular momentum that it cannot fall in directly, but rather must form an accretion disk. The theory of accretion disks makes specific predictions about the radiation spectrum that should be emitted by a disk with a given accretion rate, around a black hole of given mass. Whereas the spectrum of an X-ray binary is predicted to peak, as observed, at X-ray wavelengths, the accretion disks in AGNs are predicted to peak at ultraviolet wavelengths. How does this prediction square with observations? Much of the ultraviolet wavelength band is difficult to observe directly because it is absorbed by intervening clouds of dust and gas. Still, it appears that the spectra of quasars do indeed exhibit an "ultraviolet bump," which could be the radiation emitted by an accretion disk.

But an accretion disk alone cannot explain all the features of AGN spectra. This is hardly surprising, given that galactic nuclei display activity

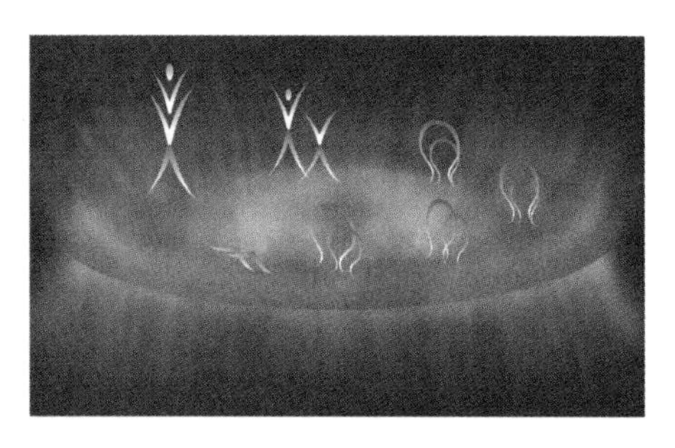

*Because the inner parts of an accretion disk move faster than the outer parts, lines of magnetic force are sheared and stretched. The magnetic field strengthens until it forces its way out of the disk plane. The resultant loops of magnetic field can flail around and dissipate energy, rather like the magnetic arcs sometimes seen on the Sun. Magnetic fields may also provide the main friction (or viscosity) that removes angular momentum and allows the material in a disk to swirl inward toward the black hole.*

in many forms, in all wavebands from radio to gamma rays. Unfortunately for us theorists who prefer simple all-encompassing explanations for phenomena, it seems that many different physical processes contribute to the emission from AGNs. The infrared radiation probably comes from dust, which has been heated to approximately 1000 degrees Celsius by absorbing some of the ultraviolet radiation produced close to the black hole. The X-rays may come from a hot, turbulent corona that forms a kind of "atmosphere" above the disk, much as the solar corona produces X-rays from a region that extends well beyond the cooler chromosphere. Radio and gamma-ray emissions probably come from jets of gas that are accelerated parallel to the disk's rotation axis. A theme that is coming to dominate many aspects of AGN modeling is that of "reprocessing." Radiation emitted in one waveband (say, X-rays) can be intercepted by the disk, for example, and re-emitted in a modified form. Thus, X-rays can be reprocessed into ultraviolet radiation, and ultraviolet into optical and infrared. These complex transformations of energy from one form to another make it much more difficult to track down the "primary" sources of radiation.

The central engines may nevertheless be similar in all the most highly active galactic nuclei. Accretion power and "flywheel" power could both to various degrees contribute to an AGN's primary power output. That output would then be reprocessed in a variety of ways, depending on details of the galactic environment. Conditions around black holes are extreme, but the relevant physics is known, and the key problem is at least well posed: Given a black hole with a particular mass and spin, in a given environment, how much power is derived from accretion, how much extracted from the hole's spin, and in what forms do these respective luminosities emerge? Such calculations play the same part in the modeling of AGNs that nuclear physics plays in the modeling of stellar structure and evolution. The evidence that black holes have anything to do with AGNs is circumstantial, but the same is true for other cherished beliefs in astrophysics: before the detection of Solar neutrinos, the evidence that stars are powered by nuclear energy was "merely" circumstantial, but the theory of stars was nonetheless considered to be established well beyond "reasonable doubt." However, the confrontation of theoretical models with observations, indirect even for stars, is admittedly much more ambiguous for AGNs. In stars the energy percolates to the observable surface in a relatively steady and well-understood way; in AGNs, on the other hand, it is reprocessed into all parts of the electromagnetic spectrum over a huge range of spatial scales, in a fashion that depends in a poorly understood way on the particular environment offered by the host galaxy.

Nor would it be that simple to devise a test that *refutes* the black hole paradigm. One might imagine an observation—for instance, the discovery that all the central stars and gas were orbiting very slowly—that disproves

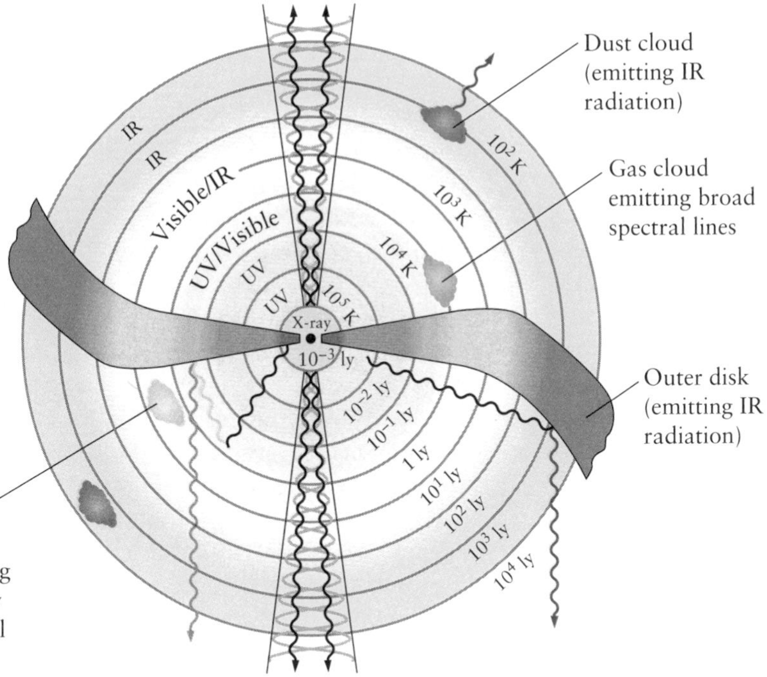

*A possible model for an active galactic nucleus, according to Sterl Phinney and his collaborators. It is drawn "logarithmically" (i.e., each equal increment of radius represents a factor of 10 in distance) so that the outer regions can be depicted on the same diagram as the central black hole, which is 100 million times smaller. Energy comes out of the innermost regions in the form of ultraviolet and X-ray radiation and in high-speed jets of gas. Outer parts of the accretion disk are cool enough to contain dust, which would be vaporized closer in. The dust absorbs radiation from the hotter inner regions and re-emits the energy at infrared wavelengths. If the outer disk is "warped," as drawn, a larger fraction of the radiation from the inner region is intercepted and "reprocessed." Radiation could also be reprocessed by gas clouds distinct from the disk.*

the existence of $10^8$ solar masses at the center of a radio galaxy whose power output seems to demand an energy source of that mass. Or one might search for regular pulses on a timescale of a few minutes, proving that the source is too small to be the expected black hole. But these examples are rather wishful thinking. A clean-cut refutation, leading to the abandonment of some theory, happens only rarely in astrophysics: many untenable models have persisted unrefuted for a long time. A cynic might argue that they have survived only because they don't go beyond generalities, or else because their proponents have been adept at replacing or modifying faulty parts to keep shaky old models "roadworthy." Such a cynical attitude is not necessarily justified, and to explain why we must digress briefly into methodology.

The way we are told science is done is this: the data suggest a model, and the model suggests further tests, whereby the model is either refuted or refined. Such a simple scheme is realistic in fields like particle physics, where the fundamental entities may be reducible to a few basic constants and equations. But other sciences deal with inherently complex phenomena, and no theoretical scheme can be expected to account for every detail. In geophysics, for instance, the concepts of continental drift and plate tectonics have undoubtedly led to key advances; but they cannot be expected to explain the shape—the precise topography—of the continents. What we should aim to do, in our attempts to understand quasars, is to focus on those features of the data that genuinely test crucial ideas—such as the efficiency of energy generation or the compactness of the central mass—and not be diverted into measuring or modeling something that is accidental or secondary.

What does the future hold for quasar research? If the past is any guide, the way is unlikely to be paved by theorists. Astrophysics is not like "gravitational physics"—the study of general relativity—where strong theoretical development has been driven by the utilization of new mathematical techniques. In the astrophysics of the last 25 years, theorists have been followers more often than leaders. In complex phenomena like AGNs, too many processes *could* occur, and picking the likely winners in advance of observations has proved too difficult.

In fact, progress in understanding quasars and other distant phenomena has mainly been driven by the advent of new techniques, new instruments, and engineering innovations. This is hardly a break with astronomical tradition—in some ways, astronomy has always been "big science." It was the first science to be professionalized and to use elaborate equipment (even though there has always been a strong amateur interest). One can see this as far back as the late eighteenth century by contrasting Herschel's telescope, a massive and elaborate construction, with the small-scale equipment that would have been used by his famous contemporaries

Lavoisier (a chemist) and Cavendish (a physicist), either of whose "laboratories" would have fit on a kitchen table. In the mid-nineteenth century, the principal impetus to astronomy was provided by the development of the spectroscope and of photography, at a time when the main theoretical tool was still Newtonian celestial dynamics. Only thereafter did people come to realize that it might indeed be feasible to study the constitutions of celestial objects.

Hubble's great book of 1936, *The Realm of the Nebulae,* concludes with these words: "With increasing distance our knowledge fades and fades rapidly. Eventually we reach the dim boundary, the utmost limits of our telescope. There we measure shadows, and we search among ghostly errors of measurement for landmarks that are scarcely more substantial. The search will continue. Not until the empirical resources are exhausted need we pass on to the dreamy realm of speculation."

The bright and beautiful photographs of galaxies seen in popular books give a misleading impression. Galaxies are barely detectable above the telltale glow of the night sky, even at the darkest sites on Earth or from space, and only long exposures reveal them at all clearly. Our view of the extragalactic Universe is still a sadly dim and blurred one. But the search continues, as more powerful telescopes and detectors are deployed. Orbiting observatories sensitive to gamma-ray, X-ray, ultraviolet, and infrared radiation, and radio telescopes on the ground, have all contributed fundamentally to our understanding of AGN phenomena.

The pace of technological breakthrough has often far outstripped the ingenuity of the theorists and phenomenologists who puzzle over the interpretation of data. Some of us have wasted a good deal of time trying to develop various models to explain fragmentary data, when only a year or two later the observations had improved so much that they showed all prior ideas, or maybe all but one, to be nonstarters. Sometimes the quality of data improves so dramatically that some fairly obvious interpretation suddenly emerges, and a consensus is quickly established.

We devote the entire next chapter to what is perhaps the most remarkable manifestation of black holes in active galactic nuclei—cosmic jets. The discovery and interpretation of jets is the fruit of a new technology—radio astronomy—that, during the last thirty years, has given us a fresh picture of the Universe. The most prominent objects in the radio sky are very different from those that dominate in optical photographs, and the relevant radiation is emitted by quite different mechanisms. Thus, we can view the story of jets not only as the tale of a great scientific discovery, but also as a case study in how a new technology can revolutionize our understanding of a subject.

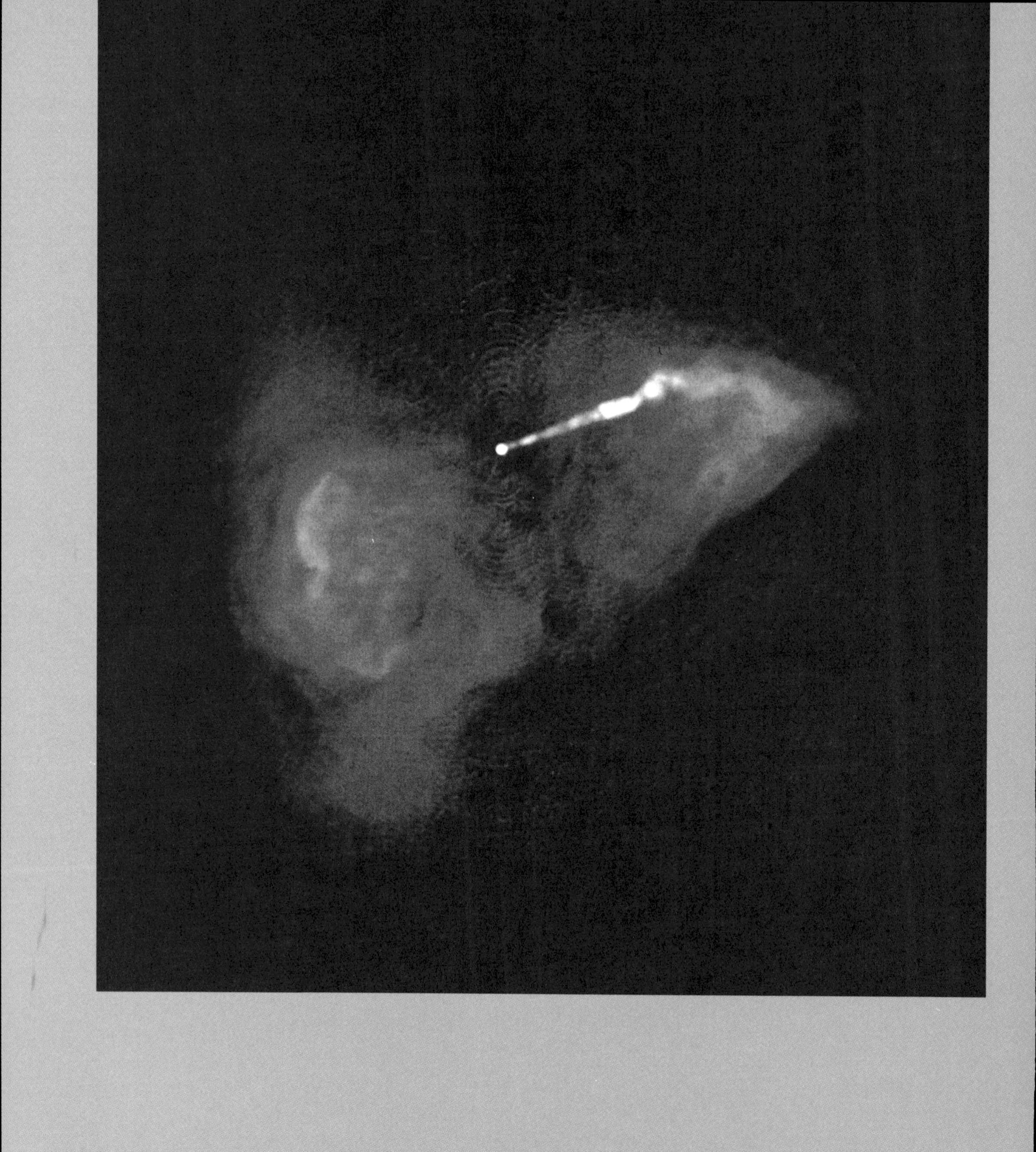

# *Chapter* 6

# Cosmic Jets

*The M87 radio source, as imaged by the Very Large Array. Although only one jet is apparent, astronomers have found evidence for a "counterjet" on the opposite side of the nucleus, which is very faint but nevertheless could carry a similar amount of power. Energy flowing through the jets inflates the cocoon, which glows at low intensity due to synchrotron radiation. Jets of comparatively low power, such as this one, often do not produce the intense hot spots that characterize the most powerful sources, but rather seem to merge smoothly into the cocoon.*

Narrow, fast streams of gas emerge from deep in the nuclei of radio galaxies and often project millions of light-years into space. Those galaxies are suspected to harbor black holes at their centers, and yet such jets of gas are just about the last thing you would expect a black hole to produce. Aren't black holes supposed to be (at least in popular imagination) "cosmic vacuum cleaners," sucking in everything within their reach? It seems obvious that gas being drawn toward a black hole can become hot and luminous just before it falls in, thus accounting for the luminosity of an accreting black hole. But the idea that much of the infalling gas can be "turned around" and propelled outward, at 99 percent of the speed of light or more, strains our credulity. Rather, it *would have* strained our credulity had it been proposed before the advent of radio astronomy. As has happened so many times in the past, a new technology has turned our view of the Universe on its head.

Since their discovery in radio galaxies, jets have proven to be a common phenomenon in the Universe. They often occur when gas with a lot of angular momentum swirls deep into a gravitational field, as, for example, at the center of an accretion disk. We might think of them as celestial waterspouts. They form in all kinds of systems: radio galaxies, X-ray binaries, and even ordinary stars in their infancy. So they are clearly not a manifestation of black holes alone. Yet the jets that apparently do form close to black holes carry the signature of the extreme conditions surrounding their birth—nowhere else in the Universe do we see evidence of matter propelled at such enormous speeds, sometimes to within a fraction of a percent of the speed of light.

## Radio Astronomy and the Prediction of Jets

To be precise, radio astronomers *re*discovered jets from active galactic nuclei. As long ago as 1917, Heber Curtis had noticed a jet of visible light emerging from the nucleus of the large elliptical galaxy M87, which lies in a cluster of galaxies in the constellation Virgo. In his 1918 publication, "Descriptions of 762 Nebulae and Clusters Photographed with the Crossley Reflector," he notes that "a curious straight ray lies in a gap in the nebulosity . . . apparently connected with the nucleus by a thin line of matter. The ray is brightest at its inner end." He did not follow up the observation, nor did anyone else. Not to fault Curtis too much, we should recall that in 1918 it had not yet been established that M87 was a distant galaxy! It took another sixty years, and the development of radio astronomy, before jets were definitively established as commonplace structures in the Universe and their true natures became apparent.

*Karl Jansky, working at the Bell Telephone Laboratories in Holmdel, New Jersey, discovered radio waves from the Milky Way Galaxy in 1931.*

Radio astronomy has its roots in the telecommunications industry. In 1931, Karl Jansky, an engineer with Bell Telephone Laboratories in Holmdel, New Jersey, built a sensitive antenna to track down radio waves that were interfering with transatlantic telephone calls. He noticed that one of the sources of noise varied consistently with a 24-hour period; moreover, its appearance and disappearance coincided with the rising and setting of the Milky Way on the sky. Clearly, the source of these waves was not terrestrial, nor even located in the Solar System, but was tied to distant regions of our Galaxy. Jansky's announcement that he had discovered radio waves coming from thousands of light-years away startled the astronomical community and made the front page of the *New York Times*. But curiously, it did not trigger a rush of astronomers eager to study the Universe with this radically new tool.

A few years later Grote Reber, an amateur radio enthusiast from Wheaton, Illinois, built a dish-shaped receiver in his garden specifically to look for cosmic radio signals. His maps, the first radio "sky survey," were published in *The Astrophysical Journal* in 1942. Along with the radio waves from the Milky Way, Reber found localized patches of radio emission at a few other places on the sky. One of these lay in the direction toward the constellation Cygnus, where we now know that the brightest radio galaxy, Cygnus A, is located. Radio astronomy was primarily the realm of amateurs until after World War II, when a large cohort of physicists and electrical engineers (many of whom had worked on the development of radar) re-entered civilian work at universities and government research laboratories.

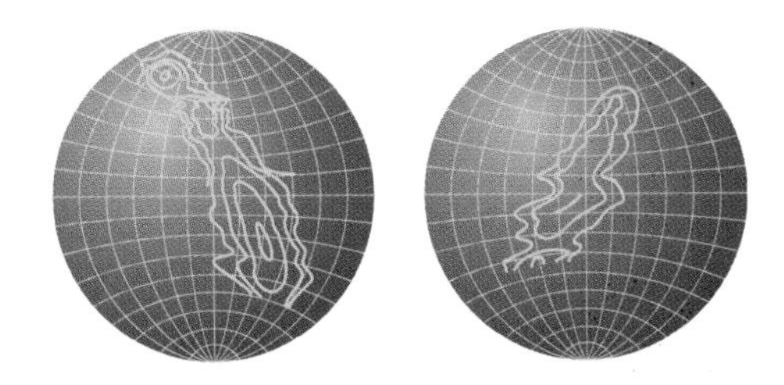

*Grote Reber's maps showed the Galactic center, the supernova remnant Cassiopeia A, and the extragalactic source Cygnus A.*

The earliest radio telescopes, essentially scaled-up versions of Jansky's antenna and Reber's single dish, were severely hampered by poor resolution. As we remarked in Chapter 5, the fineness of detail (the "resolution") observable with any telescope is fundamentally limited by the ratio of the size of the aperture to the wavelength at which the observation is being made. The higher this ratio, the better the resolution. In fact, the smallest angle on the sky that can be measured by a telescope, in radians, is roughly the reciprocal of this ratio. Radio waves have long wavelengths, ranging from centimeters to many meters. By comparison, the wavelengths of visible-light waves are less than one-millionth of a meter. Therefore, in order for a radio telescope to obtain the same level of resolution as an optical telescope, its "dish" would have to be between 10,000 and a million times larger! Building a single dish several kilometers across is clearly impractical, but radio astronomers have been able to exploit a powerful technique called *interferometry,* which enables a spread-out "array" of small dishes to simulate the resolution of a single large dish.

*After working on radar during World War II, Martin Ryle pioneered new types of radio telescope. He invented the technique of "aperture synthesis," whereby an array of modest-sized dishes can, in combination, provide images of the radio sky as sharp as could be produced by a single dish several kilometers across. The "Cambridge surveys" played an important role in the identification of radio galaxies and quasars, as well as yielding the first significant evidence against the steady-state cosmology.*

We can think of a source of radio waves as sending out ripples (consisting of electromagnetic fields) into space, much as a rock dropped into a lake sends out circular ripples of rising and falling water. A telescope effectively obtains information about the patterns of objects on the sky by measuring the directions from which the ripples are emanating. Now, if you were dozing on a rowboat in the lake and a single ripple of water came by, you would feel the boat rise and fall, but would have little clue as to the source of the ripple. But *two rowboats,* spaced sufficiently far apart on the lake, would be able to ascertain the direction of the rock by timing the rise and fall at their respective locations, and comparing notes. In particular, if the rowboats were equidistant from the source, they would rise and fall simultaneously. In a telescope, the analogous "comparing of notes" is accomplished by adding together in real time the (positive or negative) displacements of each wave, a process called *interference.* This is what happens, for example, when the radiation reflected off different portions of a mirror or radio dish are collected together at the focus. If all parts of the dish were "equidistant" from the source, all the added displacements would be either positive or negative, resulting in a large net signal.

Of course, in radio astronomy we are not dealing with a single ripple, but rather with a train of periodic ripples, spaced by a fixed wavelength. If the ripples are close together, the peaks and troughs are well defined, and timing them is easy. But if the ripples are long, gentle swells, it is difficult to determine precisely when the peaks and troughs pass by. To compensate for the difficulty of perceiving ripples of longer wavelength on a lake, the rowboats would have to be spaced farther apart. Similarly, a telescope observing at a longer wavelength has to have a larger aperture in order to obtain the same angular resolution as a telescope observing at a shorter wavelength.

To address the resolution problem without building gigantic dishes, radio astronomers resort to a clever "trick." The key to achieving high resolution is to compare signals that are collected across a sufficiently long "baseline"—that is, from points that are far enough apart to discriminate the direction from which the waves are coming. It is not necessary to merge signals from every point within that baseline. In other words, to get the resolution of a 10-kilometer dish, it will suffice to mix the signals from smaller dishes that are separated by 10 kilometers. Radio astronomers can simulate the resolving power of a single 10-kilometer dish by placing a number of much smaller dishes in an array that spans 10 kilometers and using the array as an "interferometer."

But how many smaller dishes are needed, and how should they be arranged? This question must be considered carefully because directional measurements by an interferometer are much more accurate in certain directions than in others. On a lake, we can pin down the direction to the rock most accurately if the rowboats happen to be roughly equidistant from it to begin with. In a telescope, the ability of, say, a north–south pair of dishes to resolve a pair of compact sources is greatest if the sources are also lined up in a north–south direction on the sky. Moreover, a pair of dishes with a very wide spacing will be sensitive to fine detail, but will not be able to "see" the large-scale structure of an object. Clearly, to make a detailed picture of a radio source using an interferometer, you would need many pairs of dishes, with many different orientations and separations—and before long you'd be back to trying to build a single dish 10 kilometers across.

It was Martin Ryle, of Cambridge University, who in the 1950s conceived a brilliant way out of this difficulty. In retrospect, it was unbelievably simple. He realized that, as the Earth rotated, the baseline of each pair of dishes would rotate with it! And as the baselines rotated, the orientation of each baseline would change with respect to sources on the sky. In effect, an array with just a few pairs of dishes having various separations could simulate a single, large dish that was much more "filled in." To build up an image of a radio source one simply had to wait and let the Earth do the work.

Throughout the late 1960s and the 1970s, a series of increasingly large interferometers were built, particularly at the University of Cambridge, at the Westerbork Observatory in the Netherlands, and at the National Radio Astronomy Observatory (NRAO) in West Virginia. These culminated, in the late 1970s, in the construction of the Very Large Array (VLA) near Socorro, New Mexico. The VLA consists of 27 linked radio telescopes,

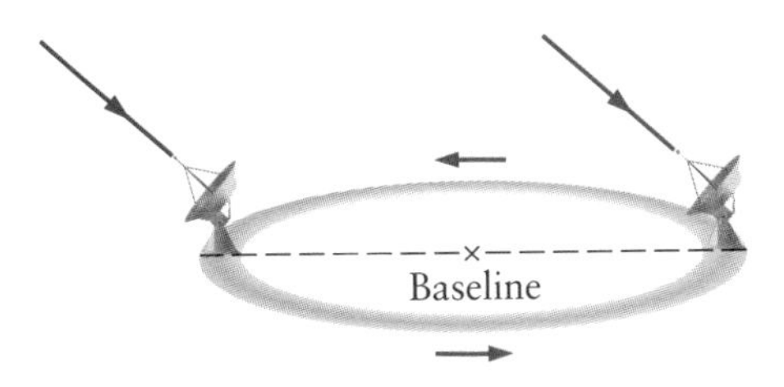


*As the Earth rotates, the small dishes trace out an annulus whose diameter equals the baseline separation. After combining the signals from these dishes for 12 hours, astronomers can "synthesize" the resolution of a reflecting dish with the size and shape of the annulus. By changing the separation of the two telescopes and repeating the process, one can in effect "fill in" the aperture of a single dish whose diameter is the maximum separation. To speed up the procedure, arrays of several dishes are used, rather than just two.*

each 25 meters in diameter, distributed in a Y-shaped array spanning 40 kilometers. The VLA can uncover features that subtend an angle on the sky as small as one-tenth of a second of arc. This is the same as the angle subtended by a dime at a distance of 20 kilometers and is comparable to the resolution of the Hubble Space Telescope. We have already discussed in Chapter 5 how, in an interferometry experiment performed as early as 1953, Jennison and Das Gupta had established that the bulk of the radio emission from Cygnus A was coming not from the galaxy itself, but from two giant lobes placed symmetrically on either side of the galaxy. The new interferometers revealed this symmetric double structure to be a common feature of powerful radio galaxies. One was forced to conclude that whatever was powering the lobes had been ejected symmetrically from the galaxy.

The more sensitive radio telescopes detected additional structure. At the outer extremities of the lobes in the brightest double radio sources (the ones like Cygnus A), they often found "hot spots," compact regions emitting particularly intense radiation. Many of the brightest sources also showed tails or bridges of low-intensity emission extending backward from the hot spots toward the center of the source, where there is usually a compact focus of radio emission called a "core." Most double radio sources are associated with an elliptical galaxy (as Cygnus A is) or with a

*The Very Large Array (VLA), near Socorro, New Mexico, uses 27 dishes, arranged in a Y-shaped pattern, to construct radio images that are as finely detailed as those produced in visible light by the Hubble Space Telescope.*

quasar; in such cases the core is invariably found to coincide with the galaxy's optical center.

This wealth of new detail challenged the ingenuity of theorists attempting to explain double radio sources. Initial hypotheses that the lobes were huge blobs of superhot magnetized gas, ejected in gigantic explosions, fell out of favor when it was realized that the energy would seep away as the blobs expanded, leaving too little behind to explain the observed radio emission. Some source of energy had to be replenishing the lobes so that they could continue to emit radiation. Could the hot spots themselves provide energy for the lobes, perhaps from embedded black holes flung out of the galaxy by a gravitational "slingshot"? Ingenious schemes were proposed, but an explanation for the high degree of symmetry and alignment proved elusive. It was the detection of the central cores that provided the crucial clue.

The presence of radio emission from the core suggested that there was continuing activity at the center of the galaxy. Suppose there were some kind of conduit through which this energy could be resupplied continuously to the lobes. This would solve the problem of how the energy losses in the expanding lobes were replenished and perhaps explain the symmetry as well. But what form would this energy supply take, and how could it travel with such geometric precision across so many thousands (or, in some cases, millions) of light-years of intergalactic space? Various possibilities were tried—superintense beams of electromagnetic radiation, for example—but nothing seemed to work very well. Then, in the early 1970s, researchers at Cambridge University proposed an answer that provides the basis for all current theories of radio galaxies. The lobes had to be powered by twin streams of fast-moving gas, created in the nucleus of the galaxy and propagating in diametrically opposite directions into the lobes.

To understand how the lobes and hot spots are created, one has to recognize that no galaxy exists in a vacuum. Just as the space between the stars in a galaxy is filled with an *interstellar medium* of gas and dust, so is the space between galaxies filled with an *intergalactic medium*. However tenuous this intergalactic gas may be, it will ultimately resist the free expansion of gaseous jets emanating from the galaxy. It is the presence of this intergalactic medium that leads to the observable structure of a radio source. After activity begins at the center of the galaxy, the jets propagate outward through the galaxy rather like the jet of water that emerges from the nozzle of a hose. The cosmic jets pass first through the interstellar medium and then through the intergalactic medium. As the jets progress outward, the density of the surrounding matter declines from roughly one hydrogen atom per cubic centimeter to perhaps one per million cubic centimeters. Still, the advancing jet must push the matter out of the way; hence the end of the jet moves slower than the gas flowing inside the jet.

As a result, energy accumulates at the end of the jet; this is the likely interpretation of a hot spot.

The flow of gas in the jet is thought to be supersonic: that is, the speed of the gas is faster than the speed of a sound wave in the gas. The latter is much greater than the 300 meters per second attained by a sound wave in air, and may be as great as 60 percent of the speed of light. The speed of the gas is still more tremendous, in some cases reaching to within a fraction of a percent of the speed of light. As the gas approaches the hot spot, however, it decelerates suddenly. Since no sound wave can propagate upstream to warn the gas to slow down, the abrupt deceleration causes a shock wave to form across the jet. The effect of the shock is important. Before the jet reaches the shock wave most of the particles are streaming in unison, and most of the jet's energy is in the form of ordered kinetic energy. Passage through the shock converts much of this ordered energy into two forms: the energy of relativistic electrons (electrons moving randomly at speeds near the speed of light) and the energy of a magnetic field. You may recognize these as the two ingredients that together yield synchrotron radiation. It is entirely natural that the most intense radio emission in a double-lobed radio source should be generated where the jet is decelerated by the action of the surrounding gas.

After splattering at the hot spots, the jet material has nowhere to go but back toward the galaxy. Thus it inflates the large lobes that can be seen in radio images. In fact the observed lobes are merely the brightest

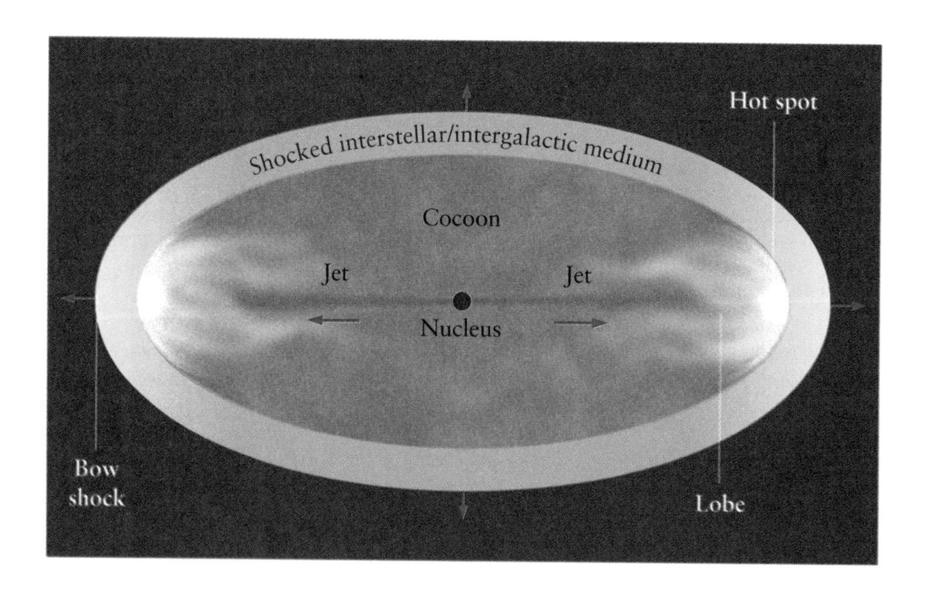


*Schematic diagram showing the structure and mechanism of strong double radio sources. Most of the radio emission is produced in the "hot spots" and lobes, by jet material that has splattered against the interstellar or intergalactic medium.*

 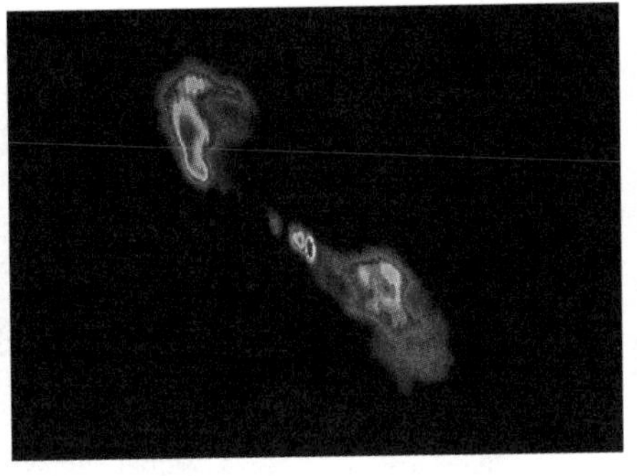 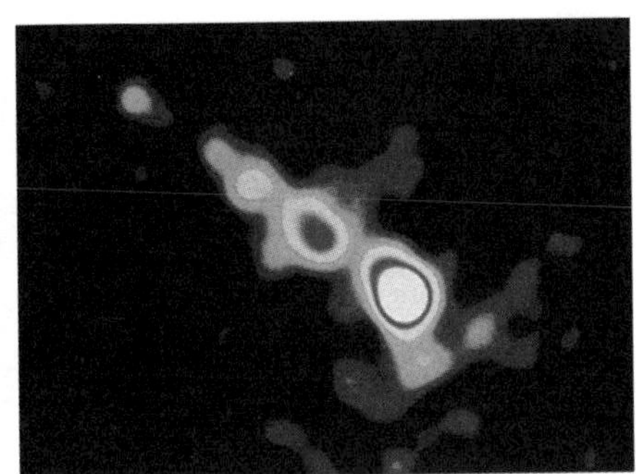

*Centaurus A (also known as NGC 5128), a nearby galaxy in the southern sky, exhibits huge radio lobes stretching several degrees across the sky, as well as a jet on smaller scales (center and right). The energy content of the giant lobes (not shown) is colossal. Cen A was probably once a very powerful radio source; it has gradually faded as the lobes have expanded into the surrounding intergalactic medium.*

parts of a much larger, cigar-shaped bubble that envelops the entire source, called a "cocoon." Each bit of jet material spends a relatively short time (roughly 10,000 to a million years) in a hot spot before moving into the lobes or cocoon, carrying with it the relativistic electrons and the lines of magnetic field. Most of the energy of the radio source is therefore contained in the lobes and cocoon, which probably accumulate gas for 100 million years. The energy content of the lobes is prodigious, as was first realized by Geoffrey Burbidge in the 1950s. In a typical source, it must be about the amount that would be liberated if the mass of a million stars were converted entirely into energy.

## Skywriting

By 1978 the theoretical arguments for the existence of jets were compelling, but astronomers were becoming uneasy. Heber Curtis's lone photograph notwithstanding, no one had yet seen any direct evidence of gaseous streams connecting the cores to the lobes of double radio sources. But this quickly changed when the VLA came on line. In image after image, the faint traces of the energy conduits were made visible by the VLA's superior sensitivity. Soon it became obvious that jets, while ubiquitous, assumed a wide variety of appearances. In some of the brightest sources, such as Cygnus A, the jets were exceedingly faint, dead straight, and

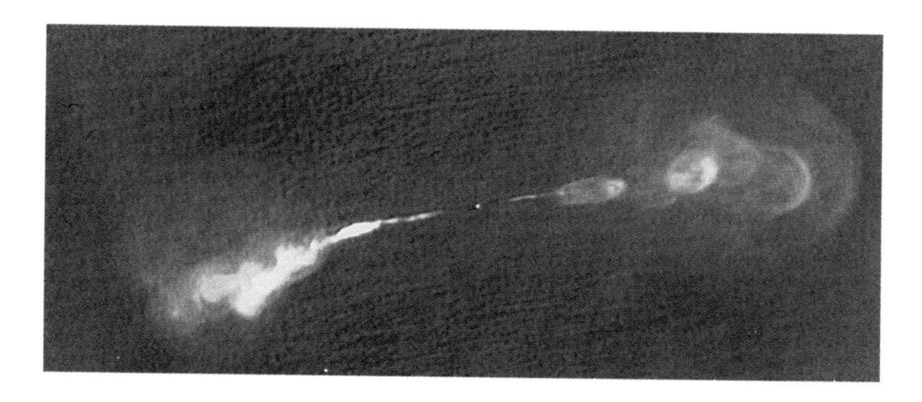

*This radio image of Hercules A vividly suggests the turbulent disruption of a pair of jets. The radio structure extends to a much larger scale than the galaxy itself.*

sometimes little more than a hair's width across on the image. Other jets appeared broader and brighter, but these had a tendency to "fizzle out" without producing an intense hot spot. Often the jets appeared only on one side, even when the lobes and hot spots they fed were decidedly symmetrical.

While these observations seemed to fit the prediction that jets would spurt out from the core, there was no direct evidence that the jets really represented streams of flowing gas. All we had were streaks on the radio sky that looked as if they *could* be streams of gas. Therefore, acceptance of the nature of jets by most observers gained ground slowly. In retrospect, this is not surprising. Astronomers by nature tend to be conservative, and are much more trusting of quantitative evidence than of theory and attempts to draw inferences from "morphology"—or the appearances of things on the sky. (Almost the lone exception to the latter bias among eminent twentieth-century astronomers was Fritz Zwicky—co-inventor of the concept of neutron stars—whose treatise *Morphological Astronomy* is a classic of astronomical iconoclasm.) Therefore, many are very reluctant to admit the existence of a new phenomenon unless forced to by overwhelming observational evidence.

Our insistence on such evidence represents a healthy form of "checks and balances" because the acceptance of any new phenomenon in astronomy reverberates through our entire "worldview." We operate under the assumption that the Universe is so vast that any phenomenon we manage to discover is almost certain to be commonplace. This philosophical principle dates back to Copernicus, who argued in the fifteenth century that the Earth was not a unique kind of body in the Solar System. Indeed, the controversy over quasar distances, discussed in Chapter 5, arose from a similar conservatism—it was easier for some to accept that redshifts could be misleading than it was to accept the incredible energy outputs of quasars.

The debate over the jet hypothesis reached a watershed in late 1978, when an X-ray binary in our own Galaxy, called SS 433, was found to be producing a pair of narrow jets. Unlike the jets from radio galaxies, the SS 433 jets were cool and dense enough that their composition and velocity could actually be measured, by spectroscopy. The jets were found to consist of gas with an ordinary chemical composition—mainly hydrogen—and the gas was seen to be moving at a quarter the speed of light. Here was an incontrovertible signature of the nature of jets in at least one object, albeit one that was very different from a radio galaxy.

The discovery of the SS 433 jets paraphrased, in a wry way, the discovery of the distances to quasars fifteen years earlier, when Maarten Schmidt noticed that the mysterious emission lines of quasars were ordinary emission lines of hydrogen that had been redshifted. The object's moniker pegs it as a member of a catalogue compiled by C. Bruce Stephenson and Nicholas Sanduleak, listing stars that show strong spectral emission lines. But in 1978, Bruce Margon and his collaborators at UCLA noticed that some of the lines in the SS 433 spectrum did not correspond to the wavelengths produced by known chemical elements. So far the story is reminiscent of quasars, and we know how that puzzle was resolved. But to make matters truly incomprehensible, the wavelengths of the lines in SS 433 changed with time! Eventually, Margon and his team figured out that the spectrum could be resolved into *three* sets of familiar spectral lines: one close to the correct wavelengths, one with a big redshift (that is, moving away from us at high speed), and one with a large blueshift (moving toward us at high speed). Margon reported his initial results at the Ninth "Texas Symposium" on Relativistic Astrophysics in Munich, Germany. (In keeping with the penchant of astronomers for nondescriptive terminology, the biennial Texas Symposium is seldom held in Texas. Recent venues have included Jerusalem, Chicago, Berkeley, and Brighton, England.) Margon's announcement was great fodder for journalists. Newspapers reported on a "mysterious dying star" that "seems to be coming and going at the same time," while Margon speculated to the press that "this is some kind of star that's in some terribly weird kind of trouble."

After about a year of observations the changes in wavelengths were found to conform to a clear pattern, which George Abell and Margon were able to fit to an extraordinarily simple model. The X-ray binary SS 433, they argued, must be producing a pair of identical, narrow jets, with speeds fixed at 26 percent of the speed of light, which precess around a fixed axis in space once every 164 days. In other words, SS 433 is a kind of cosmic lawn sprinkler, with a geometry like that shown in the figures on the following page. But the real test of the precessing jet model would be to see the traces of the jets as they moved across the sky. Only 16,000 light-years away, SS 433 is close enough that the motion of the jet material

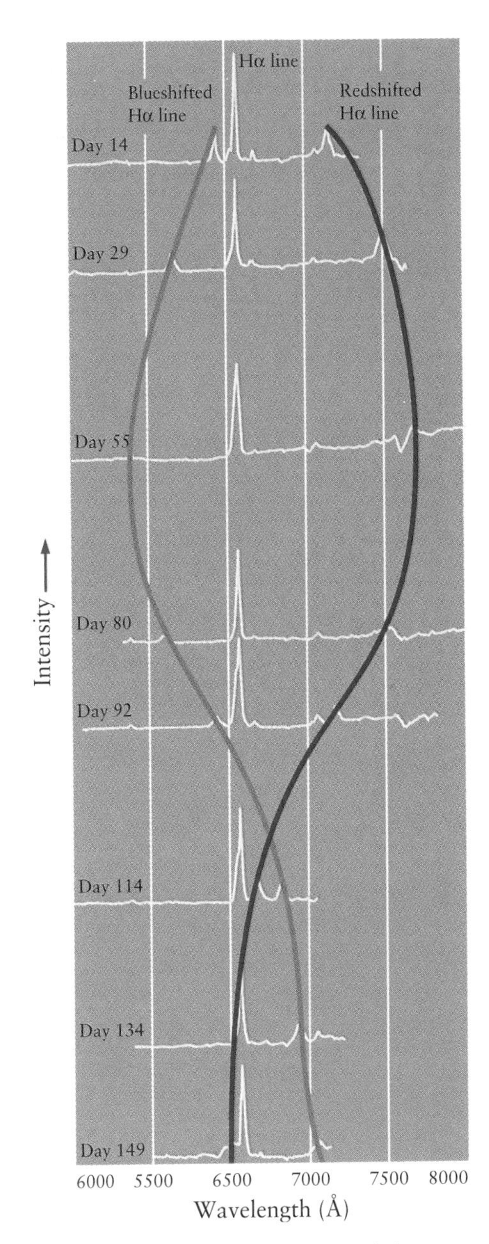


*This diagram shows spectra of the X-ray binary SS 433 taken at eight different times. Curves connecting the spectra show how the observed wavelength of a particular spectral line, emitted by hydrogen, changes from one observation to the next.*

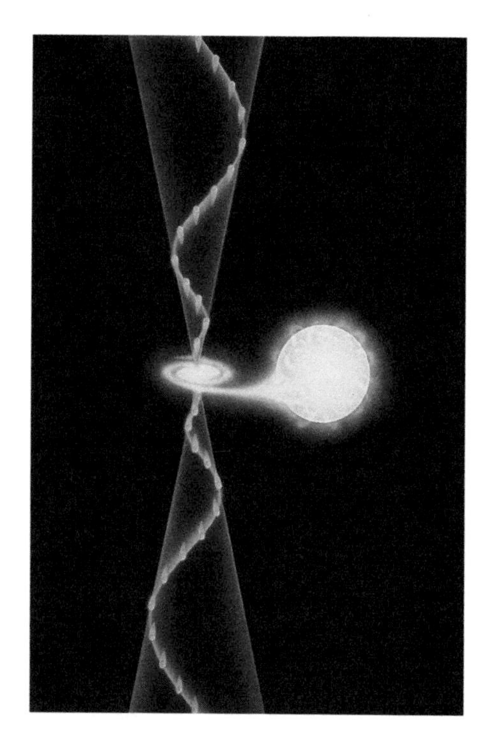

*Individual "blobs" of gas ejected from SS 433 travel in straight lines, but as the direction of ejection precesses around a cone, the blobs emitted at different times should trace out a widening corkscrew pattern, as illustrated here in an artist's impression. Robert Hjellming and Kenneth Johnston observed the predicted pattern using the VLA.*

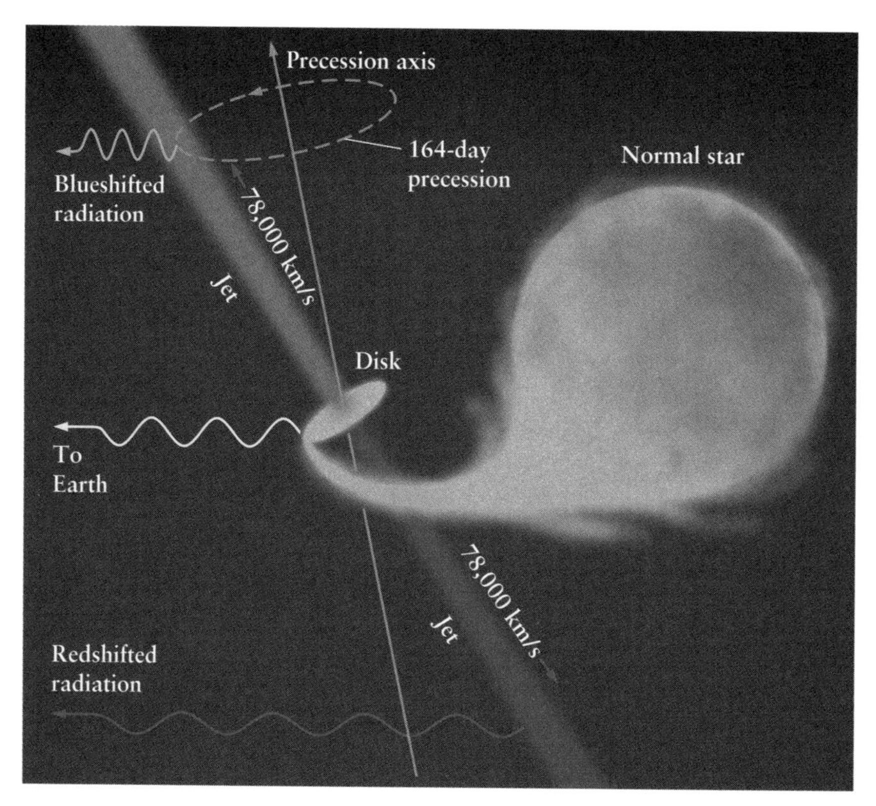


*The spectral lines of SS 433 come from twin jets of gas that emerge, in opposite directions, from a disk around a compact object, which is in orbit around a normal star. As the jets swing steadily around a cone every 164 days, the component of velocity directed along our line of sight varies. Because of the consequent change of the Doppler shift, the observed wavelengths of the spectral lines trace out a pattern that repeats every precession period. We see both a redshifted and a blueshifted line, with wavelengths that change with time.*

should be detectable within a couple of weeks by means of a telescope with the angular resolution of the VLA. If the jets are really swinging in a wide cone once every 164 days, then we might expect to see a widening corkscrew march outward from the center, with one new turn appearing every five months. An observation like this would be a theorist's dream; indeed, many of us pinched ourselves when Robert Hjellming and Kenneth Johnston produced a "movie" of the SS 433 jets behaving exactly as predicted! Moreover, SS 433's jets seem to have produced a pair of "lobes." SS 433 lies in the center of a radio nebula, called W50. Such nebulae are usually found at the sites where stars have exploded as supernovae, and they typically have a roughly circular shape. But W50 is different. It has "ears" that are exactly aligned with the jets. Presumably, the "ears" of radio emission are places where the jets have bored most deeply into the surrounding interstellar gas. Thus, the analogy between SS 433 and giant radio galaxies seems to be a solid one.

Once we accept the idea that jets are gaseous streams, we confront a series of tricky theoretical problems. As was the case with the black hole paradigm for quasars, even after we understand the general concept of jets we must still answer many questions about how they behave in detail. To take one example, the observed jets seem amazingly stable against breakup. In extreme cases they seem to maintain their integrity even as they traverse millions of light-years. But everyday experience with fluid streams suggests that a jet should be highly unstable as it bores its way through the interstellar and intergalactic medium. Anyone who has handled a flexible garden hose will be familiar with this phenomenon. If the

*SS 433 lies at the center of this radio source, called W 50. The circular shell-like structure is typical of supernova remnants, but the two "ears" are unique. They are believed to have been produced by the pressure of the precessing jets impacting on the interstellar gas. Hjellming and Johnston's observations confirm that the axis of the precession cone lines up with the axis of the ears.*

*Three-dimensional computer simulations are beginning to reveal the complex nature of jet flow. Four "snapshots" of a supersonic jet (red) plowing into stationary gas (violet) illustrate the growth of the cocoon (blue), which consists mainly of backflow from the head of the jet. Turbulence inside the cocoon disturbs the jet's steady flow, and leads to some mixing between the jet gas and the cocoon gas (yellow). Nevertheless, because of the densities and velocities specified in the calculation, the jet retains its integrity. The fifth image shows a (two-dimensional) simulation of a flow that is not so lucky. It becomes violently unstable and loses its integrity entirely.*

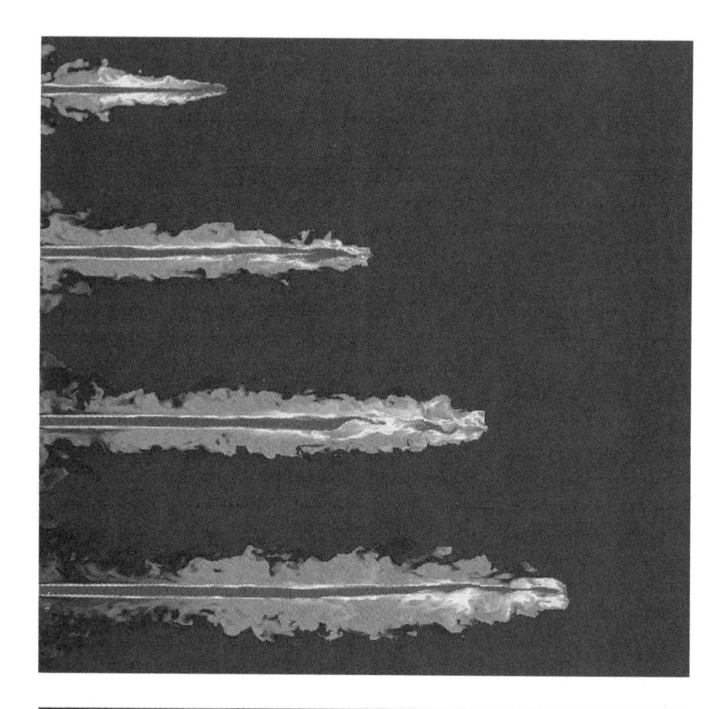

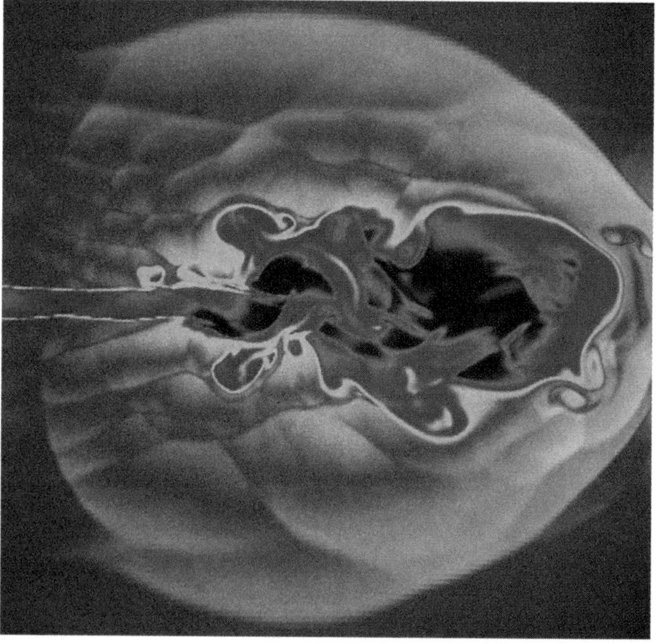

hose is laid on the ground in a curve and the water is turned on, the hose will start to writhe. This kind of instability occurs whenever two fluids slide past each other: it is responsible for making flags flap in the breeze and whitecaps rise on wind-whipped lakes. We might expect it to occur along the interface between a jet and the surrounding gas.

Clues to why cosmic jets might be much more stable than jets in the laboratory are now coming from computer simulations, which we can use to perform "numerical experiments" under conditions completely unattainable on Earth. These have already uncovered some unanticipated properties of supersonic flows that may be observable in images of radio galaxies. But high-resolution three-dimensional computations incorporating electromagnetic effects are only now becoming feasible, and simulations that are capable of handling the huge range of densities and the high speeds thought to exist in jets are still beyond the capabilities of supercomputers.

Many jets trace out large bends on the sky. In some cases these bends may simply be manifestations of the instability described above. In a cosmic jet, however, not all bends are produced in a random manner. Consider the sources called "radio trails." At low resolution the radio emission from such a source appears not to straddle an optical galaxy as it does in a typical double source but to extend in a long curve on one side of the galaxy. A high-resolution image shows why. Two jets emerge from the center of the galaxy, but instead of terminating at hot spots the jets bend steadily to one side so that they both join up with the long, curving trail of radio emission.

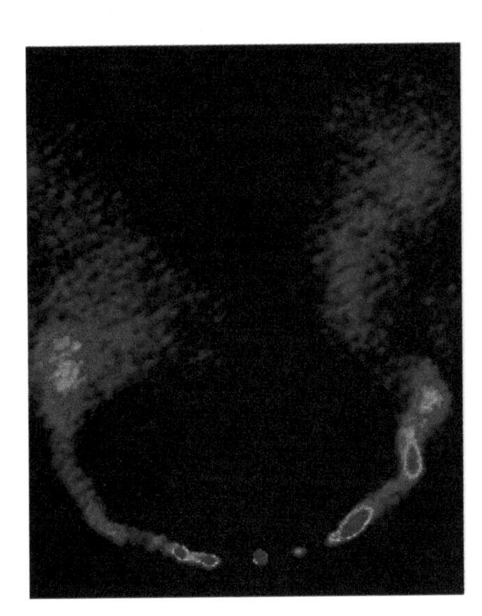

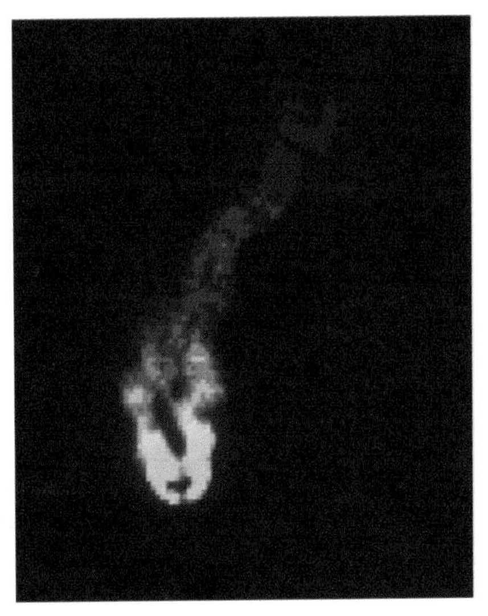

*NGC 1265, the prototype of a "radio trail," is a low-powered radio galaxy in the Perseus Cluster of galaxies. It is moving through the gas in the cluster at several hundred kilometers per second. The high-resolution image (left) shows a normal pair of oppositely directed jets, squirting out of the nucleus at right angles to the galaxy's direction of motion through the intracluster gas. When these jets penetrate beyond the galaxy's interstellar medium (which moves with the galaxy) they are swept into a "radio trail" (low-resolution image) that delineates NGC 1265's orbit.*

What happens in radio trails is that two comparatively low-power jets are formed at the galaxy's center and are then blown sideways by the intergalactic medium as it rushes past the galaxy at speeds that may exceed a thousand kilometers per second. One is reminded of smoke that emerges from a chimney and then is caught in a high wind. The fact that radio trails are usually found in rich clusters of galaxies supports this interpretation of their morphology, for the space between galaxies in such clusters is filled with comparatively dense, hot, ionized gas. The jets feel a wind because the galaxy itself is moving at high speed through this intergalactic gas, under the gravitational influence of all the other galaxies in the cluster. What is most extraordinary about radio trails is that their jets can be bent through almost a right angle without losing their integrity. In the images on this page, a numerical simulation of a jet in a crosswind shows how this can happen. If jets can survive this extreme bending, it is clear that they are not always unstable. After the jets merge with the trail, their gas presumably slows down until its velocity matches the local velocity of the intergalactic medium. By observing the length of the trail and given the galaxy's velocity through the background gas, one can measure the age of the source. In a typical case the age turns out to be 300 million years.

A somewhat different explanation has been advanced for radio sources such as 3C 449, in which the jets show several sharp bends but are otherwise basically linear. Since the bends closely mirror each other on opposite sides of the nucleus, these sources are called *mirror-symmetric*. In

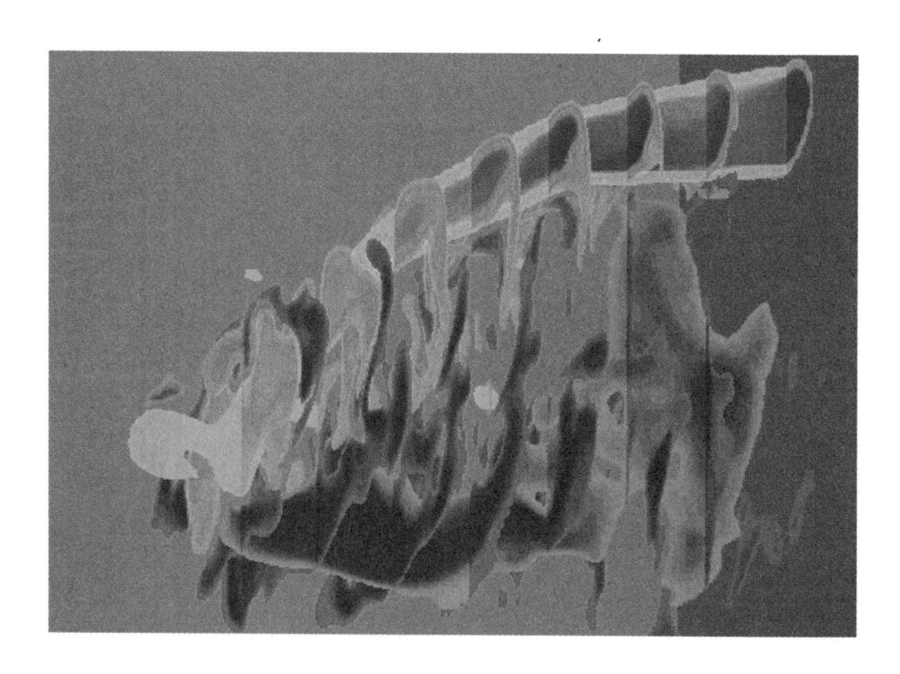

*A series of slices showing how the cross section of a jet evolves as it is swept aside by a crosswind. The jet comes in from the right and the wind blows from the top of the frame. These three-dimensional computer simulations, by Dinshaw Balsara and Michael Norman of the National Center for Supercomputer Applications, support the idea that jets can be bent by nearly 90 degrees in radio trails without losing their integrity.*

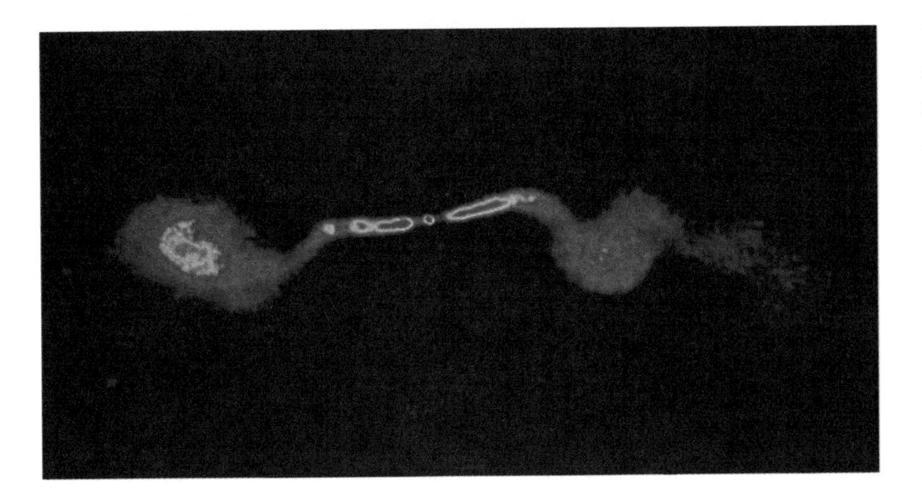

*The radio jets in 3C 449 display mirror symmetry, which is believed to result from the orbital motion of the galaxy about a companion.*

general these sources are not found in rich clusters of galaxies, but they usually have close companion galaxies. It is likely, therefore, that we are witnessing the galactic version of a binary star system, with the radio galaxy and its companion in orbit around each other. If the velocity of the gas in the jets is not much greater than the orbital velocity, the orbital motion of the galaxy will be "written" on the sky in the form of bends in the jets. In a sense the bends are an illusion: the gas in the jets does not flow through them. Imagine a photograph of someone watering a garden by swinging a hose. The water hanging in the air has a zigzag pattern, but a given drop of water in the pattern is moving in a ballistic arc. Analogously, a given particle of gas in a jet follows a straight path. The shape of the bends in a cosmic jet can be used to infer the trajectories of the orbiting galaxies. Note that orbital motion will automatically produce a mirror-symmetric pattern on the sky, exactly as we see in objects like 3C 449: if one jet bends to the right, the other jet must also bend to the right at about the same distance from the center of the galaxy.

A third class of "bent" radio sources exhibits *inversion symmetry.* In this case a bend to the right in one jet corresponds to a bend to the left in the other. Here the source of the jets is thought to be precessing like a top, just like SS 433 but on a much longer timescale. Again the individual bits of gas in the jets move in straight lines, but the pattern that should be observed at any one time is that of a corkscrew. As yet there is no widely accepted explanation for the cause of the precession, but it is likely that it arises at the very source of the jets (in contrast to the other types of bends, which arise from environmental conditions on large scales). Therefore, such precession could be an important clue to the conditions at the center that give rise to the jets.

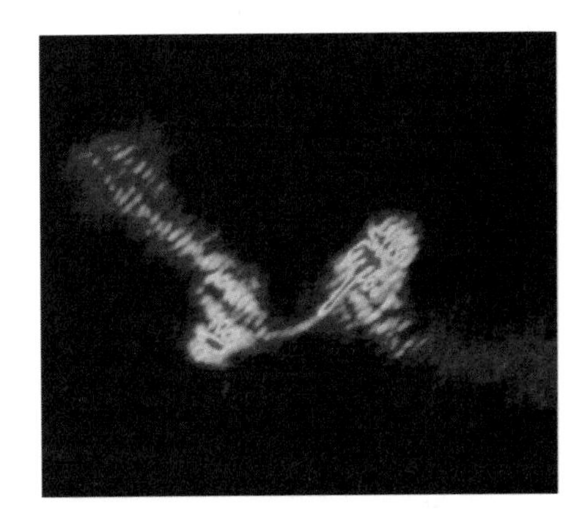

*The inversion symmetry seen, for example, in this radio image of the galaxy NGC 326, could arise if the jets precess. This pattern is analogous to the corkscrew pattern observed in the stellar-mass object SS 433, although the precession timescale would be tens of millions of years rather than 164 days.*

## Approaching the Speed Limit

If the properties of jets on large scales seem surprising enough, their perceived nature becomes even more exotic as we explore closer to the nucleus. But on this exploration even the powerful Very Large Array fails us, since its angular resolution is limited to one-tenth of an arcsecond, or about 300 light-years at the distance of Cygnus A. If there is a black hole at the center of Cygnus A, its event horizon must be more than a million times smaller! Fortunately, radio astronomers have devised instruments that can resolve details up to *a thousand times finer* than the VLA.

As we explained earlier, the resolution of maps made by radio interferometers is limited by the separation of the individual radio dishes: the greater the separation, the greater the resolution. To perform interferometry one must be able to combine the signals from the individual receivers. Across separations as great as 40 kilometers, like those between the VLA dishes, underground wave guides can link the dishes in real time. Direct linkage with the required precision is unwieldy for much larger arrays, although Great Britain's "MERLIN" array of radio telescopes (headquartered at Jodrell Bank, near Manchester) achieves resolutions about 10 times higher than the VLA by using real-time microwave transmissions to link antennae spread over several hundred kilometers.

In the technique called *very-long-baseline interferometry* (or *VLBI*), which was developed simultaneously with the construction of the modern generation of linked radio interferometers, the receivers can be on different continents. The resolution that can then be achieved is correspondingly greater. Features as small as a ten-thousandth of a second of arc can be distinguished—at this resolution, you could read a newspaper headline from across the Atlantic. The angular scale that can be resolved corresponds to one light-month at the typical 300-million-light-year distance of the radio galaxy NGC 6251 and 3 light-years at the distance of the farthest quasars. This great resolution is achieved at a cost, however. The individual telescopes cannot be linked in real time; hence the radio signals from the individual receivers must be recorded separately (but simultaneously) and compared later, using a special-purpose computer called a *correlator.* This procedure entails a loss of information, so that one has to work harder (and observe longer) to obtain images of comparable quality to those made with shorter-baseline interferometry. Until recently, the opportunities to make high-sensitivity observations were few, since VLBI measurements were made on an ad hoc basis by combining data from radio telescopes (built for other purposes) around the world. But the newly completed Very Long Baseline Array, spanning the United States from Hawaii to Puerto Rico, allows observations to be made full-time and should routinely yield high-quality maps.

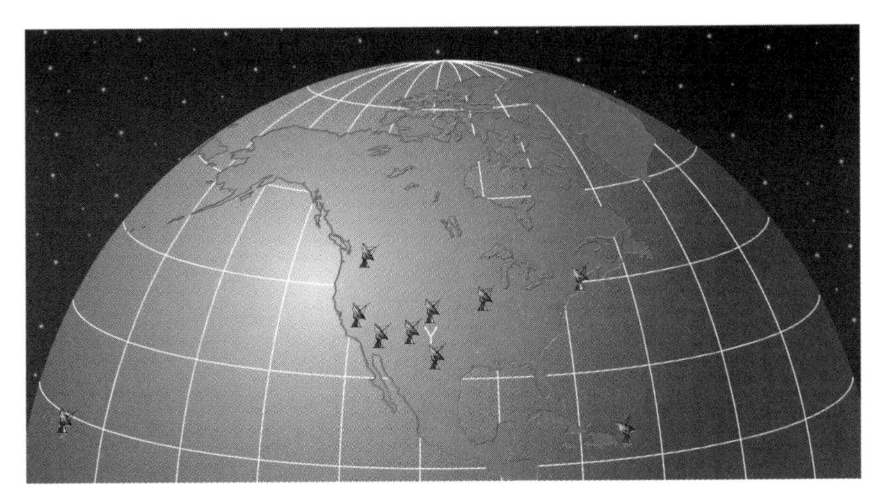


*The Very Long Baseline Array (VLBA) consists of 10 radio dishes extending 8000 kilometers. Operated as an interferometer, the array serves as a "telescope" with the same resolution (though not the same sensitivity!) as a single dish the size of the Earth. The* Y *denotes the location of the VLA, which provides additional data, with lower angular resolution but greater sensitivity, to supplement the VLBA.*

Very-long-baseline interferometry has been notably successful in probing the cores of radio sources. It has shown that they usually have features that remain unresolved. These features must therefore be smaller in angular size than a thousandth of a second of arc (corresponding typically to a few light-years). In contrast, the hot spots in the extended lobes generally have no small-scale structures. VLBI has also been successful in studying a separate class of objects called compact radio sources, usually found at locations where optical telescopes reveal distant quasars. In each compact radio source, the radio flux comes mostly from a small core, not from extended lobes. Some additional radio emission is also generally detected at a low level from an irregularly shaped region surrounding the core. The compact radio sources are often found to vary in intensity on timescales ranging from months to years. Are these compact sources related to the big double radio sources? We shall return to this question later.

By comparing high-resolution VLBI observations with broader images made using linked interferometers, we can construct an "exploded view" of a radio galaxy showing its structure on different angular scales. With these composite images we can follow jets all the way from the largest scales of the radio lobes into the innermost few light-years of the galactic nucleus. The figure of 3C 120 as seen on different scales provides a striking example. The inner region around the nucleus typically exhibits a single jet emerging from an unresolved point of intense radio emission. This has been dubbed a "core-jet" structure. In many instances the single jet points straight in the direction of one of the source's radio lobes, although the jet sometimes curves away from this direction on the smallest scales. Although only a single jet is visible, it is still possible that an oppositely directed jet exists but is too faint to be seen. If an undetected jet does point in the opposite direction, it must be at least 10 times fainter than the one that is visible. The fact that a jet appears on these smallest scales provides the strongest evidence that the lobes are indeed supplied with energy through the jets. The activity in the core must have been continuing for tens of millions of years for the lobes to have accumulated their energy from these jets. The continuous alignment of the jet all the way from the nucleus to the lobe indicates that its direction has remained constant over that time.

When we turn to the broader picture of the radio emission obtained by linked interferometers, we often still find only a single jet, on the same side as the jet seen in the small-scale images of the core. This one-sidedness often persists out to very large distances from the nucleus, more than a million light-years in some extreme cases. Because linked interferometry is more sensitive than VLBI, it is possible to place even more stringent limits

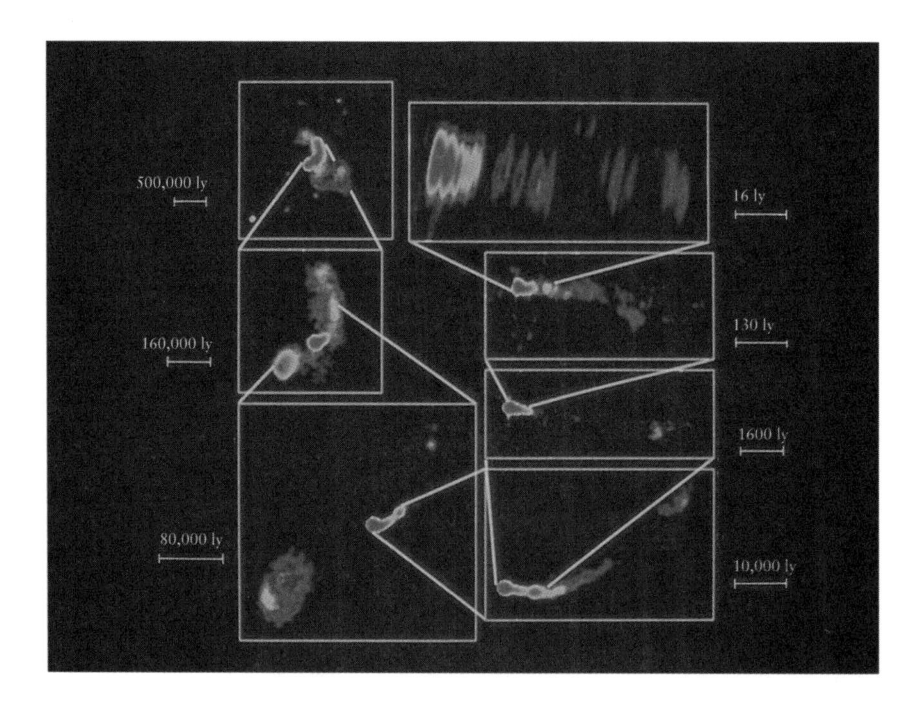


*By combining images made with different radio telescopes, Craig Walker and his colleagues have traced the structure of 3C 120 on all scales from 16 light-years to half a million light-years.*

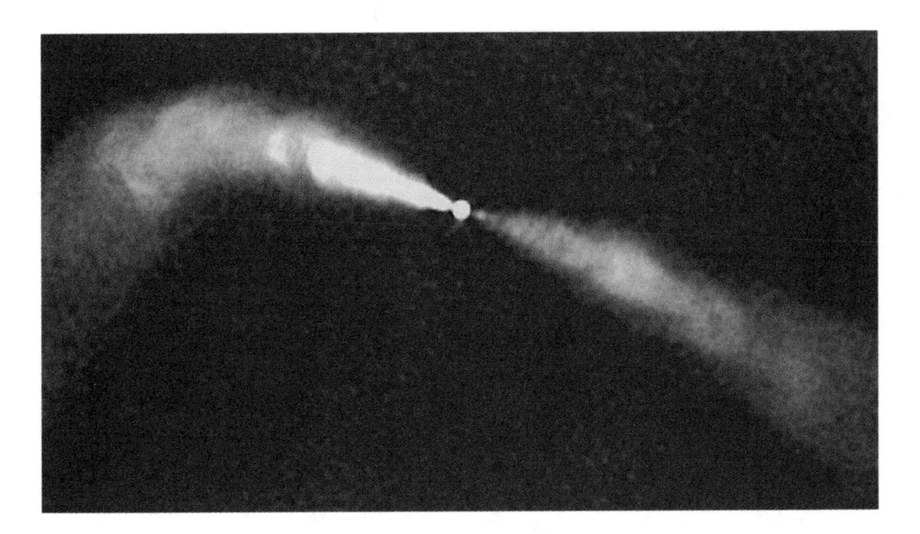

*The two oppositely directed jets emerging from the nucleus of the radio galaxy 3C31 are not exactly in the plane of the sky. We observe more intense radiation from the jet directed toward us because of the Doppler effect. The large apparent asymmetry implies jet speeds more than 90 percent that of light.*

on the maximum possible brightness of any jet on the opposite side of the galaxy from the detected jet. In M87, for example, there are some indications for the existence of a counterjet, but it must be at least 150 times fainter than the main jet.

Does the detection of only a single jet mean that all the energy is coming out on one side? Perhaps, but this can't be true for all time. Since the lobes are roughly symmetric, the energy output averaged over time must be essentially the same on both sides. To avoid this objection one might suppose that the jet flip-flops from side to side on a timescale that is relatively short compared to the roughly 100 million years it took to create the lobes. But the continuity of one-sidedness from small to large scales argues against the flip-flop theory, since one would expect to see sources that have flipped between the time the large-scale structure was created and the time we are viewing the jet on small scales.

Another possibility is that the jets are similar, except that one of them doesn't radiate much synchrotron radiation while some kind of instability, friction, or internal dissipation renders the other one much more luminous. The asymmetric friction theory is harder to assess, but the case in favor of it is not strong. Why would the friction be so different on the two sides? If the reason lies in differences between the background gas on the two sides, then one would have to explain why this difference is maintained over such disparate scales. Surely the interstellar medium a few light-years from the nucleus must be very different from the intergalactic medium at a distance of a million light-years. Why, then, should the asymmetry always favor the same side?

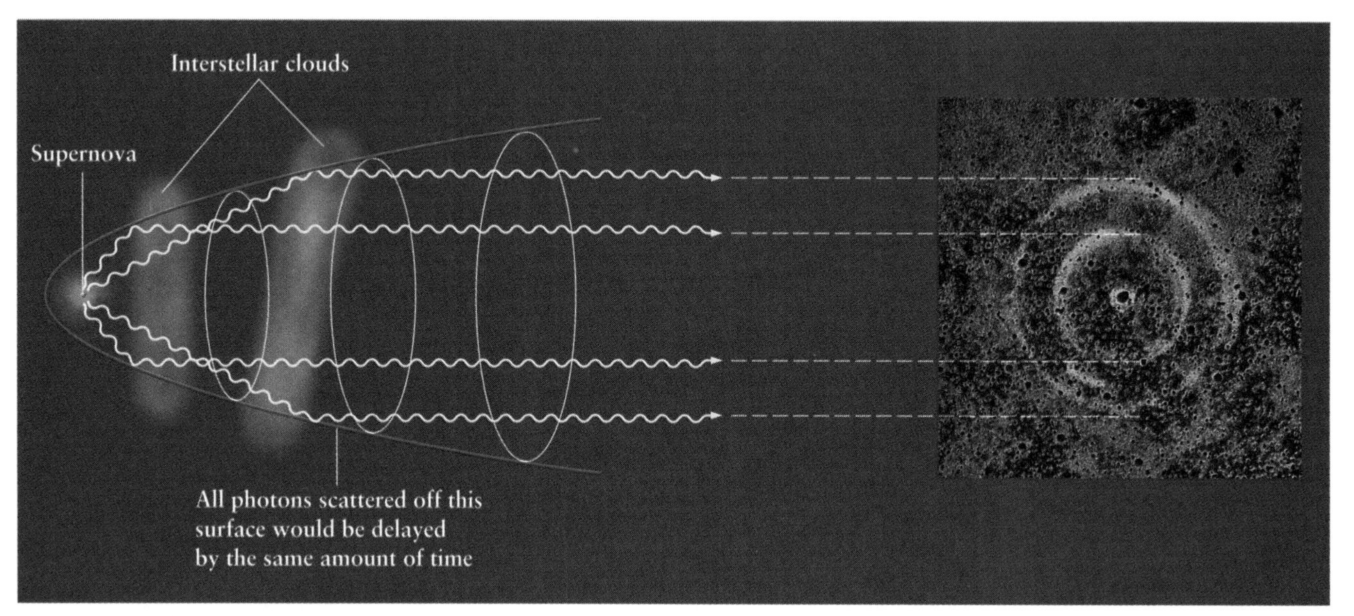


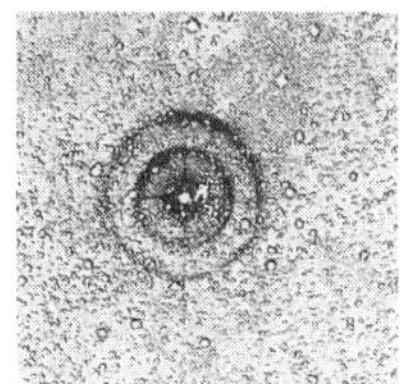

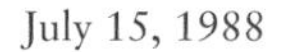

July 15, 1988

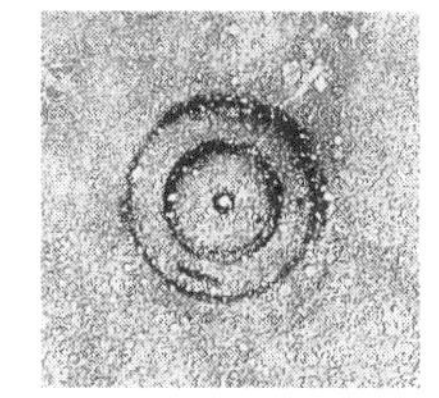

February 6, 1989

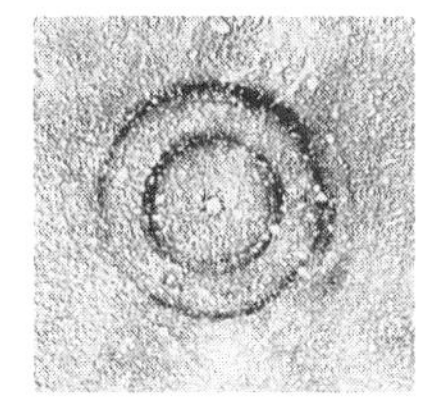

August 25, 1989

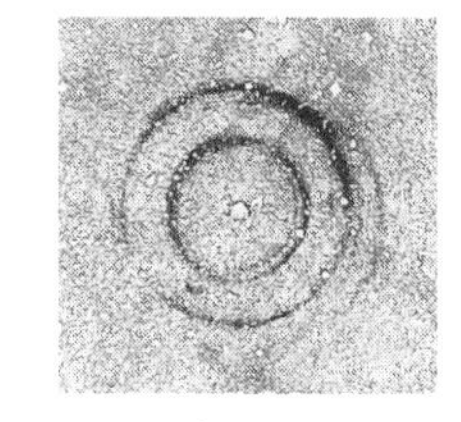

November 21, 1989

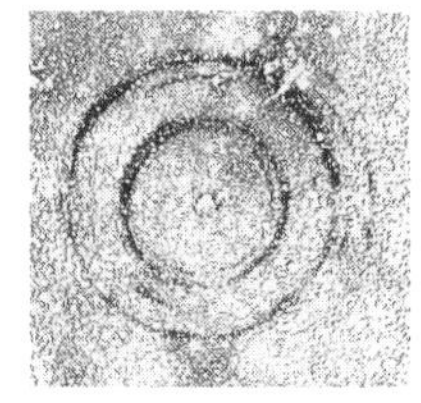

December 14, 1990

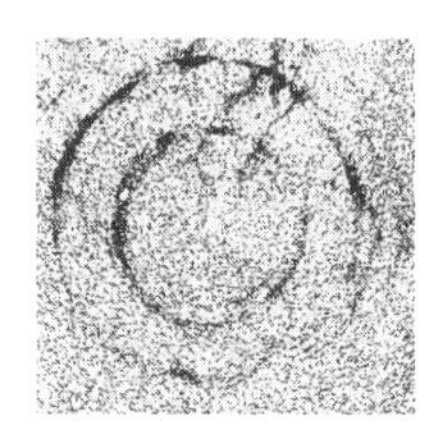

February 3, 1992

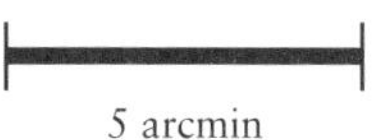


*In the years since Supernova 1987A exploded, two rings of light have been visible around it; their radii appear to be expanding at several times the speed of light. This observation, an example of a "light echo," does not mean that Einstein's theory of relativity is being violated. The apparent super-light speed results from the geometry responsible for producing the rings. The flash of light from the explosion is being scattered off dust particles that are concentrated in two clouds lying in front of the supernova in the Large Magellanic Cloud. A similar geometric arrangement leads to apparent "superluminal" expansion of compact radio sources.*

Of all the debates over jet physics, few have proven so lively as the argument over what causes the "sidedness" of jets. Above we tried to give a flavor of some of the arguments that were made early on in the debate. But now we must bring in another, even more surprising observational trend. It suggests a simpler, albeit more exotic, explanation for one-sidedness.

It turns out that the small-scale structure of jets that can be mapped with VLBI often changes from year to year, in a systematic way. From time to time bright features in these jets—often called "knots" or "blobs," for want of a better term—appear to form close to the unresolved core and then to march outward with time. Taking into account the distance to the radio source, as measured using the redshift of the galaxy, we can estimate the apparent speed at which the "blobs" move outward on the sky. Surprisingly, the features move at a speed that seems to be in excess of the speed of light; this phenomenon is therefore known as *superluminal expansion.* Does this superhigh speed of expansion mean that a basic tenet of physics—that nothing can travel faster than light—must be overthrown, or that redshifts cannot be used to determine the distances to radio galaxies? In other words, are we back to requiring "new physics," as some of our colleagues had suggested during the early debates over the nature of quasars? Most astronomers are convinced that the answer is a resounding "no." In fact, superluminal expansion can be explained very simply within the context of standard physics.

Astronomers do not contend that these "blobs" are really moving faster than light. Instead, it is thought that we are witnessing an illusion. The bright patch appears to be moving faster than light because the radiation from the patch and the radiation from the core of the galaxy, which serves us as a benchmark, travel along paths of different lengths and therefore take different amounts of time to reach us. The simplest geometric arrangement requires that the patch be moving toward us at a small angle to the line of sight between the radio source and the Earth, with a speed slightly less than the speed of light. Radiation from the core always takes the same amount of time to reach us, whereas the time it takes for radiation to reach us from the patch decreases with time. This makes the patch *seem* to separate from the core at an apparent rate that can be far greater than the true rate and indeed greater than the speed of light. The important point is that superluminal expansion does not imply that blobs in radio jets are traveling at speeds faster than the speed of light, but does require that they travel at speeds *very close* to the speed of light. For example, the superluminal expansion at 10 times the speed of light observed in 3C 273 requires blob speeds exceeding 99.5 percent of the speed of light.

Once we accept the idea that the speeds of gas flow in jets are *relativistic* (that is, very close to the speed of light), the puzzle of one-sidedness

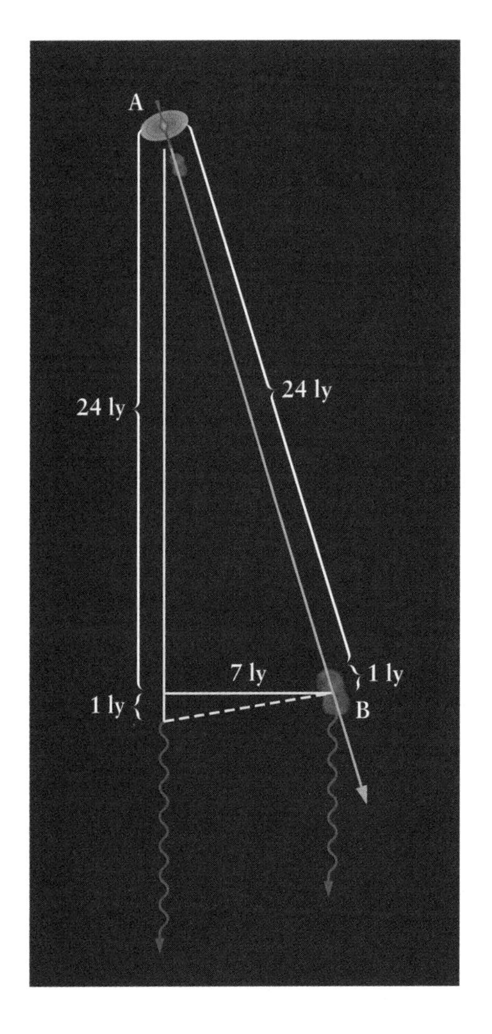


*The geometry of superluminal expansion. If the blob in this example moved exactly at the speed of light, it would appear to have a transverse speed of 7 times the speed of light. It would actually take the blob 25 years to move from point A to point B, but light emitted when the blob starts out at point A would reach us only one year before the light signaling its arrival at point B. It would therefore* appear *to make the trip in only one year! Nothing can, of course, move at exactly the speed of light. However, this example shows that an object moving at 96 percent* $(\frac{24}{25})$ *of the speed of light would appear to move 7 light-years transversely in just over 2 years, implying a transverse speed nearly 3.5 times that of light.*

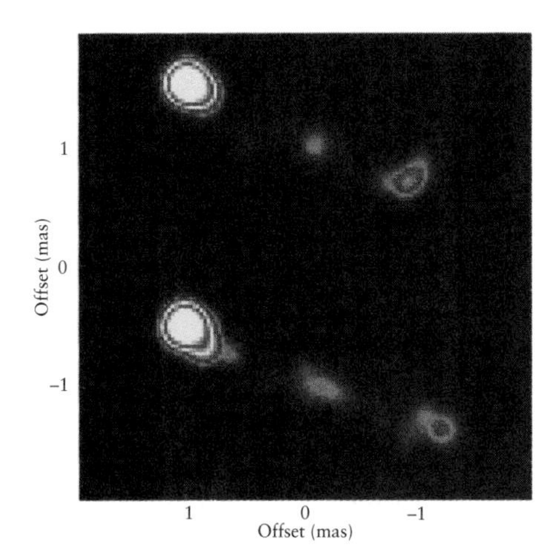


*High-resolution radio images of quasars typically show a bright core and a one-sided jet with a blobby structure. This pair of images of 3C 279, made nine months apart by Ann Wehrle and her collaborators, reveals noticeable changes in the structure of the jet. The (blue-green) "blobs" are marching outward, while a new feature has appeared near the core. The blob near the end of the jet has "moved" about 0.2 milliarcseconds, corresponding to more than 2 light-years at the distance of 3C 279. This kind of "superluminal expansion" is observed in many compact radio sources; its explanation requires the existence of outflows pointed nearly at us with speeds only fractionally less than that of light.*

is immediately solved. To explain why, we must introduce the phenomenon of *relativistic aberration.* Suppose a hunter wants to shoot a duck when it is directly above. If the shotgun is aimed vertically upward, it must be fired before the duck passes overhead. The shotgun pellets will travel vertically upward to the duck, which will have moved forward to meet them (the hunter hopes). Now consider all of this from the vantage point of the duck. It sees the hunter moving toward it, and the pellets, instead of moving vertically, have a horizontal component in their motion. To put it another way, the duck sees the pellets travel along a slightly aberrant trajectory that is inclined toward the direction in which the hunter is moving with respect to the duck.

The same thing happens with photons, the "particles" of electromagnetic radiation. A blob of hot gas that radiates photons equally in all directions will appear to be shining preferentially along its direction of motion. When the blob is moving relativistically (that is, almost as fast as the photons), the effect is so pronounced that it is sometimes called the "headlight" or "beaming" effect. Take the case of a blob of gas in a radio jet whose relativistic motion gives the illusion of superluminal expansion at an apparent speed of five times the speed of light. Here roughly half of the photons that the gas emits will be radiated within a cone facing into the direction of its motion with an opening angle of only 5 to 10 degrees. Furthermore, each of the photons within the cone will have been made more energetic by a Doppler-shift decrease in its wavelength (since the energy of a photon is inversely proportional to its wavelength). The net result is striking. If an observer is in the cone, the source will look brighter than it would if it were stationary, typically by a factor between 100 and 1000. Conversely, if the observer is well outside the cone, the source will be effectively invisible. The fact that the jet is seen only on one side of the galaxy most likely indicates that its flow is relativistic; unless the axis were precisely in the plane of the sky, the Doppler beaming effect would brighten the jet on the approaching side.

These ideas have motivated astrophysicists to develop so-called unified models for radio galaxies. The basic idea is that radiation sources of a standard kind, if they emit in a directional ("anisotropic") way, can have very different appearances depending on how they are oriented relative to our line of sight. Suppose most radio sources consist of a pair of relativistic jets emerging in opposite directions from the heart of a galactic nucleus. If the jets are more or less aligned with the direction from the source to the Earth, we detect the jet that is pointed toward us; the other jet is invisible. We therefore see a bright, one-sided jet, foreshortened by the small viewing angle. The jets end in a pair of extended radio lobes, which radiate photons roughly equally in all directions because the relativistic flow is drastically slowed when the jets pass into the hot spots. However, because we are viewing the source "end-on," the lobes appear to us as a relatively faint halo of extended emission that is seen to surround the strongly beamed compact source. This is the most likely explanation of the compact sources described earlier: they are probably those big double sources that are seen nearly end-on.

In the radio galaxies that show extended lobes, the jets are presumably traveling at a large angle with respect to our line of sight. For the most powerful of these sources, the ones such as Cygnus A, the jets are highly relativistic and extremely faint. For sources somewhat less powerful, the jets are mildly relativistic. Nevertheless, the aberration of their radio emission is sufficiently pronounced that only the jet on our side of the galaxy's center is bright enough to be detected. For weak sources such as 3C 449 the jets are significantly slower than the speed of light on the large scales observed by linked interferometers such as the VLA, so that both jets are equally visible. We note, however, that even in these symmetric-jet sources the jets appear one-sided on small VLBI scales, so that we are apparently looking at jets that start out at near-light speeds and are slowed down by their interactions with the interstellar gas.

The ultimate test of a good scientific theory is its ability to account for observational data that were not used to arrive at the model. "Unified models" for radio galaxies pass this test with flying colors. One generic prediction of unified models is that the jet we see in a one-sided source should be the one on the near side of the galaxy. Since galaxies contain interstellar gas, this means that the radio lobe on the "jetted" side of the galaxy should have much less gas in front of it than the lobe on the "unjetted" side. Several years after unified models were proposed, Robert Laing

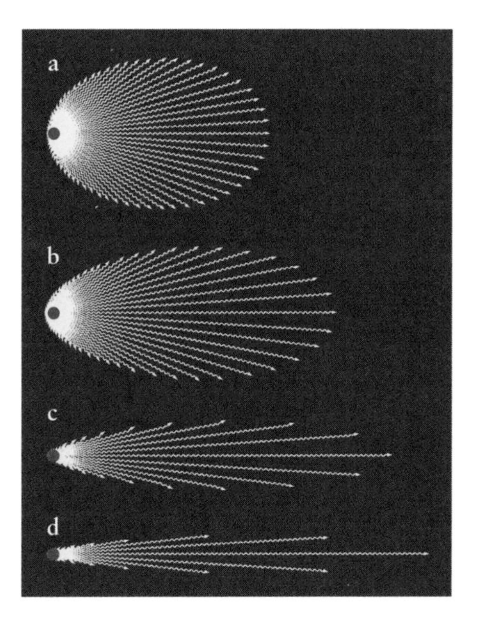


*When an object moves at nearly the speed of light, relativistic aberration concentrates the radiation it emits into a "forward cone" around its direction of motion. In* a, *the emitter (a cloud that intrinsically radiates equally in all directions) is moving toward the right at half the speed of light. In* b *it moves at 75 percent the speed of light, in* c *at 94 percent, and in* d *at 98 percent the speed of light. The emitter is rendered almost invisible except from the forward direction. The shape of each pattern shows how the intensity varies with the angle of emission. Seen from directly in front, the emitter appears brighter than if it were stationary by factors of 7, 30, 440, and 3100, for the four cases* a *to* d *respectively. (These enormous enhancements are, of course, not shown to scale in the figure.) Aberration leads to a strong selection effect in favor of detecting relativistic jets pointing almost toward us—precisely the ones that manifest apparent superluminal motions.*

and Simon Garrington, of Cambridge University, found exactly this effect in nearly every source they measured.

Another correlation that neatly falls into place concerns the alignments of small-scale jets observed with VLBI. Anthony Readhead, Marshall Cohen, and their colleagues at the California Institute of Technology discovered that in certain radio galaxies the jets align neatly with the lobes. These are precisely the galaxies dominated by extended radio emission—that is, whose emission comes mostly from double lobes. In these sources, any misalignment is by no more than a few degrees. In contrast, the VLBI jets associated with compact radio sources—in which the core dominates the emission—are noticeably curved, so that they are seriously misaligned with the faint extended structure of the lobes, typically by angles of 30 degrees or more. This correlation is easily explained by unified models. One does not expect the jets to be precisely uniform and straight—doubtless they exhibit some natural bending and inhomogeneity. In compact radio sources, which we are viewing end-on according to unified models, the bending is exaggerated because our line of sight is almost straight along the jet. Moreover, we would expect superluminal motion to be most pronounced in the compact sources, as indeed the observations seem to show.

A third correlation concerns the variability of radiation emitted by AGNs. When we were discussing the black hole paradigm for AGNs in Chapter 5, we used variability to place upper limits on the size of the energy-emitting region. But some kinds of variability are incompatible with this simple interpretation. Observers using the large single-dish radio telescope in Bonn, Germany, have found that compact radio sources often vary with extraordinary rapidity, sometimes doubling in brightness in a matter of hours. Even more striking are results obtained with the *Compton Gamma Ray Observatory,* a NASA satellite that has been orbiting the Earth since 1991. Some compact radio sources appear to be emitting a large fraction of their total luminosities in the gamma-ray portion of the spectrum and, moreover, these gamma rays sometimes fluctuate within a matter of days. If the timescales on which either the gamma rays or the radio waves vary were assumed to represent the light travel times across the emitting regions, the sources would be so tiny that the radiation could not escape. But unified models offer a way out. Just as relativistic motion nearly along the line of sight can lead to the illusion of superluminal expansion, it can also lead to apparent variations that are much faster than the true variation. Therefore, the emitting region can be much larger than suggested by variability, and the escape problem goes away.

The "Doppler beaming" interpretation of extragalactic radio sources accounts for so many diverse observations that it is unlikely to be seriously wrong. Thus, radio astronomy leads us to some profound generalizations about the nature of activity in the nuclei of galaxies. Those AGNs that produce intense radio emission—radio galaxies, BL Lac objects, and a subset of quasars—invariably do so through the production of jets, which by every indication appear to start out with speeds exceedingly close to the speed of light. Where do these jets come from? Even VLBI cannot resolve the sources of jets. But what we can say is that jets are already well formed on scales that are smaller than those accessible to VLBI. In the case of the nearby radio galaxy M87, which is "only" about 50 million light-years away, this means scales smaller than 0.03 light-year, or less than 100 times larger than the likely size of the black hole.

## Making a Relativistic Jet

Hidden from direct observation by its small scale, the process of jet formation in AGNs is very much the playground of theorists. We think we know something about the environment in which a jet forms, and we can see the end product as the jet emerges from the nucleus. But everything else about the process must be deduced from first principles. What elements must go into a theory of jets? First, there must be a source of energy that can propel gas to high speeds. Second, there must be a mechanism that splits this energy flow into two equal branches (since jets come in pairs) and gives these flows a specific direction. Third, there must be some process that confines the twin flows into the narrow streams we see. We can be reasonably confident that the jets are initially propelled along the rotation axis of the "central engine." But the natures of the first and third elements are much more uncertain.

Creating a flow that approaches the speed of light is especially challenging. The easiest way to accelerate gas to high speed is to push it with some form of pressure. If you heat gas to a high temperature in a confined space and then let it go, the pressure will accelerate the gas as it is released, like air being released from a balloon. If all the thermal energy goes into acceleration, this process is called *adiabatic expansion.* The final velocity reached by the accelerated gas will then be the typical random particle velocity—roughly equivalent to the sound speed—that the gas had when it was hot. But heating and releasing gas turns out *not* to be a good way to attain speeds close to the speed of light. To reach relativistic speeds by this kind of mechanism, one would have to heat the gas to such a high temperature that the thermal energy per particle would greatly exceed the rest-mass energy ($E = mc^2$) per particle. Gases at such high temperatures tend to be unstable. There is so much energy present that it is very easy for some of the energy to be turned into new particles! This energy is then not

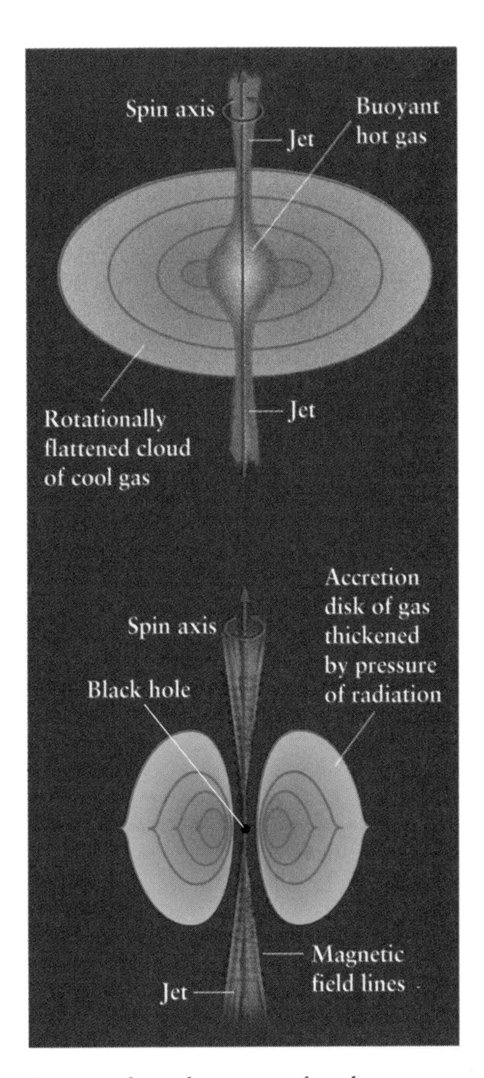


*Proposed mechanisms whereby material can be collimated into two oppositely directed jets in the vicinity of a black hole.*

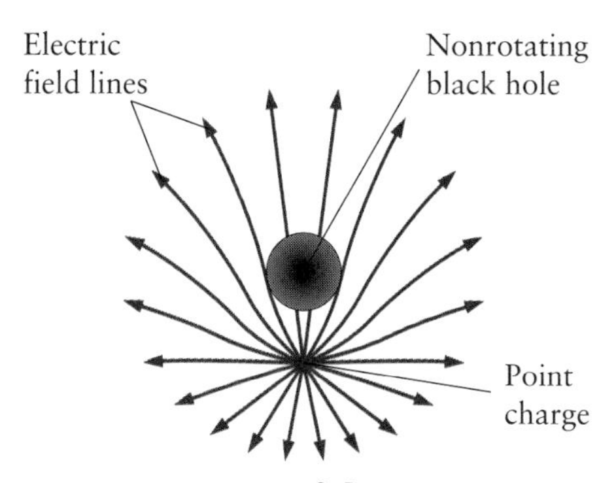


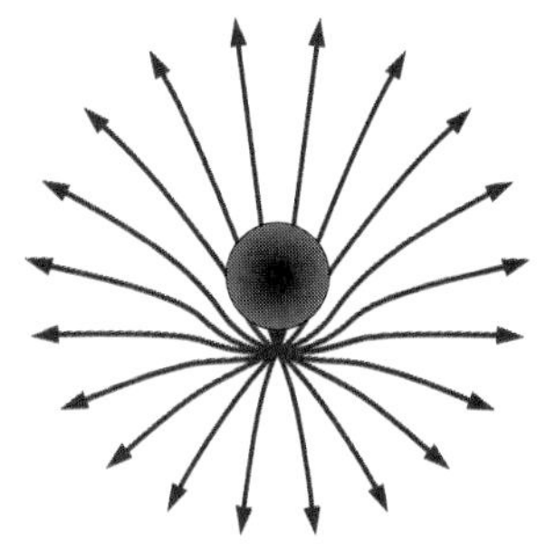


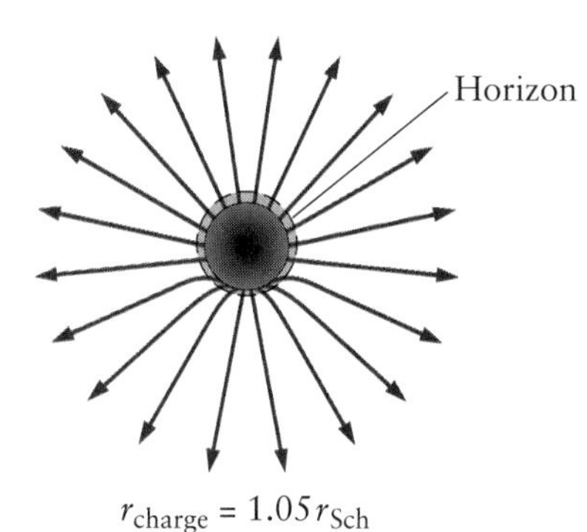


*If an electric charge is lowered toward a (Schwarzschild) black hole, the lines of force become distorted. As the charge gets close to the Schwarzschild radius (or "horizon") the distortion becomes so great that the lines of force appear to emanate from the center of the hole. It is as though the horizon were a conducting sphere over which the charge spreads itself uniformly. If the point charge is falling into the hole, the apparent spreading occurs on the free-fall time. This makes it appear as though the horizon has an electrical resistance of about 100 ohms.*

available to help with the expansion. Such hot gases also tend to cool very rapidly, which means that they radiate away their energy (or even emit other escaping particles such as neutrinos) before it has a chance to contribute much to the acceleration.

Another way to accelerate gas in AGNs is to push on it with radiation. But the relativistic aberration effect discussed earlier prevents radiation pressure from accelerating gas to the highest speeds that have been observed. Suppose you are riding on a blob of gas that is being accelerated by a beam of radiation coming from behind. Realistically, the light rays pushing on the gas cannot be perfectly aligned; they will have at least a small spread in direction. As the gas approaches the speed of light, the light rays on the edge of the beam will appear as if they are coming at you from the front! These rays *oppose* further acceleration and will generally limit the maximum speed.

In addition, pushing on a gas with radiation cannot yield a final jet flow with more energy than was contained in the radiation to begin with. Unless the central radiation source can be very well hidden from view, any jet source powered by radiation pressure should also be very luminous. Yet strong radio galaxies, such as Cygnus A and M87, are not luminous enough. To energize the extended radio lobes, the jets would have to have more power than is available in the radiative luminosity of the nucleus.

Is there a mechanism that could accelerate a powerful outflow of gas to relativistic speeds without producing an intense source of radiation? It is difficult to see how such speeds could be attained using only the gravitational energy released by infalling matter. An accretion disk may dispose of some of its energy by creating a wind, but such an outflow is unlikely to reach the speeds observed in jets. But there is another source of power that is available to be tapped: the rotational energy of a spinning black hole.

When this possibility was first discovered by Roger Penrose of Oxford University, it came as a shock, since people had thought that *nothing* could be extracted from a black hole. But this is not quite true. The real law governing the growth of black holes requires only that the *surface area* of a black hole's horizon can never decrease. Since a spinning hole has a smaller surface area than a nonrotating hole of the same mass, there is no reason in principle why the rotational energy could not be extracted. But how does one do it in practice?

Astrophysicists have proposed plausible mechanisms for extracting the spin energy of a black hole; these mechanisms depend on exploiting the remarkably close analogy between a black hole and an ordinary electrical conductor. This analogy can be illustrated for a nonrotating, or Schwarzschild, black hole (discussed in Chapter 1) as follows. Imagine observing the electric field around a point charge that is gradually moved toward the hole. When the charge is far away from the hole, the electrical lines of force stick out radially from the charge. But as the charge is moved closer to the horizon, the lines of force get progressively more distorted: they "wrap around" the hole so that eventually they appear to emanate radially from the center of the hole instead of from the charge. It is as though the charge had spread itself over the hole's "surface," just as would happen if the "surface" were an electrical conductor.

A spinning, or Kerr, black hole behaves like a spinning conductor. Suppose that the gas approaching a spinning black hole contains a magnetic field. Like any spinning conductor in a magnetic field, the "surface" of a Kerr black hole will develop a voltage difference between the equator and poles. This means that a black hole spinning in a magnetic field behaves like a battery, analogous to the battery in a car or flashlight. In electrical engineering parlance, this kind of battery—a conductor spinning in a magnetic field—is called a *unipolar inductor.* The only difference between a black hole battery and a car battery is that the voltage drop across the poles of a black hole battery in an AGN would be around $10^{15}$ volts, instead of the more familiar 12 volts!

To extract energy from a battery, you have to hook it up to a circuit. In the case of a black hole battery, one "wire" should be attached to the pole and the other to the equator. In an AGN, the role of the "wire" is played by the ionized gas that surrounds the black hole, since this gas is an excellent conductor. The amount of energy extracted from the battery then depends on the relationship between the electrical resistance (or "impedance") in the circuit and the internal resistance inside the battery. The power output is maximized if the two resistances are equal, a condition called an *impedance match.* Now it turns out that a black hole is not a perfect conductor. No matter what its mass, it behaves as if it has a resistance of about 100 ohms, roughly the same resistance as the light bulb in a flashlight. But what is the external impedance? If the effective circuit closes at some significant distance from the hole, as we expect, then the impedances will automatically be roughly matched. Energy, drawn from the hole, accelerates charged particles and ultimately propels the jet. As the energy is extracted, there is a drag on the spin of the hole, slowing it down.

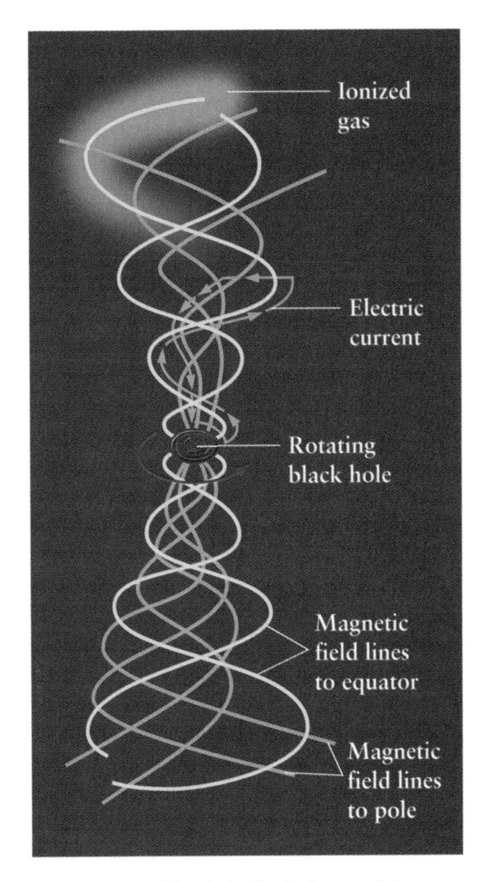


*A spinning black hole behaves like a spinning conductor. If the hole is surrounded by conducting gas carrying a magnetic field, then a voltage difference is set up between the poles and the equator, and the hole can act like a battery. The electric current driven by the voltage drop can extract energy from the spin of the hole, which can be transformed into outflowing jets.*

*Roger Blandford has contributed several key ideas on the nature of active galactic nuclei. In particular, he studied the formation of jets, and with Roman Znajek showed how the spin energy of a black hole embedded in a magnetic field could be converted into an outflowing jet.*

In this sense, we might think of the spinning hole as a kind of flywheel, storing rotational energy for later use.

Conditions are naturally favorable for energy extraction when the hole is spinning rapidly and the infalling gas contains a moderately well ordered magnetic field. Do we expect a supermassive black hole to be spinning? It may, of course, have "original spin," because it formed by the collapse of a rotating gas cloud or supermassive star. Moreover, any black hole in an active galactic nucleus can be "spun up" after a sufficiently long time by accreting material with angular momentum. A third way to obtain a rapidly spinning black hole, in shorter order, is from the merger of two holes with comparable masses. In that case, a sizable fraction of the orbital angular momentum ends up in the spin of the merged hole. It is more difficult to predict what the magnetic field will be like in the gas surrounding the hole. In accretion disks, powerful instabilities tend to amplify the field to a sufficient *strength* that efficient energy extraction could occur. But it is not known whether the *structure* of the field will favor energy extraction. If the field is too chaotic, with reversals on very small scales, then the impedances may not be well matched, and energy extraction will be inefficient. There is certainly no observational evidence that the conditions for jet formation are satisfied in all cases. Indeed, only a small fraction of AGNs produce powerful jets.

Curiously, in cases where the galaxy hosting the AGN can be identified by type, the "radio-loud," jet-producing nuclei seem to be in ellipticals, whereas the radio-quiet nuclei are in spirals. We think of the spiral vs. elliptical classification as referring to the gross properties of galaxies, whereas the question of whether or not to form a jet is decided on scales not much larger than the horizon of the black hole. How the innermost nucleus "knows" in which kind of galaxy it resides is one of the principal puzzles in AGN research.

The huge voltage drop across a "black hole battery" may actually be responsible for *creating* the matter that flows along a jet. In the presence of such a large voltage, any electron is accelerated to such a high energy that it creates a "cascade" of particles of matter and antimatter—mostly electrons and positrons—from the vacuum of space. One could even say that the black hole "short circuits" and compare the creation of particles to the energy released in a bolt of lightning when the voltage between a thundercloud and the ground "shorts out." So many particles are created in such a cascade that they are easily able to neutralize any unbalanced electromagnetic forces. One can then describe the electromagnetic properties of the ionized gas in a very simple way, using a viewpoint first introduced by the Swedish physicist Hannes Alfvén. We can think of the magnetic field lines as being part of the fluid, in the sense that when the fluid moves, the mag-

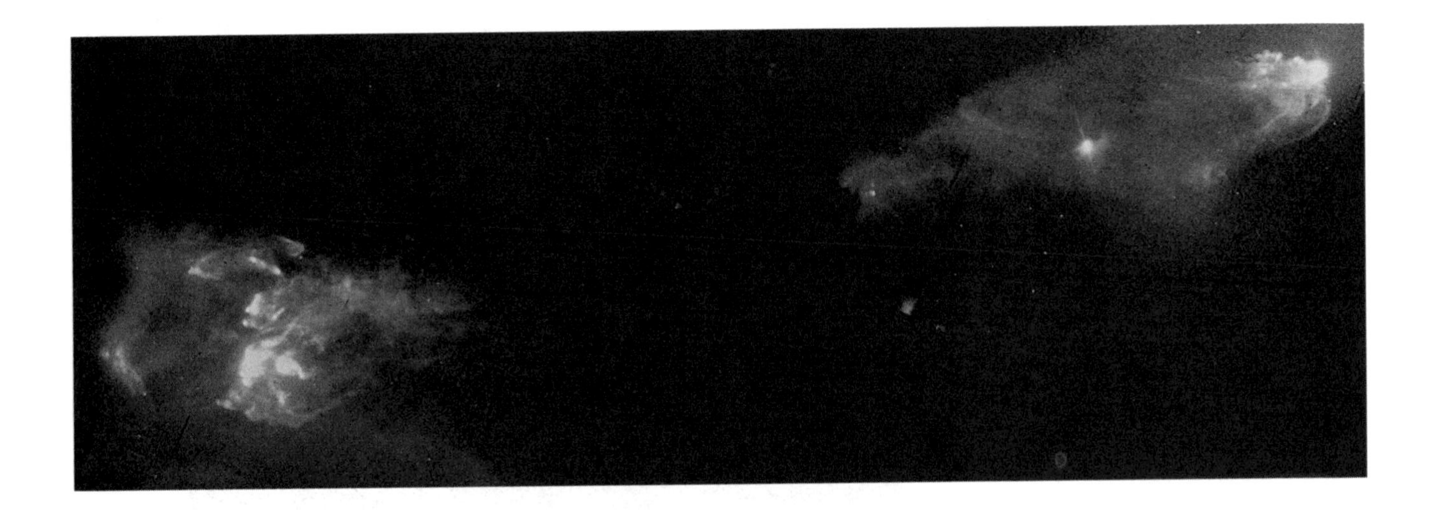

*Despite its eerie resemblance to a giant double radio galaxy, this is a visible-light image of the region around a young star, made using the Hubble Space Telescope. The protostar itself is hidden behind a dusty cloud, but its jets penetrate the cloud and create a pair of lobes spanning more than a light-year. Theorists believe that magnetic propulsion and focusing mechanisms, similar to those operating in AGNs, can also explain protostellar jets.*

netic lines of force must move with it—and vice versa. The embedded magnetic field also gives a fluid added resiliency. If the field lines are crammed closer together, they will try to spring back; if they are gently curved, they will try to straighten out; but once curved into a closed loop, they will try to contract like a stretched rubber band.

This tendency of a loop of magnetic field to contract could explain why jets are so narrowly focused, or collimated, even on the smallest observable scales. The spinning black hole effectively grabs onto the magnetic field lines and rotates them. Particles of ionized gas outside the black hole are then forced to rotate along with the field lines, and consequently feel a centrifugal force that flings them outward. Beyond a certain distance, however, the magnetic field is not strong enough to force the particles to corotate. The gas particles lag behind, bending the magnetic field lines backward. As the field lines become more and more twisted about the rotation axis, they begin to exert a strong pinching force that may be responsible for the initial focusing of the jet. Under certain circumstances the coiled magnetic field can also act like a compressed spring, helping to accelerate the gas still further.

The unipolar inductor mechanism is especially attractive because it seemingly explains both the acceleration and the collimation of jets using a single mechanism. But detailed models of relativistic jets based on this mechanism have yet to be produced. The calculations that have been done so far make very idealized assumptions about the distributions of both magnetic field lines and matter close to the black hole. No one has yet attempted to model self-consistently the creation of particles at the base of the jet. Furthermore, the initial collimation produced by magnetic forces is

*Nova Scorpius, located within our Galaxy, is an X-ray binary that occasionally expels jets that appear to move superluminally. It almost certainly contains a black hole of 5 to 10 solar masses that behaves like a miniature (and speeded-up) version of the supermassive holes that energize jets in quasars and strong radio galaxies. The systematic outward motions along the jets are clearly apparent, even over timescales of weeks.*

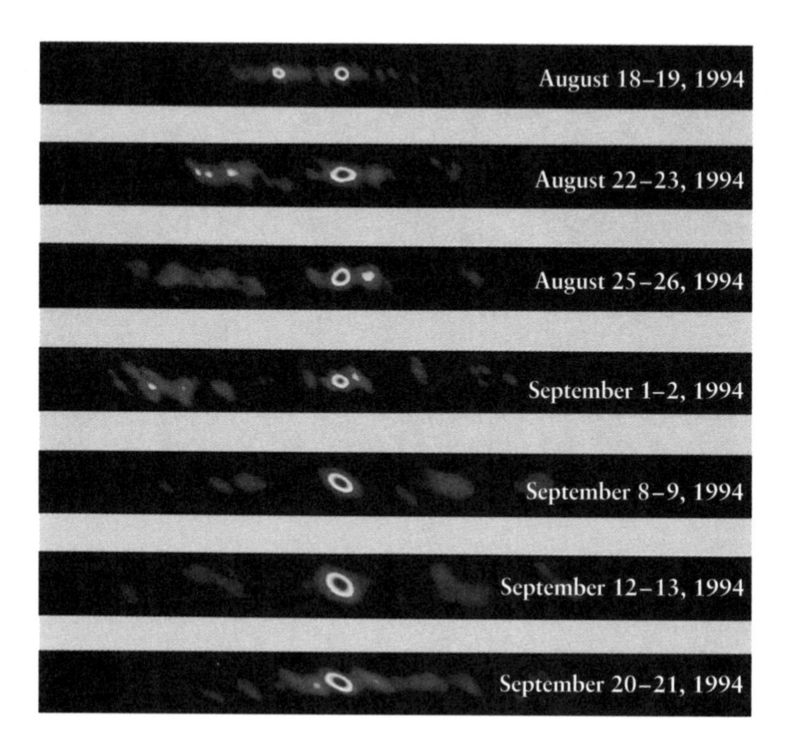


unlikely to be solely responsible for keeping the jet in a narrow beam farther out in the galaxy. Once away from the nucleus, the jet flow is more likely to be confined by the pressure of the surrounding gaseous medium.

The same electromagnetic mechanisms that have been proposed to explain jet production and collimation in active galactic nuclei, of which the unipolar inductor model is just one example, can be generalized to explain jets that form in other astrophysical situations. We now realize that jets are ubiquitous in the Universe. Even the swirling gaseous disks within which stars like the Sun are forming produce strikingly narrow jets, which have been traced over several light-years in length. As is the case for the relativistic jets in AGNs, both hot gas and radiation pressures seem inadequate to propel these "protostellar jets" to the observed velocities. Once again, speculation centers on magnetic forces as both the driving and collimating agents.

Protostellar jets are very different from their AGN counterparts. Their velocities are "only" about 100 kilometers per second ($\frac{1}{3000}$ that of an AGN jet), and they are made out of gas of ordinary composition (stripped off the protostellar disk), rather than electron–positron pairs. Indeed, they may even contain molecules. Yet they can be understood according to the same general principles as the jets from spinning black holes: a gaseous disk rotates magnetic field lines, material is lifted off the disk and flung outward by centrifugal force, a twisted field pinches the incipient jet, and so on.

Magnetic forces may also propel the jets in SS 433 as well as the rapidly variable jets that have recently been discovered shooting out of several X-ray binary systems in our Galaxy. The latter resemble miniature versions of the jets produced by extragalactic double radio sources. They may result from similar processes but are energized by a black hole of around 10 solar masses, rather than one of many millions. Since the jets develop over timescales proportional to the hole's size (and therefore to its mass), observing one of these nearby objects for a year may help us to understand what happens to a giant radio source over its lifetime of many million years. The concept of the magnetic propulsion of jets applies so widely that it has acquired the cachet of universality, a quality that astrophysicists find almost irresistible.

Where should we place jets in the scheme of phenomena in the Universe? Their production is not restricted to black holes. They appear to be a rather general phenomenon, seen on all scales from single protostars to the span of a cluster of galaxies. Maybe they are apt to form anytime a flattened, rotating system manages to inject some extra energy into the material near the rotational axis. But the jets from active galaxies are remarkable in that they transport energy at more than 99 percent of the speed of light. These *relativistic* jets may be the calling cards of supermassive black holes. Theory tells us that the jet power could come directly from the hole's spin, its extraction catalyzed by a magnetic field around the hole. The grandest manifestations of activity in galaxies—the vast radio structures, sometimes stretching millions of light-years across the sky—may therefore be a direct outcome of processes governed by Einstein's theory in its most extreme limit, operating deep in the galactic nucleus on one-ten-billionth the radio lobes' scale.

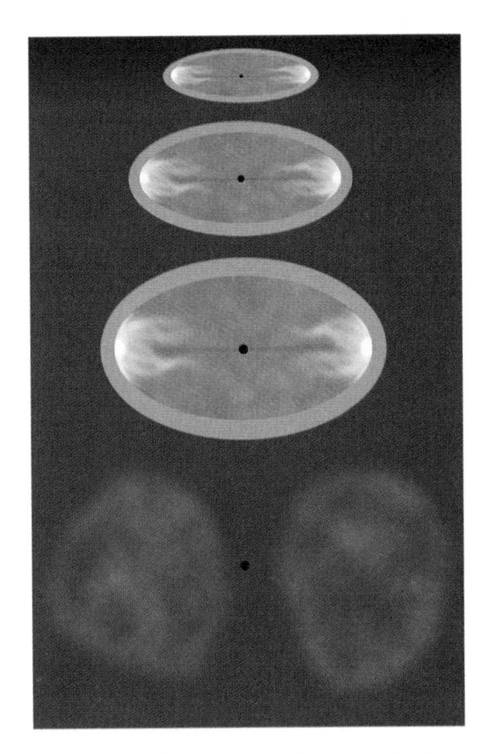

*Schematic illustration of stages in the lifetime of a strong radio source. The jets force their way outward through the interstellar (and then intergalactic) medium. When activity in the nucleus dies out, the extended lobes and cocoon expand and fade, eventually merging with the surrounding gas.*

*Chapter* 7

# Black Holes in Hibernation

*This Hubble Space Telescope image shows the central parts of a rich cluster of galaxies as it was when the Universe was only two-thirds of its present age. Spiral galaxies were much more common, compared to ellipticals, than in present-day clusters, and encounters between galaxies were more frequent. Collisions and near-misses among these gas-rich galaxies could have triggered recurrent episodes of activity in their nuclei and built up the masses of central black holes. In today's calmer galactic environments, most of these massive black holes would be quiescent.*

Select a nearby galaxy at random and observe its nucleus, and there is only one chance in a hundred that you will detect moderately vigorous activity that can be attributed to a massive black hole. The chance of stumbling on a powerfully active galaxy—a quasar or giant double radio source—is even smaller. But this does not mean that massive black holes occur only in the nuclei of a small fraction of galaxies, for activity requires not only a massive black hole but also a supply of "fuel." To power a quasar or luminous Seyfert nucleus by accretion there must be more than a solar mass per year of gas flowing into the black hole. Radio jets powered by a black hole flywheel might survive on a smaller yearly mass budget, but a substantial gas supply is nevertheless needed to maintain the magnetic fields in the hole's vicinity. And, without more vigorous episodes of accretion, the hole's rate of spin will slowly decline.

A solar mass of gas per year is a sizable fuel supply, even by galactic standards. It is unlikely that the nucleus of a galaxy maintains this level of gas inflow for more than a small fraction of its lifetime. Perhaps there is a time early in the galaxy's life when the stars and gas in the nucleus become so closely packed that some kind of runaway catastrophe occurs, dumping matter copiously into the black hole. From time to time, the galaxy might merge with a smaller, gas-rich galaxy or swallow an intergalactic gas cloud—we see examples of such encounters. For a while following such an event, matter might pour into the nucleus at a rate fast enough to power a quasar. But such episodes of violent activity most likely represent a short-lived phase in the life of a galaxy. When the fuel supply dwindles, the activity dies away but the black hole does not disappear. Black holes in "hibernation"—massive black holes now starved of fuel, and therefore quiescent—are now being discovered in nearby galaxies, and one may even be present in the center of our own Milky Way. These objects can provide clues to the enigma of quasars and remind us that even the benign-looking galaxies around us could reawaken and turn violent once again.

## The Rise and Fall of Quasars

When the Universe was only 10 to 20 percent of its current age, or about 2 billion years old, quasars were far more common than they are now. The bright quasar nearest to us, 3C 273, is more than a billion light-years away. Although it is hundreds of times more luminous than a galaxy, its remoteness renders it a thousand times too faint to be seen with the naked eye. In contrast, a hypothetical astronomer living during the "quasar era" would find the nearest quasar "only" about 25 million light-years away, and it would be just bright enough to be seen with the naked eye on a dark night (i.e., comparable in brightness to a 5th-magnitude star). We might have expected quasars to be closer at that time because the entire Universe had not expanded as much since its origin in the big bang. But the effect is much larger than can be accounted for in this way: quasars were a thousand times more common then, relative to galaxies, than they are today.

It seems unbelievable that we can tell what the Universe was like billions of years ago, but there's nothing very mysterious about this. When we observe a galaxy or quasar at high redshift, we are not only seeing out to a very large distance—we are also looking backward in time. The light reaching the Keck telescope on the summit of Mauna Kea in Hawaii tonight might have left a quasar when the Universe was a small fraction of its present age. So, to find out what the Universe was like when it was much younger, we need "merely" to catalogue the mix of objects found at distant redshifts. Carrying out such a program, however, is hard, painstaking work. Oftentimes, the objects being surveyed are close to the limits of detectability, and there is always the problem of "incompleteness": the likelihood that some objects that should be in the catalogue will be misidentified or otherwise missed. In the early days of quasar research, the samples were not systematically obtained, and it was a time-consuming procedure to measure redshifts. But by now, many thousands of quasars have been catalogued, their luminosities and redshifts recorded.

What we find is that quasar activity peaked sharply about 2 or 3 billion years after the big bang. At that time galaxies were three times closer together, on average, than they are today. But quasars were 30 times closer together. Does this trend continue at earlier epochs? Apparently not. Astronomers have discovered very few quasars at redshifts much larger than

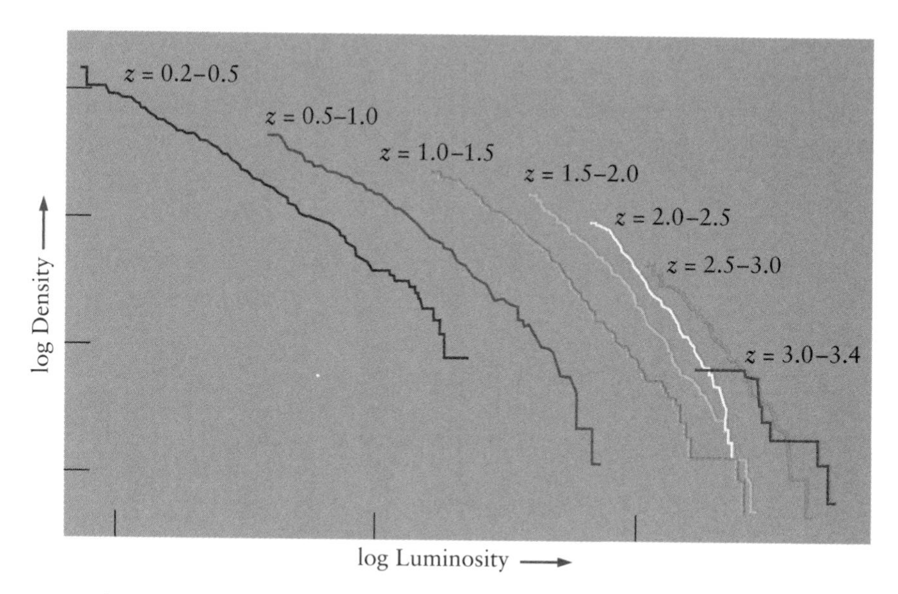


*Data from the Large Bright Quasar Survey, by Paul Hewett, Craig Foltz, and Frederic Chaffee, show how the "luminosity function" of quasars changes with redshift. On the axes of this log–log plot, each tickmark corresponds to a factor of 10 in either the luminosity or the density of quasars in space. From each curve one can read off the density of quasars brighter than a certain luminosity, for a specified range of redshifts,* z. *The plot clearly shows that quasars of a given luminosity become more common with increasing redshift. The fact that the curves are flatter at the highest redshifts implies that very luminous quasars were relatively more common at earlier times, compared to less luminous ones.*

the redshift at the quasar peak. Of course, very distant quasars are fainter and more difficult to detect. But this only partly explains the paucity of quasars observed in the very distant past: at larger redshifts—and thus earlier times—the quasar population genuinely thins out. Presumably, we are seeing the Universe as it was before most quasars had formed. A few quasars have been found with such high redshifts that their light was emitted when galaxies were six times closer together than they are now. This means that the wavelength of the light has been "stretched" by a factor of 6 between emission and reception: the visible light reaching us was emitted in the far ultraviolet part of the spectrum. When we observe the remotest quasars, we are "seeing" the Universe as it was little more than one billion years after the big bang. It seems, though, that most quasars appeared on the scene only after the Universe had been expanding for 2 or 3 billion years. They were relatively common for a similar period of time, and then, like the dinosaurs, they more or less became extinct.

What does this pronounced evolution in the quasar population signify? Quasar activity presumably can't start until at least some galaxies have condensed from the expanding Universe and developed dense nuclei where matter can be fed into a black hole. During the course of a galaxy's evolution, its gas gradually becomes "tied up" in stars; quasar activity might decline because less gaseous fuel is available as galaxies age. Intense activity could then restart only if the nucleus suddenly acquired a fresh supply of fuel—for example, through a merger with a neighboring gas-rich galaxy.

But fuel supply is only half the story. The level of activity also depends on the mass of the black hole. We do not know whether massive black holes form concurrently with galaxies, or predate them. Are they the "seeds" around which galaxies condense? How much does the black hole grow during the period of quasar activity? Until we know more about galaxy formation, and about the relationship of quasars to galaxies, there seems little hope of understanding the life cycle of a quasar theoretically.

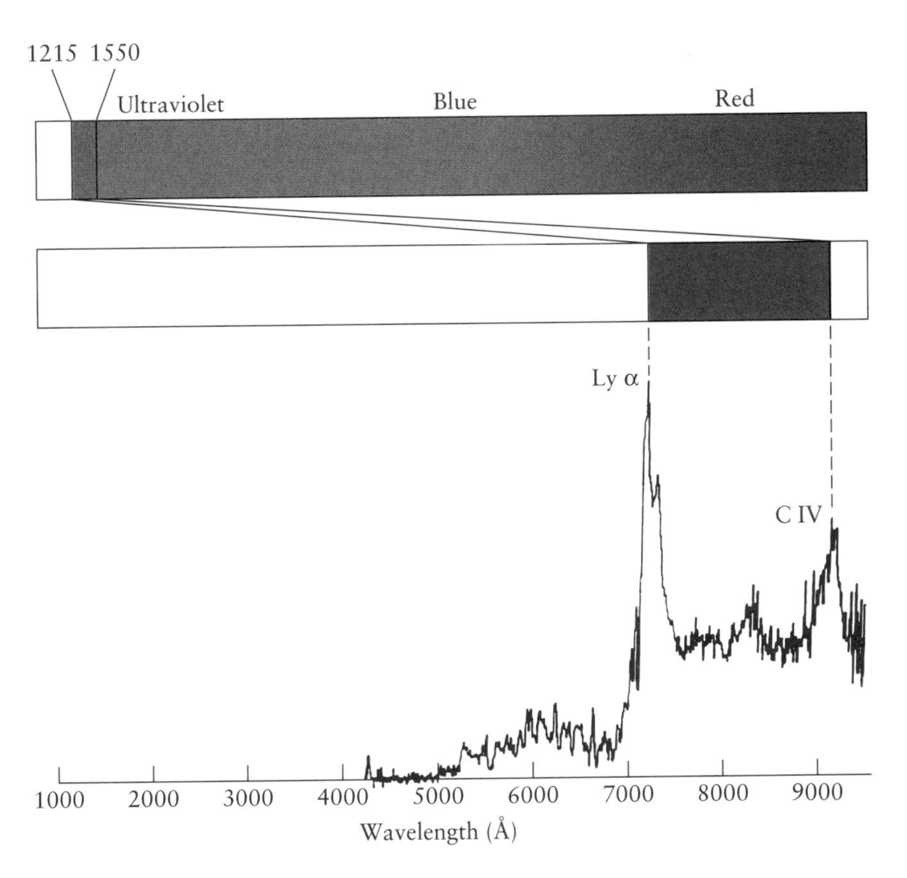


*The quasar PC 1247 + 3406 has a very high redshift of z = 4.897. The ultraviolet lines of hydrogen (Lyman α, emitted at 1215 Å) and triply ionized carbon (C IV, emitted at 1550 Å) are observed in the red part of the spectrum. Light recorded in this spectrum left the quasar when the Universe had expanded only to one-sixth of its present scale, about one billion years after the big bang.*

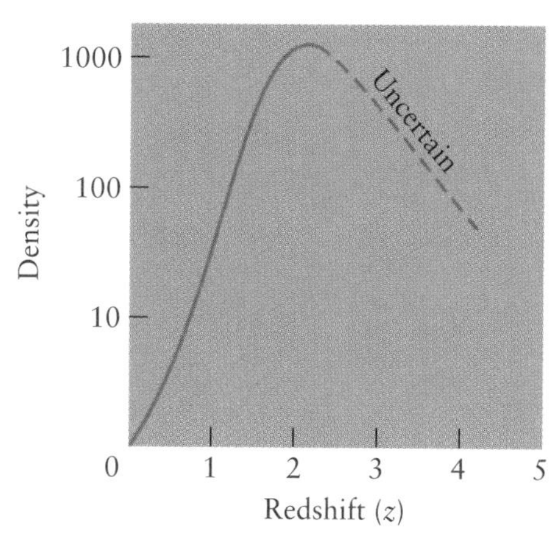


*This graph shows how the abundance of quasars has changed with redshift. The vertical scale shows the density of quasars in space relative to their local (z = 0) density. Changes in density resulting simply from cosmic expansion have been removed before plotting the diagram. The pronounced peak that remains shows that quasars were most common at a redshift of 2 to 3.*

We must rely instead on empirical clues. For this reason, the quest for ever more distant quasars is very important: their existence very much earlier in cosmic history would embarrass theorists who argue that galaxy formation didn't start until the Universe was more than a billion years old. If the theorists are right, larger telescopes would obviously offer a clearer and more detailed picture of early quasars, but wouldn't reveal quasars at much greater distances. We may already be seeing the first quasars that ever formed, in which case looking still farther out (and farther back into the past) would merely probe a featureless pregalactic dark age.

The dramatic rise and fall of the quasar population sends a clear message: most galaxies that were ever active were active in the remote past. Their remnants—massive black holes in the centers of galaxies—must therefore lie about us. How common are these hibernating black holes, and how massive are they? These questions are interrelated: both depend on how long the typical quasar lives.

Quasars are rare cosmic beasts. There are now fewer than one for every 100,000 galaxies. Even during the era of peak quasar activity, when the Universe was only 2 billion years old, quasars were 100 times rarer than normal galaxies. One might infer from this that only one galaxy in 100 has ever indulged in quasar activity, and that we shouldn't expect to find remnants in more than about 1 percent of galaxies today. But this argument assumes that individual quasars live for more than a billion years. There is another possibility. A graph like the one on the facing page plots the rise and decline of the *population* of quasars: it need not represent the life cycle of a typical object. Just as in the biological case, many short-lived generations could be born, evolve, and die over the period during which the overall population rises and falls. If there were many generations of quasars, each living much less than a billion years, then we would expect to find more remnants.

The longer a quasar lives, the more massive its black hole has to be. The most powerful quasars radiate energy equivalent to several times the Sun's mass each year. The energy is released when gas swirls downward toward the hole, where gravity converts, at most, about 10 percent of the mass-energy ($mc^2$) of the infalling matter into quasar radiation. Thus, the hole would need to swallow at least 10 solar masses for each solar mass worth of energy that is radiated. Obviously, if a very powerful quasar kept shining for more than a billion years, it would end up weighing tens of billions of solar masses. Given the same lifetime, 90 percent of all quasars would weigh more than one billion solar masses. These mass estimates are lower limits, since they do not include the mass that might have accumulated in the black hole before the quasar epoch began. Since the lower limit on the mass is proportional to the total energy expended over the active lifetime, the final masses needn't be so high if there have been many successive generations of short-lived quasars.

Did one generation of quasars live for a billion years, leaving behind rare, enormous holes? Or did many generations flare up and fade during the quasar era, leaving behind less massive holes in most, or even all, galaxies? We could answer these questions if we found an independent way to measure the masses of the black holes in quasars. Unfortunately, the best we can do at present is to bracket the most probable masses between upper and lower limits. We have reason to believe that typical quasars weigh at least 100 million suns, irrespective of their lifetimes. This argument depends on the fact that the radiation produced by the quasar exerts a repulsive force on infalling gas that tends to cancel out gravity. The pressure of the quasar's intense radiation would expel everything from a quasar's vicinity unless gravity provided a strong enough countervailing force. The maximum luminosity for a given mass is the Eddington limit, which we introduced in Chapter 5. If quasars are powered by accretion, and pull in gas from their surroundings, then they must have a big enough mass for gravity to overwhelm radiation pressure. Knowing the quasar luminosity, we can calculate a lower limit to the mass. By this reasoning, the Soviet theorists Yakov Zel'dovich and Igor Novikov deduced, as early as 1964 (only one year after quasars were discovered!), that typical quasars had to weigh at least $10^8$ solar masses, and that the most powerful ones had to weigh $10^9$ solar masses. Another argument yields a lower limit for the minority of quasars that are radio sources: the magnetic fields and fast electrons in the radio lobes contain a huge amount of energy—the mass-equivalent of millions of suns. Whatever "engine" generated these lobes must therefore weigh at least $10^8$ solar masses.

One can deduce an upper limit to the black hole mass by studying the emission from fast-moving clouds of glowing gas, which are heated by ultraviolet radiation from the vicinity of the black hole. We estimate the distances of the clouds from the black hole by measuring how much the ultraviolet light disturbs the atoms (technically, we use the spectral emission lines to gauge their level of "ionization," or how many electrons have been stripped off the atoms). The clouds around the most luminous quasars are about a light-year away from the hole; in less luminous sources, they lie at smaller distances. Using the Doppler effect, we derive the clouds' velocities from the spread in wavelengths of the spectral lines. These velocities are typically 1 to 10 percent of the speed of light. Given the clouds' distances, we can calculate how massive the black hole would have to be to generate their velocities gravitationally. Actually, we do not know whether the cloud motions are gravitationally induced, but this does not affect our ability to obtain an upper mass limit. The velocities could be larger than

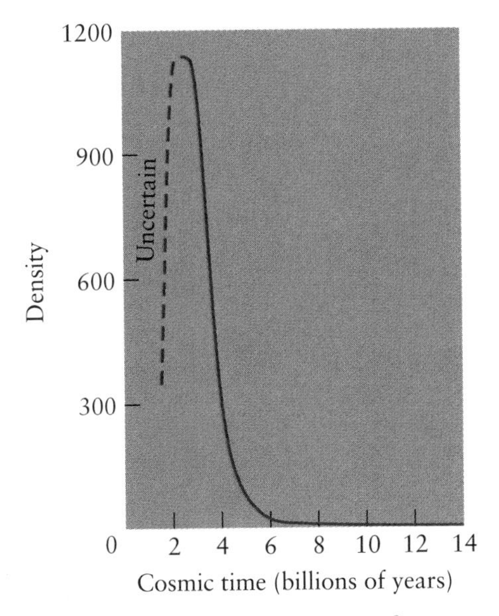


*The same data as the previous figure, except that the redshift scale of the horizontal axis has been converted to "time since big bang," assuming a standard cosmological theory; also the vertical scale is linear, rather than logarithmic, to emphasize how drastic the evolution really is. The peak of the curve represents about 2 percent of the present abundance of luminous galaxies.*

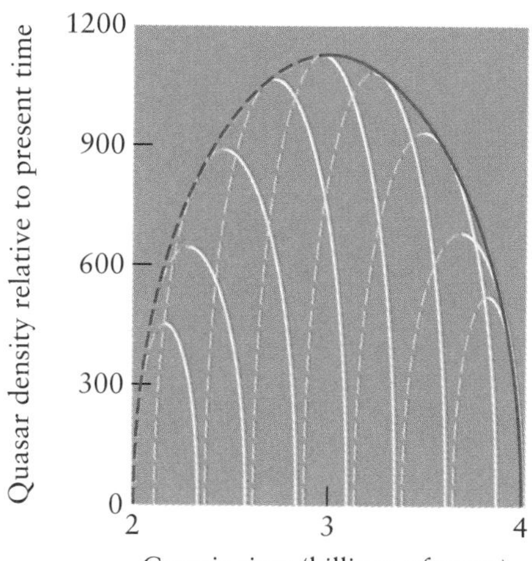


*The rise and fall of the entire quasar population is not necessarily synonymous with the life cycle of an individual quasar. The curve plotted in the previous figure, expanded here, depicts the number of quasars existing at different times, but these need not be the same objects. Individual quasars might live for much less than a billion years. If many short-lived generations of quasars emerge and die during the time over which the population rises and falls, then a large fraction of present-day galaxies could have been quasars in the past.*

the gravitational value if some other force were affecting the clouds' motions, such as radiation pressure; but they are unlikely to be smaller. One therefore gets an upper limit on the central mass.

We would be in trouble indeed if our derived upper limits were smaller than our lower limits! But fortunately they are not, and we can conclude that the black holes in most quasars probably have masses between about 100 million and a few billion suns. What does this tell us about quasar lifetimes? That quasar activity in any individual galaxy probably lasts for a total of tens to hundreds of millions of years, not billions of years. Although quasars never existed in more than about 1 percent of galaxies *at any given time,* we deduce that more than 10 percent of galaxies, and perhaps the majority, went through a quasar phase at some point in their evolution.

Like the Anasazi tribes in the southwestern United States, whose culture flourished for hundreds of years but vanished abruptly around A.D. 1300, the quasars seem to be a cosmic "culture" from an earlier era in the history of the Universe. But the descendants of the Anasazi are still among us: they are the modern Pueblo Indians. Are the descendants of ancient quasars still active in the modern Universe? Some of them could be powerful radio galaxies. These objects are rare, being found in fewer than 1 percent of large elliptical galaxies, but they do seem to be much more common than quasars at low redshifts. They do not radiate with the intensity of quasars, but their power outputs in the form of jets can be comparable. A quasar phase of activity, long ago, might have built up the mass of the black hole, and could also have given the hole spin. Perhaps the hole then lurked quiescent, the galaxy swept clean of gas, for billions of years. It took some event, maybe an encounter with a companion galaxy, to renew the infall of gas, at too low a rate to yield a quasar but at a rate sufficient to reactivate the nucleus by introducing a magnetic field to the spinning hole. This "engaged the clutch," tapping the hole's latent spin energy and converting it into jets that plow their way out to distant reaches of intergalactic space. If this is indeed what has happened in Cygnus A and M87, then these large-scale manifestations of AGN activity could offer the most direct evidence for the massive black hole remnants of quasars.

Seyfert galaxies, and lower-powered radio galaxies, could also be remnants of quasars. Seyfert nuclei, in particular, are found in about 1 percent of modern spirals. Their spectra are very similar to the spectra of quasars, but they are less luminous. Seyferts alone could account for every quasar remnant if the typical quasar lifetime were as long as a billion years. But the masses of their black holes, where these can be estimated, tend to be lower than those inferred to exist in the most luminous quasars. One should not be surprised at this. Even during the quasar era black hole masses would have been distributed over a range extending down to fairly low values. Black holes in the lower mass range would not have produced sources of radiation easily visible at high redshift but could have survived to produce lower-luminosity AGNs today.

The current generation of AGNs may thus include some of the massive black holes built up during the quasar epoch. It is also possible that the black holes in many contemporary AGNs have been built up comparatively recently. But it is very unlikely that currently active galactic nuclei account for most of the quasar remnants. There are simply not enough of them to account for all of the quasars that once existed. If successive generations of short-lived quasars placed massive holes at the centers of many, or most, contemporary galaxies—as we have argued—then most of these black holes are "hibernating," starved of fuel, and will remain inconspicuous unless accretion starts afresh.

## Quiet Holes in Nearby Galaxies

Is there any evidence that these quiescent black holes actually exist in the nuclei of nearby "inactive" galaxies? Yes—with increasing confidence, we have begun not only to detect them, but also to weigh them.

Of course, a truly quiet black hole can be detected only through its gravitational influence on its surroundings. In our attempt to observe such holes, the lack of activity can work to our advantage; the fireworks in an AGN often make it *more* difficult to detect the black hole "cleanly," that is, through its gravitational effects. AGNs contain fast, even relativistic, gas. But as likely as not, these motions are the result of nongravitational forces such as those due to radiation and magnetic fields. There are motions that would obviously be dominated by gravity, such as stellar orbits or the motion of gas in an accretion disk, but their signals are often drowned out by the intense radiation from the AGN.

We've seen that stellar-mass black holes can be detected cleanly in X-ray binaries. Measuring the Doppler shift of spectral lines from the normal star allows one to deduce a lower limit to the mass of the compact star in the binary. If this mass turns out to be larger than about 3 solar masses, one can comfortably conclude (by a process of elimination) that the compact star is a black hole. We would like to make the same kind of argument for a galactic nucleus. However, unlike an X-ray binary, where only two bodies are orbiting each other, in a galactic nucleus there are millions of stars in orbits whose properties (orientation, orbit radius, velocity, etc.) differ from star to star. We can only measure an average of the stellar orbits, projected through the star cluster. To make matters worse, the angular resolution is never adequate to isolate the region very near the black hole, so we must infer the hole's presence from trends observed on much larger scales. Nevertheless, several methods of analysis have led to convincing evidence for the existence of massive black holes in quiescent galactic nuclei.

Stars in a galactic nucleus move in random directions, like stars in a globular cluster. Each star in a galaxy responds to the sum-total gravitational force from all the other stars, as well as from any other matter that happens to be present (such as a black hole). The stars swarm around with a mean speed $v$ such that the disruptive effect of kinetic energy balances the tendency of gravity to pull them all together at the center. This average speed can be different at different distances from the center, but its variation with radius follows distinct patterns in galaxies with and without black holes. In a galaxy without a central massive black hole, this speed levels off toward the center at no more than a few hundred kilometers per second, and may even decline for stars that spend most of their time near the very middle. But a black hole changes things markedly. Stars that venture within a certain distance of the hole move much faster in response to the extra gravitational pull. Thus $v$ would increase toward the center.

*A massive black hole in the center of a galaxy has a "sphere of influence" within which the hole's gravity severely affects the motions of stars. A star entering the sphere of influence will be accelerated and deflected toward the hole. Other stars may be trapped in elliptical orbits confined within the sphere of influence. The black hole would therefore signal its presence by causing a concentration of fast-moving stars around it. A star on an almost radial orbit may get so close to the hole that it is swallowed or torn apart.*

The radius of the black hole's "sphere of influence" in the nucleus is determined by how close a star has to get before the hole's gravity overwhelms that of all the other stars combined. This is just the radius within which the escape velocity from the black hole's own gravity (or, almost equivalently, the orbital speed of stars gravitationally bound to it) becomes a large fraction of $v$. This sphere of influence has a radius proportional to the hole's mass and inversely proportional to the square of $v$. It is millions of times larger than the size of the hole's horizon, but it still only subtends an angle on the sky of a few arcseconds, even in the nearest galaxies.

The extra gravity contributed by the black hole alters the motions of stars in a second, more subtle way. Suppose we were to isolate each star and look at the shape its orbit traces out as it makes one loop around the nucleus. Some orbits would be nearly circular, staying roughly at the same radius, while others would be cigar-shaped, swooping between large and small distances during each circuit. If we consider the *mix* of orbits (how many stars are on near-circular orbits, how many cigar-shaped, etc.) outside the sphere of influence, we would be unlikely to find a marked preference for one type of orbit over another. But inside the sphere of influence, the story is quite different. The extra pull of the black hole would distort many of the near-circular orbits into cigar-shaped ones, so that there would be many more of the latter.

While this sounds like an esoteric technical point, it has a big effect on the appearance of the galactic nucleus. The distorting of the orbits increases the number of stars found close to the black hole at any one time, relative to the number that would be present if the mix of orbits were random. This extra concentration of stars within the black hole's sphere of influence, called a "cusp," can be observed as a sharp enhancement in the starlight at the very center of the galaxy. If the black hole has been present long enough for the stars in the cusp to interact gravitationally with each other on a one-on-one basis (which might take as long as a billion years), then the cusp can deepen still further as more stars become trapped within the sphere of influence.

Thus two telltale signs in the population of randomly moving stars could give away the presence of a quiescent black hole. The first is a central "blip" in the starlight, signaling the enhanced concentration of stars within the hole's sphere of influence. Such "blips" have been searched for, and found, but there is always an ambiguity in that the nonstellar light from a weak AGN—which might result from accretion onto a central black hole of relatively low mass—would be hard to untangle from the starlight of the cusp. A second, less ambiguous, signature would be evidence that the stars nearest the center are moving anomalously fast. Of course, ordinary stars cannot be detected individually in these galaxies. The sharpest details that a ground-based optical telescope can resolve have angular sizes of around an arcsecond, roughly the same size as the typical black hole's sphere of influence. An arcsecond corresponds to a region 10 light-years across at the distance of a nearby galaxy such as Andromeda, and such a region contains large numbers of stars. Even the Hubble Space Telescope can do only a factor of 10 better. Nevertheless, one can take optical spectra of the combined light from such a region and find absorption lines characteristic of stellar spectra. The redshift and breadth of these lines tell us about both the organized and random motions of the stars.

Nearby galaxies obviously offer the best prospects for discerning the signatures of a quiescent black hole. The giant elliptical galaxy M87 is a radio source with an unusual optical jet (the one photographed by Heber Curtis in 1917) as well as weak extended radio lobes. The presence of these features indicates some central activity, though at a power level far below that of quasars. M87 is therefore a good candidate for having a central black hole in relative quiescence. In 1979, Peter Young, Wallace Sargent, and their collaborators at the California Institute of Technology claimed to have found evidence for a mass of 3 billion suns in its center. This claim was made on both of the grounds described above: a central peak in the starlight was attributed to the enhanced concentration of stars gathering around a central dark mass; and the typical random velocities appeared to rise from 300 kilometers per second to more than 500 kilometers per second as one observed on smaller and smaller scales, as would be expected if a central mass were present.

The credibility of this claim was debated throughout the 1980s. Part of the central peak in the light is now believed to originate not from stars

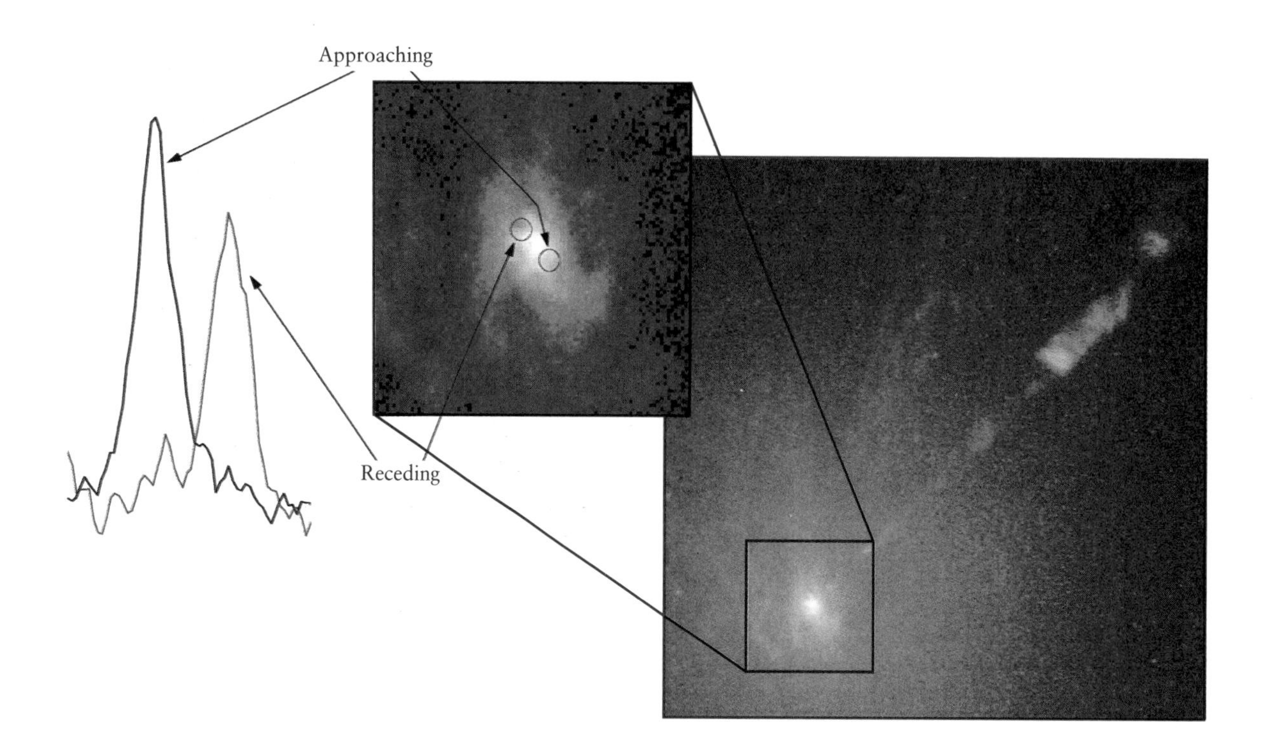


*The Hubble Space Telescope has revealed a disk of glowing gas orbiting the nucleus of M87 at a distance of 60 light-years. A systematic pattern of redshifts and blueshifts indicates rotation at speeds of roughly 750 kilometers per second. The central mass required to keep gas in orbit at this speed and distance is 2.4 billion solar masses. Peter Young and collaborators derived a similar black hole mass using different observational techniques in 1979.*

but from a low level of accretion onto the central black hole. In addition, interpreting the random velocities has turned out to be trickier than expected. Even in the absence of a black hole, it is possible for one kind of orbit to be favored over another, and the mix of orbits may be different in different parts of the cluster. The problem is that our information about the average stellar velocity is incomplete. When we infer the random velocity by measuring the Doppler width of a spectral line, we are only detecting the part of the velocity that is along our line of sight. We then have to "fill in" the missing part (the part parallel to the plane of the sky) using theoretical arguments. Unfortunately, if we make the wrong assumption about the true mix of orbits, we can be led seriously astray. Suppose, for example, that most of the orbits were cigar-shaped. If we looked at the edge of the cluster, we would see stars whose motion is mainly in the plane of the sky and would therefore measure a small Doppler width. However, if we looked at the center of the cluster, the stellar motion would be mainly along our line of sight, and we'd measure a large Doppler width. If we didn't correct for the unusual orbit mix, we would conclude that the velocity goes up steeply toward the center and therefore infer the presence of a massive black hole—even if none were present.

By now, most of these concerns have been laid to rest. Observations with the Hubble Space Telescope have vindicated Sargent and Young's original claim by producing sharper pictures of the cusp and by revealing a disk of glowing gas orbiting the center at a distance of about 60 light-years. Spectra of this gas tell us its orbital speed; it is then straightforward to calculate that there must be a mass of about 2 to 3 billion suns in the center to hold the gas in orbit.

M87 is not one of our very closest neighbors: it lies in the Virgo Cluster of galaxies, about 50 million light-years away. Its black hole reveals itself only because it has such a huge mass. At such long range, even a hole of 100 million solar masses would not cause a peak in the starlight discernable to optical astronomers, nor a visible quickening of stellar velocities. For this reason, astronomers searching for quiescent black holes have focused most of their attention on the nuclei of our closest extragalactic neighbors. Unlike M87, these galaxies do not currently seem "active" even at low levels; yet because they are more than 10 times closer than M87, a central mass concentration of "only" a few million suns might reveal itself gravitationally.

The most interesting and convincing case lies at the center of our nearest large neighbor in space, the Andromeda Galaxy, M31, which is about 20 times closer than M87. In the 1960s, Martin Schwarzschild (Karl Schwarzschild's son) and his Princeton colleagues had flown a small telescope, *Stratoscope,* in a high-altitude balloon to avoid the blurring effect of the lower atmosphere. They discovered that the stars in the central few light-years of M31's core had a flattened distribution, but they didn't guess that

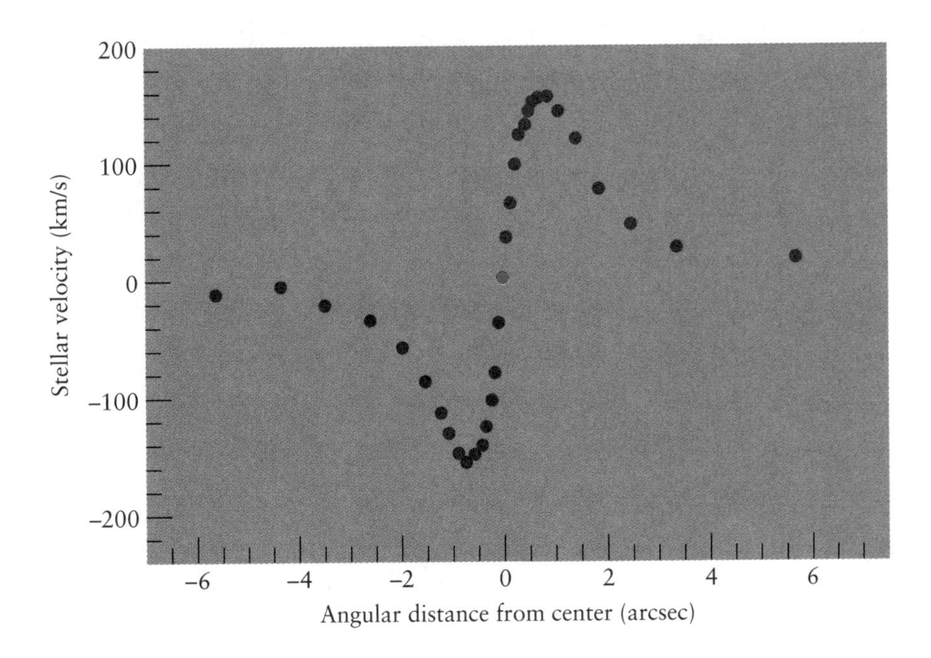


*This graph shows average speeds of stars near the center of the Andromeda Galaxy (M31). Negative velocities denote blueshifts (i.e., motion toward us), while positive velocities denote redshifts. The symmetry of the curve between redshifts and blueshifts is a signature of rotation. The sharp increase in orbital speed toward the center indicates a central concentration of matter weighing between 30 and 70 million solar masses.*

the explanation involved a supermassive black hole. In the last few years, Alan Dressler of the Carnegie Observatories, John Kormendy of the University of Hawaii, and Douglas Richstone of the University of Michigan have studied the motions of these stars. They find that the random velocities rise toward the center. But there is an additional bonus: the entire cluster of stars appears to be rotating, and it is this rotation that presumably accounts for the flattening of the distribution. The detection of rotational motion helps to remove the ambiguity about the mix of stellar orbits, which bedeviled the early work on M87. Moreover, from the rotation speeds of these stars at various radii we can tell that the central mass is highly concentrated and falls in the range of 30 to 70 million solar masses. But even this elegant result is not without its loose ends. Pictures of M31 taken by the Hubble Space Telescope have revealed more detail, but in contrast to the pictures it produced of M87 the crisper images of M31 have only confused the interpretation. These show two separate "peaks" in the light distribution. This is rather perplexing—the obvious possibility that there are two black holes seems unlikely, because two neighboring objects of such high mass would spiral together so quickly that it would be a fluke if we happened to catch the nucleus in such an unusual short-lived state.

*The central region of the Andromeda Galaxy (M31) as observed by the Hubble Space Telescope in visible light. The stars are orbiting around a central black hole with a mass of about 30 million suns; the reason for the asymmetric double structure is not understood.*

M31 is not the only nearby galaxy for which evidence of a central black hole exists. John Tonry showed in 1984 that the small elliptical galaxy M32 (a satellite of M31) harbors a central dark object of 2 to 5 million solar masses, and several other observers have subsequently confirmed his observation. The edge-on disk galaxy NGC 3115 has a nucleus in which the innermost stars are moving unusually fast, implying a central black hole of a billion solar masses. Another interesting case is the Sombrero Galaxy (NGC 4594), studied by Brian Jarvis and Pierre Dubath at the European Southern Observatory in Chile, and by John Kormendy with the Canada–France–Hawaii Telescope (CFHT) in Hawaii. Like Andromeda, this galaxy has an apparently rotating stellar core, but the central mass is about half a billion suns. The CFHT, which offers the best ground-based measurements, can resolve details down to about 30 light-years. But still better data from the repaired Hubble Space Telescope should soon firm up the evidence in all these systems.

Even if one is convinced—as we are—that each of these galaxies contains a concentrated mass at its center, the question arises whether this mass has to be a black hole. The mass certainly has to be in some dark form. There is not much light associated with the mass at the center of any of these galaxies, certainly too little to be compatible with a normal mix of stars. The object in the center of M31, whatever it may be, has a "mass-to-light ratio" at least 35 times the Sun's, meaning that at least 35 solar masses are present for every solar luminosity coming out of this region. The darkness of the mass certainly suggests the presence of a massive black hole, but there are alternative explanations. Could there, perhaps, be an unusual population of faint stars concentrated near the center? A sufficiently massive cluster composed of smaller black holes, of 10 to 100 solar masses apiece, could be "hidden" without providing much light, as could a cluster of other stellar remnants like neutron stars and white dwarfs, or a cluster of stars with individual masses below 0.1 solar mass. A cluster of dark stars, contained within a region 10 light-years across (about 1 arcsecond at M31's distance), cannot be excluded by current observations, although there is no evidence favoring such an ad hoc model, either. The sharper images that are now within the Hubble Space Telescope's capabilities do have the potential to rule out a star-cluster model. If spectra of the central 0.1 arcsecond show that the stars are moving faster still, then this result would imply that all the dark matter had to be crammed within a radius of 1 light-year. Such a small radius would rule out a cluster of many dark stars. A cluster so compact would evolve quickly, owing to stellar encounters, until some stars merged into a central black hole and the others escaped; it would therefore be implausible to find such a compact cluster *without* a massive black hole in a galaxy, like Andromeda, as old as 10 billion years.

Where else should we look for massive black holes among our near neighbors? We cannot yet predict which quiescent galaxies are likely to harbor the most massive holes. NGC 3115 and the Sombrero Galaxy, for instance, have holes at least 10 times heavier than Andromeda's. On the other hand, Kormendy and John McClure found that the nearby spiral galaxy M33 couldn't contain a central hole even a thousandth as heavy as the one in Andromeda. The black hole mass may turn out to correlate with the number of stars in the central bulge of the galaxy—if so, this might offer a clue to how the holes form. But the discovery of truly quiescent holes is likely to proceed by pure serendipity for some time to come.

The same cannot be said, though, for galaxies that show clear signs of activity, however slight. When the jets in a strong radio source switch off, the extended lobes expand and their radio luminosity gradually fades over the course of a few hundred million years. Faint but extended lobes may be a sign of past activity—and of a black hole. The nearby galaxy Centaurus A is a rather weak radio source but has giant radio lobes: it may be a dying source that, in its prime, rivaled Cygnus A in its power output. The nucleus is shrouded by dust, so we do not have the kind of optical data that Young and his collaborators obtained for M87. But Centaurus A is the nearest galaxy we have reason to think spewed out a great deal of energy in the past. It has a compact nucleus observed at both radio and X-ray wavelengths, as well as a jet; so there is some residual activity. The extremely high energy requirements of Centaurus A's lobes give us every reason to expect that a very large black hole lurks at its center.

It may well be that the largest black holes are found in elliptical galaxies, such as Centaurus A or M87. However, we have had such good luck at finding holes in random disk galaxies nearby that we might consider our own Galaxy "underprivileged" if it lacked a central black hole. Is there a massive black hole in the middle of the Milky Way? The evidence now clearly points in this direction, but the story is convoluted. Observations compiled over many years have hinted at a concentrated central mass of 3 million solar masses—only a tenth the central mass in Andromeda. This evidence came initially from the rather chaotic motions of gas streams, which are hard to interpret. Only recently has it been confirmed by observations of stellar velocities.

Studying the center of our Galaxy is not easy. The dust between us and the Milky Way's center is so thick that it completely obscures the center in optical and ultraviolet light, which are the best bands for measuring the velocities of stars. Everything we know about the Galactic center comes from observations in the infrared, radio, X-ray, and gamma-ray bands, all of which can penetrate the dust and yield considerable information about the gas but limited information about the stars. Unfortunately, gas is subject to nongravitational forces—from thermal pressure, magnetic fields, and radiation—and may not follow the ballistic trajectories of objects solely under gravitational influence; therefore, there is considerable risk in attempting to map the gravitational field by following the gas. The ambiguity is much smaller if the gas is in circular motion in a disk, but this is not the case in the Galactic center.

The complex structures at the center of the Milky Way include a compact radio source. This source is unique within the Galaxy, having a very unusual spectrum, and is the best candidate for being at the true center of our Galaxy, and therefore of being associated with a massive black hole if indeed there is one. Theorists have proposed that the observed spectrum could be characteristic of a moderately massive black hole, of anywhere from 100 thousand to a few million solar masses, accreting gas from the interstellar medium at a very low rate. It would greatly strengthen the case for a black hole if one could show that the center of gravity coincided with the location of the compact radio source.

Infrared observations have recently revealed some stars very close to the putative black hole, and Doppler measurements of infrared spectral lines have yielded the velocities of a few of them. The velocities are consistent with all the dark matter being contained within a single compact object.

Even more remarkably, Reinhard Genzel and his colleagues in Munich, using the 3.5 meter New Technology Telescope at the European Southern Observatory, have detected transverse motions in some of these stars: the stellar positions have shifted slightly over the three years for which they have been observed. The speeds along the line of sight are of course known from the Doppler effect, so we now have full 3-dimensional motions. These motions imply that the stars are orbiting a concentrated central object weighing 2.5 million solar masses. This new evidence is a far cleaner diagnostic than we formerly had from gas motions. There now seems little doubt that a black hole lurks in the center of our Milky Way. But it is, by galactic standards, a modest one. Our own Galaxy could never have been a powerful quasar.

## Crucial Clues from Orbiting Gas

The uncertain interpretation of gas motions in the Milky Way's center notwithstanding, there is now one spectacular case in which gas measurements, taken using a novel technique, offer by far the clearest evidence for a massive black hole. The spiral galaxy NGC 4258 is not completely quiescent; its nucleus contains an intensely hot emitter of X-rays. But the crucial evidence comes from something much cooler—molecules of water in gas clouds orbiting very near the nucleus. The water molecules emit powerful microwave radiation because they act collectively as a maser.

The term "maser" is an acronym, standing for "*m*icrowave *a*mplification by *s*timulated *e*mission of *r*adiation" (the acronym "laser" is the

*When a photon hits an atom or molecule that is in an excited state, the photon can de-excite it, triggering (or "stimulating") the emission of another photon moving in the same direction. This amplification process can lead to very intense radiation.*

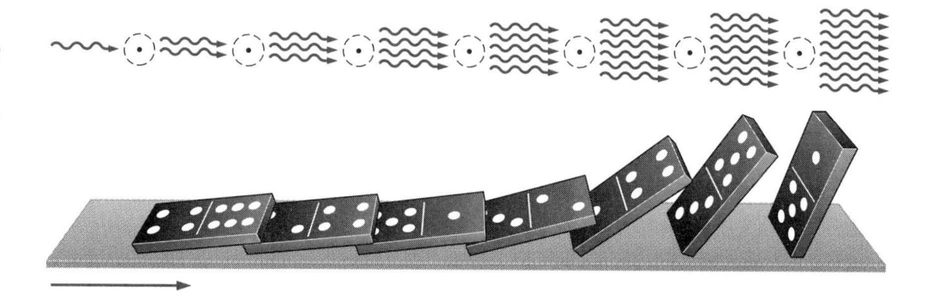

same, except that "light" replaces "microwave"). The microwave radiation in a maser is amplified to form an intense beam. The origin of the amplification lies in a sort of "domino effect." Spectral emission lines are produced when an atom or molecule is placed into an "excited state"—that is, a state of higher energy. Left alone, the excited atom or molecule will drop back into a lower-energy state, emitting a photon with a very specific wavelength. A series of these photons of specific wavelength form the radiation that creates the spectral emission line. If you have a group of molecules in the same excited state, it is impossible to predict when any one of them will emit. Therefore, this "spontaneous emission" leads to photons emerging at random, or *incoherently.* But there is a way to induce the molecules to emit sooner, and more predictably. If an intense beam of radiation with exactly the same wavelength as the spectral line passes by, it can "stimulate" a molecule to emit right away and add its photon to the beam, amplifying it. This effect becomes dramatic if you start with an anomalously large number of molecules in the excited state. Then the whole group of molecules can behave like a row of dominoes, all falling in step, and the result is a concentrated beam of radiation at uniform wavelength.

Masers and lasers require finely tuned conditions in order to operate in the laboratory (or in your CD player, or in any number of other items in everyday use). Therefore, astronomers were surprised to find that a similar process of "masing" occurred in the atmospheres of cool stars and in many interstellar clouds in our own Galaxy, leading to intense "hot spots" of emission in the spectral lines of various kinds of molecules. Exactly how the effect is triggered in astrophysical situations is uncertain. The essential point, however, is that the high intensity of the maser radiation enables radio telescopes to detect individual gas clouds in the galaxy NGC 4258; moreover, the water molecules emit a sharp spectral line with known frequency (22 gigahertz, corresponding to a wavelength of 1.36 centimeters), so that the Doppler effect can easily be measured.

Very-long-baseline interferometers yield images of the molecular clouds with angular resolution better than half a milliarcsecond. Moreover, the orbital speeds (about 1000 kilometers per second) can be measured with an accuracy of one part in 1000. The masering clouds delineate a slightly warped orbiting disk of gas, which we are viewing almost edge-on. The motions in this disk suggest that it is orbiting a central mass of 36 million suns. How do we know that these motions are gravitational in origin? Because so many individual clouds have been detected that we can actually plot a graph of orbital velocity vs. distance. The results show almost perfect agreement (to within about 1 percent accuracy) with the predicted velocity curve for objects orbiting a centrally condensed mass. This stunning agreement implies that virtually all the central mass must lie within the orbital radius of the *innermost* cloud we can detect. If more than even 1 percent of it spilled over the inner cloud radius, the fit would be spoiled. If we supposed that this mass were uniformly distributed inside the inner cloud radius, and dropped to zero sharply outside, its density would have to exceed 100 million solar masses per cubic light-year. This is already far higher than any density ever detected in a galactic nucleus or a globular cluster. But a real star cluster could never have such a sharply defined

*The center of NGC 4258 harbors a thin disk, shown here in an artist's impression, which we observe almost edge-on; the graph shows the spectrum of intensity vs. velocity, calculated from the Doppler shift. The disk contains cool molecular gas, which does not emit uniformly. Radio astronomers detect many discrete patches of emission, each coming from a region where the water molecules are acting like a maser.*

boundary. Fitting a realistic model of a star cluster to the velocity curve would require a central density *1000 times higher.* Such high densities and a total mass of 36 million suns cannot be accommodated in a cluster consisting of any known types of objects. The only plausible alternative is a single massive black hole.

The Hubble Space Telescope had already discovered an orbiting disk in M87. However, this recent evidence from NGC 4258 is much more precise, and much more compelling. It represents truly overwhelming evidence for a black hole. Radio interferometry provides even sharper images than the Space Telescope—its angular resolution is 100 times better. And the spectral resolution is 100 times better as well, so the orbital speeds can be measured very accurately. Whereas the M87 disk was measured on scales of 60 light-years, the disk in NGC 4258 is mapped on scales of less than one light-year. The evidence for a concentrated mass in NGC 4258 is

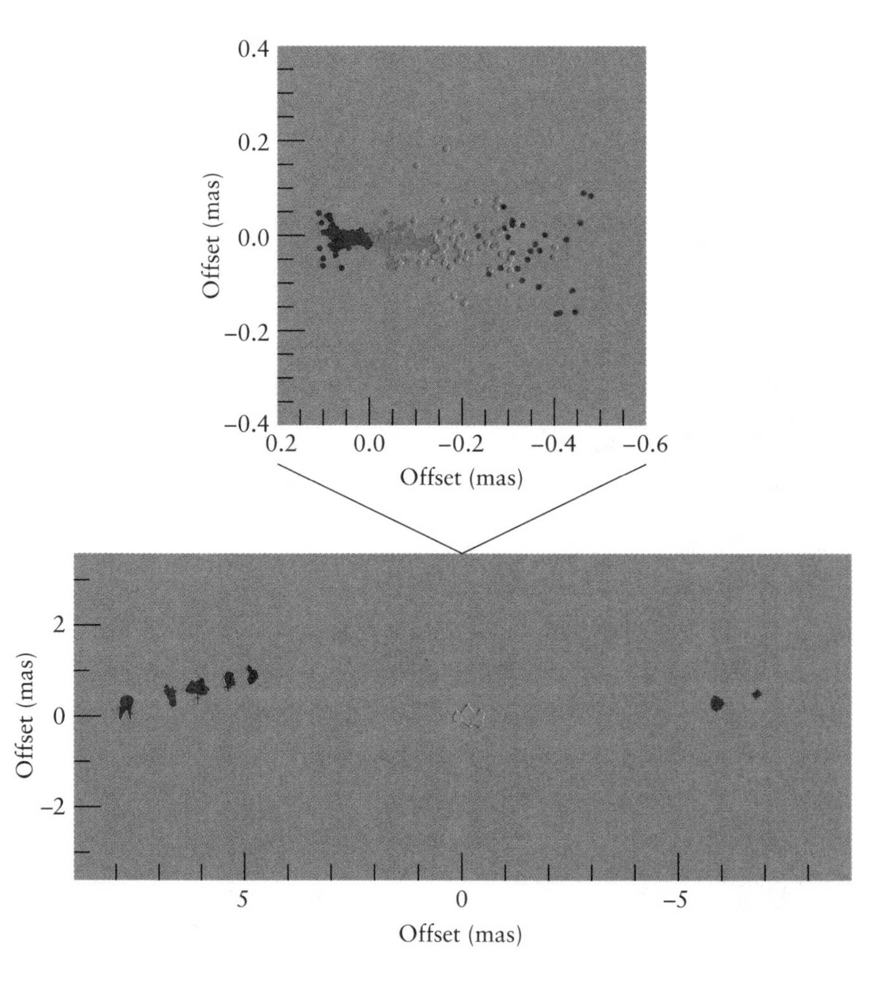


*By using the Very Long Baseline Array, radio astronomers have pinpointed the positions of the maser spots in NGC 4258 to better than a thousandth of an arcsecond (a milliarcsecond, or mas). The maser spots are clumped into three groups, which outline the foreground and extremities of a slightly warped disk, shown in the artist's impression. Measurements of the Doppler shift reveal that the disk is rotating in the sense indicated by the color-coding of the dots. The entire disk is a hundred times smaller, in angular size, than anything that could be resolved with existing optical instruments, including the Hubble Space Telescope.*

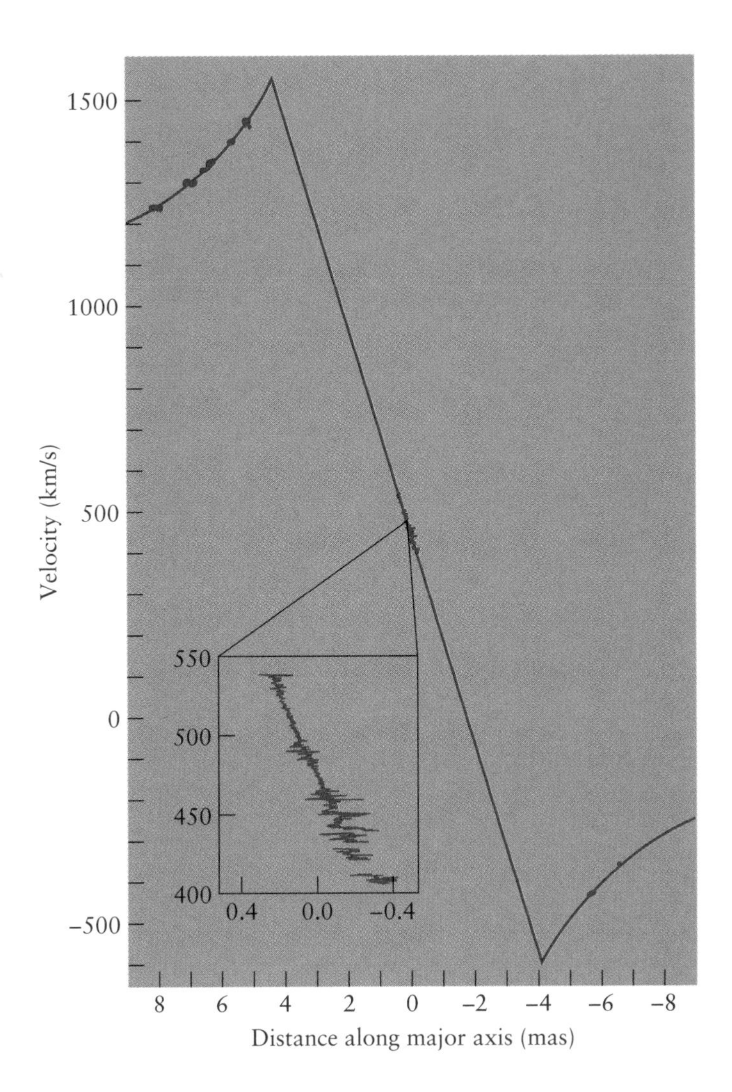


*The wavelength of the emission from each maser spot in NGC 4258 can be determined so accurately that the velocity can be inferred (from the Doppler shift) to a precision of 1 kilometer per second. The galaxy itself is moving away from us at 500 kilometers per second. When plotted against distance from the center, the velocities trace out a very clear pattern. The continuous line shows what Newton's theory would predict if the gas in the disk were orbiting under the influence of a 36-million-solar-mass central body. The disk has a sharply defined inner edge, at which the orbital speed is 1080 kilometers per second. Outside the edge, the orbital velocity falls off as the square root of the radius, exactly as expected if the gas is moving under the influence of a centrally concentrated mass.*

correspondingly stronger than the evidence assembled for the other galaxies we have discussed, most of which was acquired by optical telescopes. Moreover, NGC 4258 is the system for which it is hardest to envisage that the mass comprises anything but a single black hole.

Unfortunately, one should not be overly optimistic that other massive black holes will be discovered in this way. Maser emission travels in narrowly defined directions, which depend on the velocities of the gas clouds and their geometric relationships. In all probability we see the masers in NGC 4258 precisely because the gas motions are so well ordered and because the disk is so very nearly edge-on. A few other AGNs show strong maser emission, but it is too early to tell whether they have the precise orientation or the clean geometry of NGC 4258. If either were missing, the interpretation of measured velocities would be much more ambiguous. We may get lucky again, but we should be prepared to accept the remarkable observations of NGC 4258 as a stroke of good fortune.

## Flares from Tidally Shredded Stars?

The presence of massive black holes in ordinary galaxies isn't surprising—indeed, our estimates of quasar lifetimes (and how many generations have lived and died) imply that we should expect to find black holes in most galaxies. But before accepting this conclusion (and dismissing alternative ideas involving, for instance, compact clusters of dark stars), we must ask a further question: Can massive black holes lurk at the centers of these galaxies without revealing themselves through some detectable form of violent activity, however weak or intermittent? We are used to the idea that black holes are implicated in the most powerful emitters of radiation in the Universe and that, when accreting, they can convert mass to energy with extraordinary efficiency. But there is no sign of such activity in the Andromeda Galaxy: the output from its center is less than a ten-thousandth the output of a quasar. Can a black hole be so completely starved of fuel that it doesn't reveal its presence as at least a "miniquasar"?

We do not know how much gas there is near the black hole in the Andromeda Galaxy, or in the nuclei of other nearby galaxies showing evidence for a black hole. There is no reason why these regions might not be "swept clean" of gas. The star density, however, is much better known—after all, if the stars weren't closely packed near the center of the galaxy we wouldn't have evidence for the black hole at all. The stars trace out complicated orbits under the combined influence of all the other stars and of the hole itself. But these orbits are not "fixed," like the orbits of the planets in the Solar System. They gradually change, shifted by the cumulative effect of encounters with other stars. From time to time, these encounters would shift a star onto an orbit that brings it very close to the presumed black hole; the occasional star might even plunge right in.

A star approaching a black hole faces a risk even if it misses falling through the horizon. Since the gravitational force varies with distance from the hole, there is a difference between the force acting on the side of a star facing toward the hole and the side of the star facing away. From the star's point of view there is a "tidal force"—just like the one exerted by the Sun and Moon to cause the tides on Earth—that stretches the star along the imaginary line running from the star to the black hole. If the tidal force becomes too strong, it can overwhelm the gravitational force that holds the star together—and tear the star apart. The tidal forces near the horizons of the biggest black holes are gentler than the corresponding forces near small black holes. A star like the Sun could be swallowed by the black hole in M87 while still in one piece. However, if the hole's mass is less than 100 million suns (as those of the nearest galaxies are suspected to be), then a star like the Sun would be tidally disrupted if it came within 10 times the hole's radius. Whether a star is susceptible to this kind of destruction depends on its density. A very compact star, such as a white dwarf, could fall inside any supermassive black hole more or less intact.

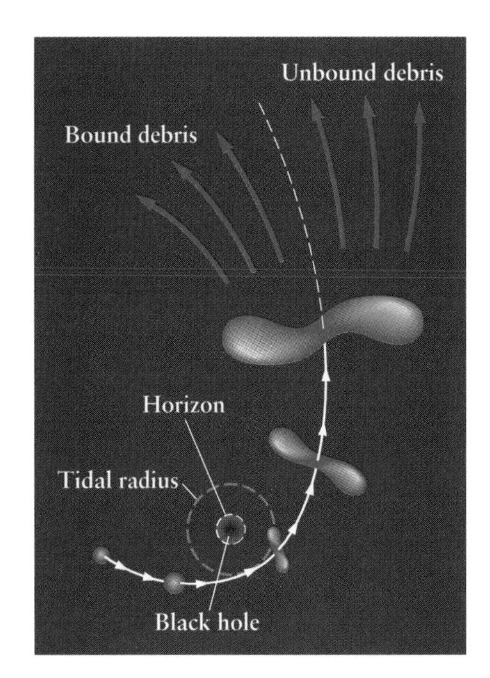


*A star approaching very close to a massive black hole can be torn apart by tidal forces. The disruption process violently distorts the star, removing orbital energy from some parts and adding it to others. The debris are stretched as they move out. Nearly half the debris will escape; the remainder will move on elliptical orbits bound to the black hole. The bound debris return to the hole after (typically) a few months, at which point they can be accreted. During this transient accretion phase, a massive hole in a quiescent galaxy could become briefly as bright as a quasar. Here the size of the star is exaggerated relative to the size of the hole and the orbit.*

Tidal forces are bound to destroy some ordinary stars from time to time in any galactic nucleus containing a massive black hole. Every few thousand years, one of the stars would pass so close to the hole that it would be torn apart. The exact rate of these events would depend on the statistics of the stellar orbits, and particularly on how quickly stars are shifted onto orbits that take them close in. A disrupted star would send off flares that could be the clearest signal of a black hole's presence. But what would we observe when such an event occurs?

When the tidal force starts to compete with a star's own gravity, the material of the star responds in a complicated way. Because the star is moving, the direction and strength of the tidal force change rapidly with time. The star is consequently stretched along certain directions, squeezed along others, and strongly shocked. This phenomenon poses a grand challenge to experts in computer simulation. Within a few years, we should have computer-generated "movies" showing us what happens when stars of different types are tidally disrupted, and what a distant observer might measure as the observational signatures of such events. Relatively crude calculations have already been done, and they are good enough to convey the essence of what goes on.

The debris would spread out onto a range of orbits. About half would remain behind, bound to the black hole, while the rest would fan out, at speeds of up to 10,000 kilometers per second, moving on trajectories that would take them away from the hole. The most conspicuous effects would result from the bound debris, which return to the hole and eventually swirl down into it. The nucleus would flare up to a brightness much greater than that of a supernova—almost to that of a quasar—but only for about a year. It is hard to calculate how much of this radiation would emerge in the visible band, rather than in the ultraviolet or even X-ray portions of the electromagnetic spectrum. An important question, still being investigated, is: How rapidly does the brightness fade? In other words, how long does it take for the last dregs of debris to be digested? The answer is important because we want to know whether the nucleus would have faded below detectable levels before the next stellar disruption occurred, 1000 to 10,000 years later. Knowing the answer, we could determine how probable it was that *no* emission should be observed at any random time. We could then estimate, statistically, how common massive black holes are in ordinary nearby galaxies. The details are still controversial—gas flow tends to be more difficult to calculate than the movements of stars, which can often be treated as points of mass responding solely to gravitational forces.

The tidal disruption of a star is so rare in any given galaxy that we are most unlikely to witness such a "flare" from Andromeda during our lifetimes (still less from our own Galactic center). But if most galaxies harbor black holes and we look at the nearest 10,000 galaxies, we would expect to catch a few at the peak of a flare, and probably rather more in a state where the effects of the most recent tidal disruption are still discernable. A highly worthwhile program would be to monitor all galaxies in the Virgo Cluster once a month, looking for the occasional stellar disruption; another option would be to study a similar number of fainter and more distant galaxies in a small patch of sky. These programs could be started right away: automated searches to detect supernova explosions, like the one being pursued by Saul Perlmutter and his collaborators at the University of California in Berkeley, could be adapted to search also for the signs of tidal disruptions. It would be surprising—indeed, baffling—if a prolonged search failed to discover any events.

The predicted flares may be the most robust test for the ubiquity of massive black holes in nearby galactic nuclei. It takes a very powerful tidal force to shred a star. If theorists are able to predict the observable signatures in sufficient detail that we can identify the events unambiguously, the disruption of stars could offer a probe of the region close to the black hole where gravity is strong. Such probes are few and far between. Most of the evidence presented in this chapter was gathered from stars and gas orbiting far away from the central mass, where gravity is still relatively weak and speeds are much less than the speed of light. This evidence is important because it implies huge dark masses in the centers of galaxies that are too concentrated to be, for instance, clusters of faint stars. But Einstein's theory describes—exactly—the astonishing and non-intuitive behavior of space and time close to a black hole. If we could only see this behavior directly, we could prove positively that these dark masses really are the black hole "solutions" of Einstein's equations. Are there any other diagnostics that can tell us about the region very close in? Can observations of galactic nuclei thereby enable us to check Einstein's theory? This is the theme of our next chapter.

*Chapter* 8

# Checking Up on Einstein

*This computer simulation shows what an accretion disk around a nonrotating black hole would look like if it could be observed sharply enough to see the detailed structure. The disk is viewed obliquely, at an angle of 30 degrees to its plane. Because of the orbital motion, the right-hand side is blueshifted and brightened compared to the (receding) left-hand side. In this rendering, the colors refer only to intensity: blue is more intense, red less intense. The "arc" of light over the black hole is caused by rays from the underside of the foreground disk that are deflected by almost 180 degrees and pass through the gap between the disk and the hole's horizon. The ring interior to this arc is due to photons that orbit the black hole once before escaping toward our field of view.*

To specialists in relativity, the black holes in binary star systems and galactic nuclei signify places where the space of our Universe has been punctured by the accumulation and collapse of large masses—collapse to smooth geometrical entities that can be described exactly by simple formulae, the "Kerr solution" to Einstein's equations. As Roger Penrose has emphasized, "It is ironic that the astrophysical object which is strangest and least familiar, the black hole, should be the one for which our theoretical picture is most complete." Black holes are, according to theory, just as "standardized" as elementary particles such as electrons and protons. All black holes of a given mass and spin are exactly alike.

It seems astonishing testimony to the power of mathematical physics that any entities of the real world can be so fully understood. This thought made a deep impression on Chandrasekhar. In a lecture given in 1975, he said: "In my entire scientific life . . . the most shattering experience has been the realization that an exact solution of Einstein's equations of general relativity, discovered by the New Zealand mathematician Roy Kerr, provides the absolutely exact representation of untold numbers of massive black holes that populate the Universe." It is now even clearer, as we've seen in earlier chapters, that massive collapsed objects indeed exist, and that the power supply in quasars and other kinds of active galactic nuclei is, ultimately, gravitational. But how confident can we be that these objects have exactly the properties that Einstein's theory predicts?

It will not have escaped the reader's attention that much of the evidence for black holes is still circumstantial. That in itself doesn't mean the case is weak. After all, until the detection of solar neutrinos the evidence that stars like the Sun are fueled by nuclear fusion was also circumstantial, yet we were persuaded by the success that astronomers had achieved in predicting the brightnesses, colors, radii, and other properties of stars with different masses and ages, and in explaining how nuclear reactions in stellar cores and supernova explosions transmute nuclei up the periodic table to account for the relative abundances of the chemical elements. Active galactic nuclei, and their miniature counterparts the X-ray binaries, provide strong evidence for very deep and compact "gravitational firepits," in which up to 10 percent of the rest-mass energy ($mc^2$) of inflowing material is converted into radiation. But that in itself tells us nothing about the detailed behavior of space and time in the domain where gravity is superstrong.

## Searching for a Sign

Could our observations of quasars and X-ray binaries give us some specific sign of the curvature of space near the black hole, some signature—in the radiation emitted by gas swirling inward toward a hole, or by jets emerging from it—that indicates whether the distortions of space and time agree with those predicted by Kerr's solution? Gas swirling into a black hole flows in a complicated way. Theorists are now computing what these flows are like, with the aim of accounting for both the radiation and the emergent jets. If the inflow is steady, the gas surrounds the hole in a configuration resembling either a thin disk or a thick donut, depending on such factors as the mass supply rate and the magnetic field. Unsteady flows—as would develop, for instance, when a star is tidally disrupted—would be much more complex, but work is under way to compute these as well. Far away from the black hole, the effects of gravity on the gas should nearly agree with Newton's theory: any deviations required by general relativity would be hard to detect. But uniquely relativistic effects should become prominent very close to the black hole. For example, according to Kerr's solution, the radius of the hole's horizon and the smallest orbit that a particle can have without falling in (which is larger than the horizon) depend on how fast the hole is spinning. The faster the spin, the closer the gas has to get before the hole swallows it, and the more energy is released (at least for gas approaching the hole in a thin disk).

If we could view the flow pattern close up, we could tell whether it was consistent with what theory predicts; if it was, we could even infer how fast the hole was spinning. A disk viewed obliquely would appear distorted by the severe bending of light. It would look a bit like an old

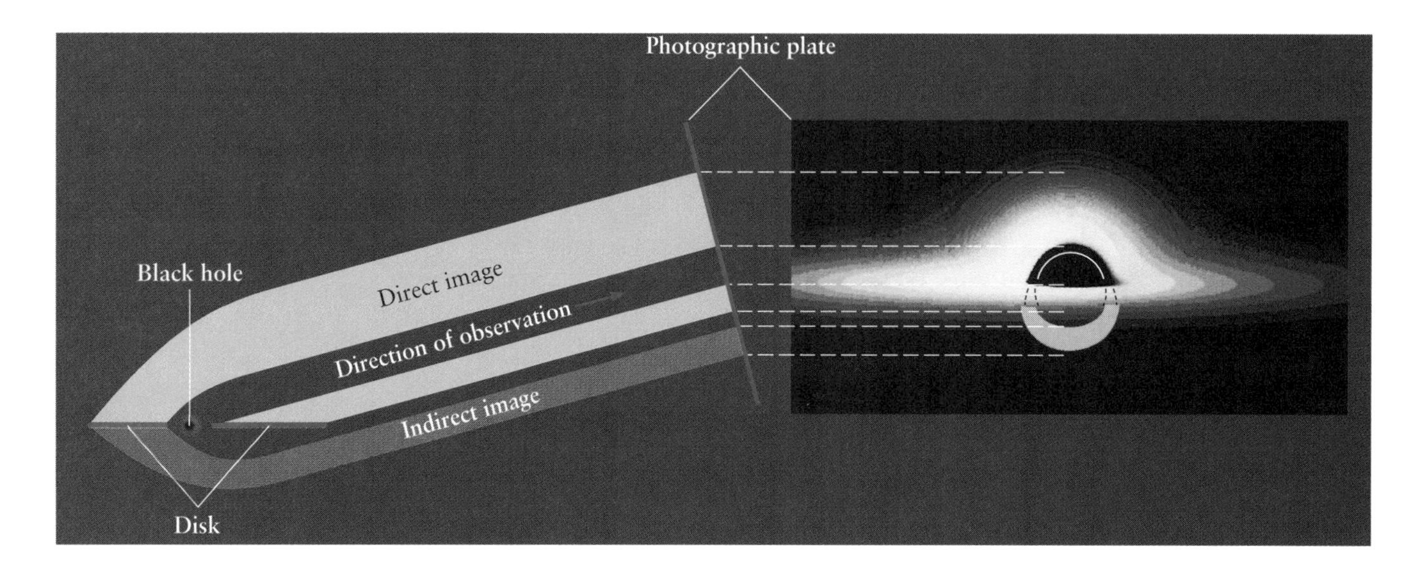


*In this figure, adapted from J.-P. Luminet, the right-hand image shows the appearance of a thin disk viewed at an angle of 10 degrees to its plane. The left-hand side shows the ray paths as they would appear viewed from the side. The bending of the ray paths as they pass over the hole causes an apparent warping of the far side of the disk, and the Doppler effect associated with the orbital motion causes an asymmetry between the left and right sides of the image. The bottom face of the disk's far side can also be observed (lower thick arc in right-hand figure). The thin upper arc hugging the hole represents the bottom face of the near side of the disk, viewed via rays that are initially directed away from us but are deflected around and over the hole. (This upper arc is seen more clearly in the chapter-opening illustration.)*

phonograph record left too long in the sun: the far side would seem to be warped upward, and the disk would appear brighter on the side where the flow comes toward us than on the receding side.

There is no chance of our obtaining images that are sharp enough to actually show the patterns in these flows, or the distortions created by the strongly curved space; no astronomical technique can achieve the needed resolution. We can nevertheless learn something just by studying the radiation emitted from the disk. We saw in Chapter 5 how optical spectra of light from hot gas in active galactic nuclei reveal Doppler shifts, which tell us how the gas is moving. The gas emitting the lines at visible wavelengths, however, lies relatively far from the black hole and moves, at most, at a few percent of the speed of light. This gas is too far away to contain a strong imprint of the black hole's geometry. The gas really close to the hole is so hot that it emits X-rays rather than visible light. These X-rays have been observed and are found to vary irregularly in intensity—active galactic nuclei display a scaled-up version of what happens in X-ray sources like Cygnus X-1 in our own Galaxy.

The ultrahot gas near a black hole is at temperatures of millions, or even tens of millions, of degrees. At these temperatures, the X-ray emission is concentrated in specific spectral lines, just like the more familiar visible and ultraviolet light from gas at 10,000 degrees. The new generation of X-ray telescopes—starting with the Japanese *ASCA* satellite, launched in 1993—carry the first spectrometers that are sensitive enough to detect the substantial Doppler shifts expected from gas swirling close to the hole. If the disk were viewed obliquely by one of these spectrometers, we would detect a broad line with two "wings," one blueshifted and the other redshifted, incorporating X-rays from the approaching and receding sides of the disk, respectively. The line would appear lopsided: the blueshifted side should be much stronger because it has been amplified by the Doppler beaming effect (see Chapter 6). Finally, despite the symmetry of the disk, the wavelength of the line "center" would appear to be offset from its normal value, toward the red, even after correcting for the beaming effect and the cosmological redshift of the galaxy. This shift comes in part from the redshift that, according to general relativity, is undergone by all radiation emitted near a strongly gravitating object, and in part because fast-moving clocks (and, therefore, vibrating atoms) appear to run slow from the viewpoint of a stationary observer. X-ray astronomers are finding very broad, asymmetrical lines, very similar to the theoretical predictions, that may soon offer a useful way to probe conditions in the inmost parts of accretion flows.

*In this computer simulation, the black hole is rotating with angular momentum equal to half the maximum amount allowed by general relativity. The axis of rotation is perpendicular to the plane of the disk, and the hole is rotating in the same direction as the accretion disk. A rotating black hole is smaller than a nonrotating hole of the same mass, and it draws the accretion disk inward toward the horizon. The arc and ring, due to photons orbiting the black hole for one-half and one full orbit respectively, are more narrowly confined, and the disk appears more asymmetric because of the hole's rotation. (The horizontal black line is an artifact of the numerical procedures used to determine photon paths near a rotating black hole.)*

Even more information can be obtained by measuring the rapid changes of the X-ray signals with time. The *Rossi X-ray Timing Explorer,* launched at the end of 1995, carries a special-purpose instrument designed to clock the arrival timers of individual X-ray photons with an accuracy of several millionths of a second. Amidst the noisy flickering thought to characterize X-ray binaries, the *Rossi Explorer* has revealed dozens of nearly periodic signals, called "quasi-periodic oscillations" or QPOs. Some of these oscillations remain remarkably steady, despite wild swings in the intensity or spectrum of the source. Do these QPOs carry information about the geometry of space close to a black hole or neutron star? Some theorists think so, but the nature of the connection is not yet clear. One possibility is that we are witnessing the ringing of the inner parts of accretion disks. The extreme curvature of space traps small oscillations near the center of the disk, allowing them to build up until the vibration is detectable. Another possible explanation is a rapid wobbling of the inner part of the disk, an effect that depends entirely on the black hole's spin.

Two of the effects we have already seen in active galactic nuclei could depend on distinctive properties of a spinning (Kerr-type) black hole. One is the emergence of jets from the very centers of AGNs. Part of the spin energy of a black hole can be tapped to power these jets. Roger Penrose first suggested, in general terms, that such energy extraction was possible; Roger Blandford and Roman Znajek later showed explicitly how the energy stored in a spinning hole could be channeled into electrical and magnetic fields, and perhaps into a fast outflowing jet, as explained in Chapter 6. The other effect is the seemingly rock-steady orientation of jets in many radio galaxies, lasting over periods of hundreds of millions of years, enough time to build the lobes in powerful sources. Such stability would be hard to understand if there were no stable gyroscope to enforce it. But a Kerr black hole is the perfect instrument to provide just this kind of stability.

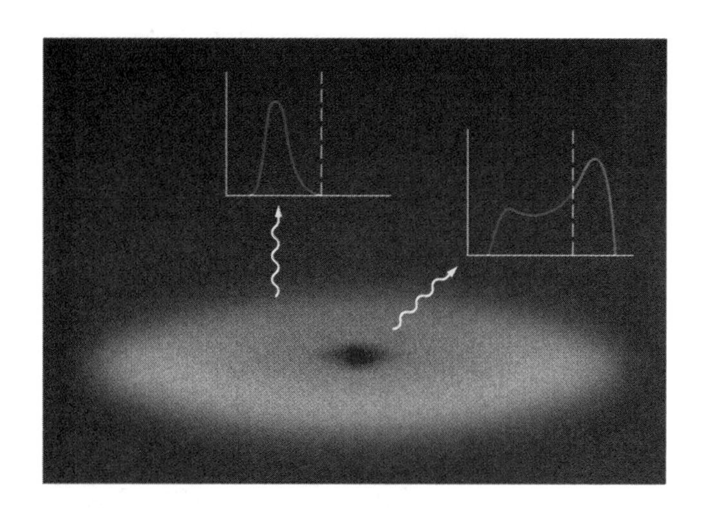

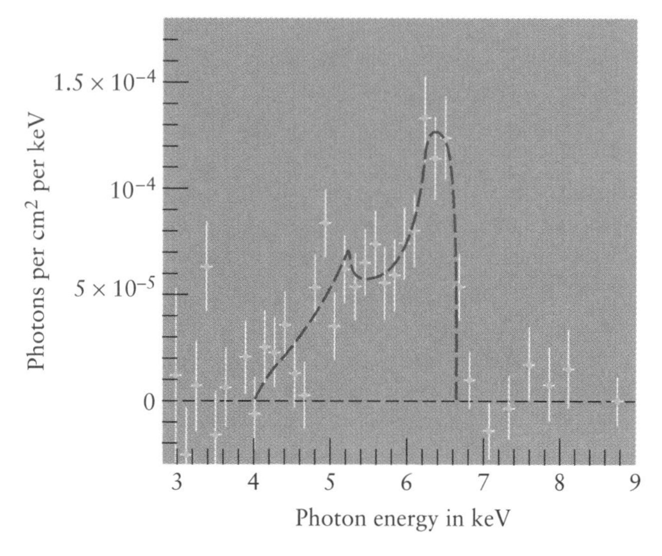


*Most of the radiation from an accretion disk comes from its innermost regions. Photons emitted with a well-defined energy—that is, in a spectral line—will be received at a range of different energies, because the Doppler shift and gravitational redshift are different for different parts of the disk. If the disk is viewed face-on, the line is redshifted: this is partly the gravitational redshift, and partly the effect of time dilation in the rapidly orbiting material. If the disk is viewed obliquely, the Doppler effect leads to two peaks, one from the approaching and one from the receding side; the blue (approaching) peak is stronger. Because there is also a gravitational redshift, the peaks are not symmetrical about the unshifted energy; the line center is shifted toward the red. The graph at right plots the strong line due to iron in the nucleus of Seyfert galaxy MCG–6-30-15 (adapted from Y. Tanaka et al.). Error bars indicate the uncertainty in each measurement; the dashed line shows what is expected.*

*Subjected to the dragging of inertial frames near a spinning black hole, the inner parts of an accretion disk become aligned with the hole's equatorial plane. This effect was discovered by James Bardeen and Jacobus Petterson in 1975.*

The direction in which a jet travels is thought to be set by the rotation axis of the swirling gas just outside the black hole. In Newtonian theory, this rotation axis does not change as the gas falls toward a gravitating body. Its direction is fixed with respect to what physicists call an "inertial frame" of reference, that is, one that is not rotating. When a physics demonstrator sets a toy gyroscope in its cage spinning in a classroom, the "inertial frame" is effectively defined by the classroom. No matter how the demonstrator tips the cage, the gyroscope will keep the same orientation with respect to the walls of the room. (In reality, a gyroscope maintains its orientation with respect to the "distant stars," slowly changing its spin axis as the Earth rotates. But this effect is seldom noticeable in classroom demonstrations.) In the same way, Newton would predict that the spin axis of the gas just outside a black hole would be the same as its spin axis far away from the hole, where it first starts drifting in. But if this were true, the direction of the jet would change every time the black hole accreted a gas cloud that approached the galactic nucleus from a different direction. (It is tempting to argue that the direction of the jet should coincide with the rotation axis of the entire galaxy, since the galaxy is supplying gas to the nucleus—but observations show that this is generally *not* the case.)

Einstein's theory predicts that gyroscopes behave quite differently in a gravitational field. According to general relativity, the "inertial frame" can actually be affected by gravity, especially in the strong gravitational field of a compact rotating "body" like a Kerr black hole. The effect is called the "dragging of inertial frames" or, sometimes, the Lense-Thirring effect, after its discoverers. A gyroscope, placed near a rotating body, will maintain its orientation with respect to the "inertial frame" in its vicinity, but that inertial frame will be rotating with respect to a distant observer. So, the distant observer will see the gyroscope's axis change with time! In a sense, the Lense-Thirring effect causes the *space itself* near a Kerr black hole to rotate in the same direction as the hole, with an angular velocity that increases as the hole is approached.

What does this mean for matter flowing toward a Kerr black hole? First, it means that nothing can fall straight into a Kerr black hole along a radius. If you dropped a rock (with zero angular momentum) into a spinning black hole, you would observe it to be deflected from its radial plunge before it disappeared beneath the horizon. No rocket engine, strapped to the rock, could develop enough thrust to stop the deflection once the rock crossed the hole's "ergosphere," within which the dragging of inertial frames becomes irresistible, as is further discussed in the box on the next page. If the rock were in a circular orbit that was not lined up with the equatorial plane of the hole, the Lense-Thirring effect would cause the plane of the orbit to wobble, or "precess," about the hole's rotation axis, as shown in the figure on the opposite page. The same effect also causes the gas orbiting the hole as part of an accretion disk to precess. Since orbits at different radii precess at different rates, the Lense-Thirring effect would tend to distort a misaligned disk and induce some degree of chaos in the flow. However, the viscosity of the disk tries to smooth it out and restore its planar structure. The competition between these effects ultimately nudges the disk's axis into alignment with the rotation axis of the black hole. Thus, the hole really does act like a gyroscope, enforcing a steady orientation on any flow pattern, whether of disk or jets, near it.

Unfortunately, the behavior of the ultrahot gas and electromagnetic

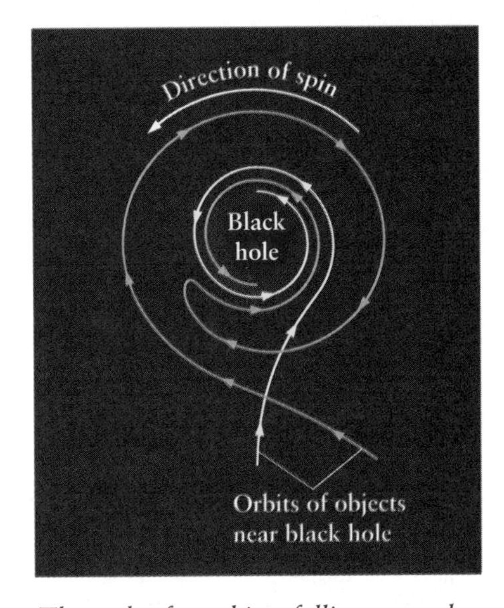


*The path of an object falling toward a Kerr black hole is influenced by the rotation of the hole, shown here as counterclockwise. The yellow particle approaches along a radial trajectory, but its path is deflected into a counterclockwise spiral. The green particle approaches on a clockwise orbit; a distant observer sees its direction reverse so that it too spirals into the hole on a counterclockwise path.*

## Spinning Black Holes

A spinning (Kerr) black hole is more complicated than its nonrotating (Schwarzschild) counterpart. Its properties depend on two characteristics—spin and mass. Moreover, the space around it is not spherically symmetrical: like any spinning star or planet, it has an equatorial bulge. Space is not only rushing into the hole but is also swirling around the rotation axis.

Two distinct surfaces are important in describing Kerr black holes. Inside the outer surface, known as the *static limit,* space swirls around so fast that even light has to rotate with the hole. Within the inner surface light is sucked inward, just as happens inside the Schwarzschild radius of a nonrotating black hole. This inner surface is therefore the *horizon.* The region between the horizon and the static limit is the *ergosphere* (from the Greek word for energy), so called because processes in this region can extract energy from the hole.

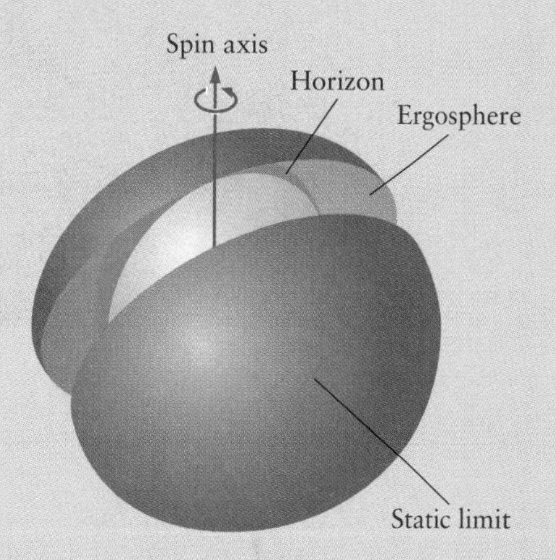


*The ergosphere of a rotating black hole is the region between the horizon and the "static limit," the oblate surface that surrounds the horizon in this exploded view, touching it only at the poles of the rotation axis. In the ergosphere no photon or particle of matter can resist the rotation of the black hole.*

It is possible for an object to orbit closer to a Kerr black hole than to a Schwarzschild black hole, so accretion disks around Kerr black holes can release more energy for a given accretion rate. The efficiency of energy extraction, measured as the percentage of mass converted to energy, ranges from 6 percent for a Schwarzschild hole right up to 42 percent for a Kerr hole with the maximum allowed amount of spin.

The nonrotating Schwarzschild solution is obviously a special and atypical case. However, what is remarkable is that Kerr's equations (though they only have one extra parameter, the amount of spin) are general enough to describe all black holes. A black hole may form by a general asymmetrical collapse, and there will be a brief phase of violent disequilibrium, during which intense gravitational waves are generated. But when the hole has settled down, Kerr's equations describe it exactly. This is the famous theorem that "black holes have no hair." Once you know the mass and spin, all other properties are fixed.

When a black hole swallows something, it gains mass and its spin generally changes. But no other traces survive of what was swallowed—this information is lost. The area of the horizon is like entropy in thermodynamics and always grows. For a Schwarzschild hole this area scales with the square of the mass. There is also a limit to how much energy can be extracted by crashing black holes together—the resultant single hole must have a bigger area than the separate premerged holes added together.

The area of a Kerr hole depends on its spin as well as its mass. It is possible for processes in the ergosphere to slow down a hole's spin and thereby actually extract energy from it, without decreasing the area of the horizon. Many astrophysicists believe that the "cosmic jets" described in Chapter 6 can be energized by the extraction of energy stored in the spin of the hole.

fields around a black hole is harder to understand than the hole itself. Although one may believe (and we do) that Lense-Thirring precession and the Blandford-Znajek process are important in AGNs, it will not be easy to bolster this belief by really compelling evidence. One would have thought that it would be easier to perform "clean" tests of general relativity when black holes rather than less compact objects like the Sun are involved, since the effects of relativity are large (rather than differing from Newtonian predictions by only one part in a million, as in our Solar System). But measurements in our Solar System achieve precisions of one part in a billion—these measurements are a thousand times better than is needed to detect the expected effects, and therefore enough to provide a test with a precision of one part in 1000. In contrast, there are no phenomena yet observed in AGNs that can tell us the properties of the central hole with a precision of better than one part in 10. But astronomers as a group possess great reserves of optimism, instilled by repeated good luck as well as the profound developments in observational techniques over the past thirty years. We will next describe some phenomena, beyond the scope of current experiments, that might soon allow us to learn something about the "heart" of an AGN and at the same time to test relativity in the "strong gravity" regime.

## Stars in Relativistic Orbits

Slowly moving bodies orbiting at large distances from a central mass follow paths that are little different from the standard elliptical orbits deduced from Newtonian theory. But general relativity does predict one small difference—there should be a very slight precession of the orbits that is missed by Newton's theory. This orbital complication is unrelated to the precession caused by the dragging of inertial frames and occurs even for orbits around nonrotating bodies. In our Solar System, the effect appears as the famous "anomaly" in the orbit of Mercury—the precession of its perihelion, or point of closest approach to the Sun—which was convincingly explained by Einstein's general theory of relativity.

Near a black hole, orbiting bodies move at a large fraction of the speed of light, and rather than showing a slight precession their orbits trace out complicated patterns. These patterns depend on the size and eccentricity (i.e., ellipticity) of the orbit; they also depend on how fast the hole is spinning because the Lense-Thirring effect tends to drag the orbit around in the direction of the hole's spin as though the space itself were spinning like a whirlpool. We have already seen that, in a galactic nucleus, stars must occasionally venture so close to the black hole that either they are swallowed whole or torn apart by tidal forces. The former is likely to have little observable consequence, while the latter is too "messy" to make a good test of relativity. Is there any chance that a star could migrate into an orbit close enough to a massive black hole that relativistic effects would become obvious, without it moving so close that tidal forces would destroy it? Could we ever observe the relativistic gymnastics of such a star?

There is certainly "room" for a star to follow such an orbit, provided that the black hole is sufficiently massive. A hole weighing as much as 100 million suns has a Schwarzschild radius twice as large as the radius of the Earth's orbit round the Sun and a tidal-destruction radius (for stars like the Sun) only a few times larger. A normal star would have to stay well outside the tidal limit to avoid suffering internal distortions that could mask purely relativistic effects. Around less massive holes only white dwarfs and neutron stars could safely populate relativistic orbits. But some galactic nuclei contain even bigger black holes, with horizons as large as the entire Solar System. Since tidal forces near the horizon are

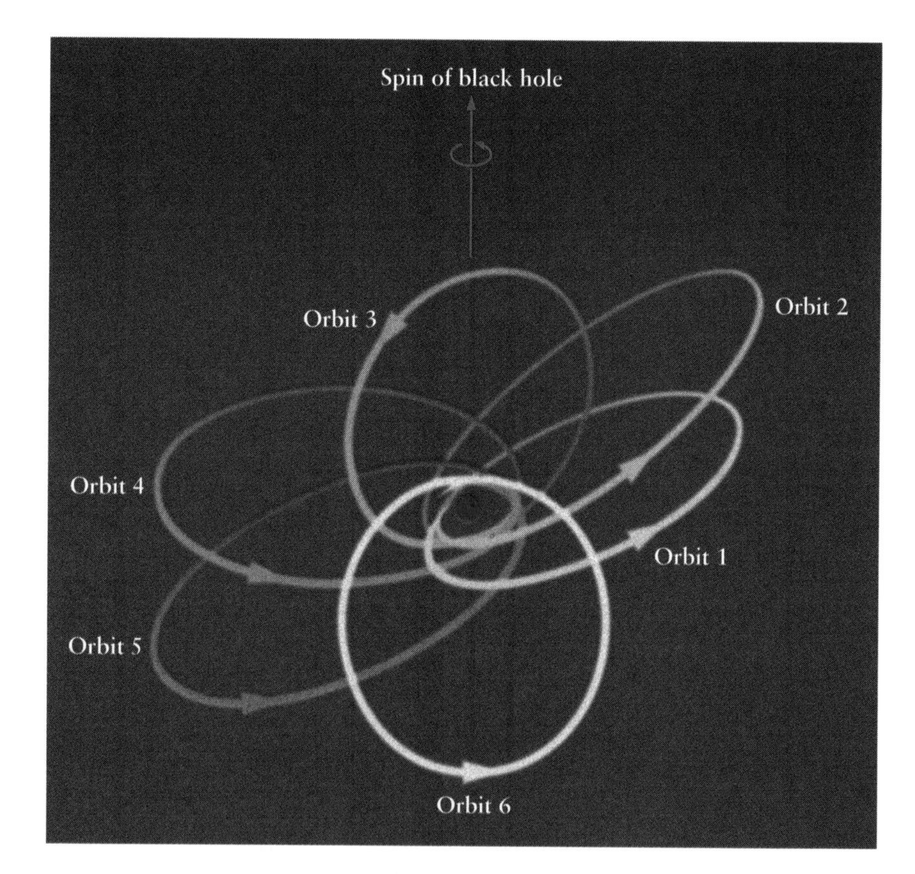


*Six consecutive orbital loops around a rapidly spinning black hole, as calculated by Kevin Rauch. The highly elliptical orbit starts out with an orientation 45 degrees away from the equator of the hole. Each successive loop can be approximated by an ellipse, but the three-dimensional orientation of each loop is different. Two effects of general relativity cause the orbit's orientation to change. First, the dragging of inertial frames, due to the hole's rotation, causes the orbital plane to precess about the spin axis. Second, the same effect that causes the precession of Mercury's orbit causes the point of closest approach to precess around the hole. When these effects occur together, they lead to the complicated orbital trajectory shown.*

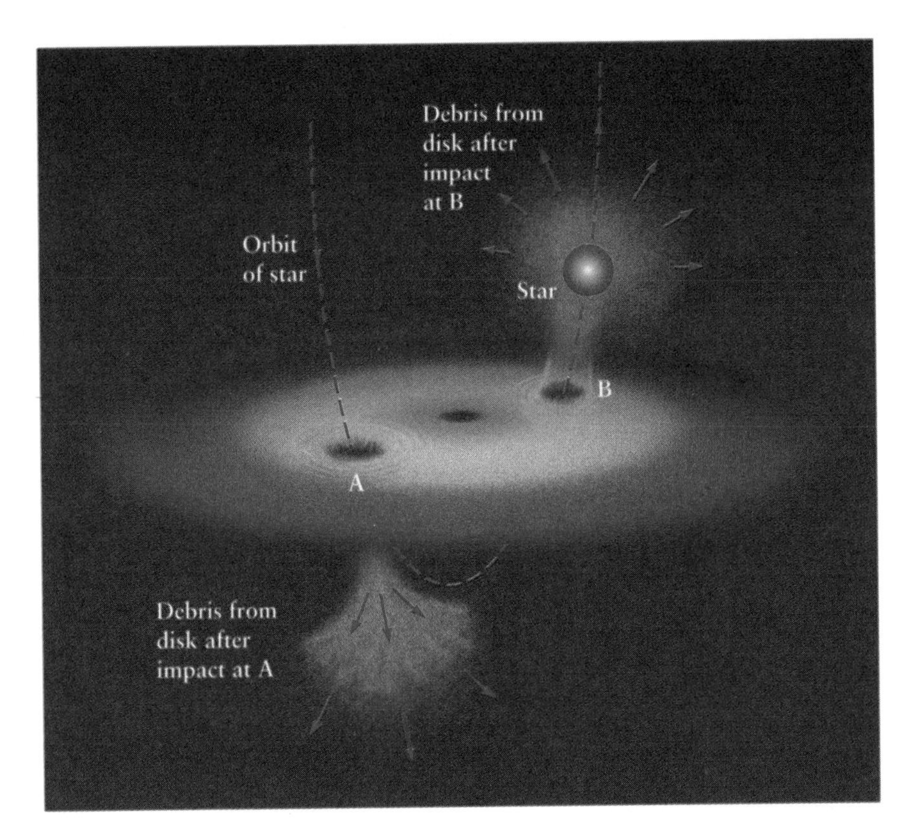


*When a star crashes into the disk it punches a hole through it, and the impacted material sprays out at high speed. If the star returns repeatedly to the disk, crashing through it twice each orbit, it will gradually lose orbital energy and may eventually be captured into a very tightly bound orbit close to the black hole.*

weaker for larger holes, an ordinary star could orbit such a hole at a few Schwarzschild radii, sustaining little damage.

To get into one of these orbits, a star would have to lose an enormous amount of energy and orbital angular momentum. How can this happen? It could undergo a series of close encounters with other stars, whereby orbital energy is transferred from one star to the other. Or it could crash, twice each orbit, into a gaseous disk. Each impact with the disk would dissipate some of the star's kinetic energy, so its orbit would gradually shrink. It is therefore possible for a star to get into the kind of orbit depicted in the figure on the opposite page. But would we "see" it, once it got there?

Such a star would be too faint to be seen directly (unless it happened to be in our own Galactic center). But information about its orbit could still be obtained by observing the activity of the AGN. A star's periodic impacts with the gas disk could modulate an AGN's power output. For instance, the debris from the impact might create an opaque screen, which would attenuate the radiation emitted by the disk for part of each orbit. Alternatively, whenever the orbiting star passed through the base of a jet, it could "short out" the electric field that accelerates the outflow. If an

AGN displayed any regular periodicity (rather than just random variations), it would be hard to explain except as the work of an orbiting star. Regular monitoring of such an object should disclose not only the basic period, but also the timescale for orbital precession—something that directly probes the curvature of space around the hole. In the early 1990s there was a claim that the X-rays from a particular Seyfert galaxy, NGC 6814, varied with a period of 3.4 hours. Disappointingly, this period turned out to belong not to the AGN itself but to an X-ray–emitting binary star that lay close to it in the sky. But that "false alarm" doesn't diminish our motive for continuing to search for periodicities, nor the chance that we will someday find one.

## Gravitational Waves

A star could persist in a relativistic orbit around a supermassive black hole for hundreds or even thousands of years. No such orbit could persist indefinitely, however. Even if there were no drag due to interaction with gas, the orbit would gradually lose energy and the star would spiral into the hole. The reason for the star's final plunge is a fundamental prediction of Einstein's theory—the emission of *gravitational radiation.*

If the size or shape of an object changes, then so does the gravity around it. The only important exception—as Newton had realized—is that the gravitational field outside a sphere stays the same if the sphere simply expands or contracts. Changes in a gravitational field cannot propagate instantaneously—if they did, they would convey information about the size and shape of the gravitating mass at speeds faster than the speed of light, which is impossible. If the Sun were to change its shape, and thereby alter the gravitational field around it, 8 minutes would elapse before the effect was "felt" at the Earth. At very large distances, this change manifests itself as "radiation"— a wave of changing gravity, moving outward from its source. In some respects the creation of gravitational waves resembles the way a change in the electric field around a group of charges leads to electromagnetic waves. For instance, a spinning bar with a charge at each end creates an electric field that is different when the bar is end-on and sideways-on. At large distances from the bar, this changing field becomes an electromagnetic wave.

There are two important differences between electromagnetic and gravitational waves. First, any gravitational wave is very weak unless large masses are involved. Atomic-scale "spinning bars"—diatomic molecules—are efficient emitters of electromagnetic waves (in the radio or infrared band, depending on how fast they are spinning). But they are hopelessly weak emitters of gravitational waves. As we emphasized in Chapter

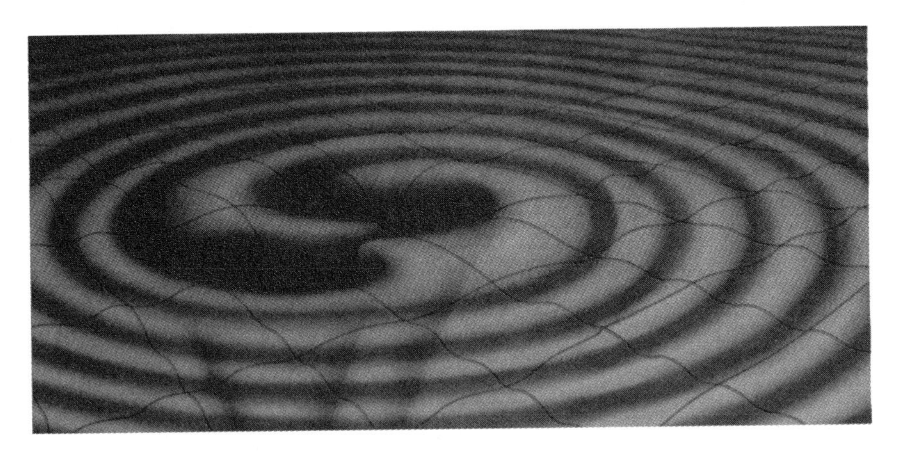

*When two black holes or neutron stars closely orbit each other, gravitational radiation will cause them to spiral together and eventually to coalesce. The waves are traveling distortions in space, illustrated here as ripples in a rubber membrane, which move outward with the speed of light. The locations of the compact objects are depicted as deep depressions in the membrane.*

1, gravity is such a weak force that even its direct effects are nearly undetectable for laboratory-size objects; *changes* in a gravitational field are far harder still to detect. Gravity becomes competitive with electromagnetism on large scales only because there are no negative "gravitational charges" (i.e., negative masses) to cancel out the positive ones, as happens, to a high degree of precision, for electricity. Large spinning dumbbells do exist in the Universe—there are huge numbers of binary stars, each containing $10^{57}$ atoms; and black holes, some of vastly greater mass, may in some instances have binary companions. These are the most straightforward sources of gravitational waves.

While the absence of negative "gravitational charge" gives gravity an "edge" over electromagnetism for very massive objects, it paradoxically also weakens the ability of an object to produce gravitational radiation. Here we come to the second important distinction between gravitational and electromagnetic waves. The most efficient way to produce electromagnetic radiation is for the "center of electrical charge" to wobble relative to the center of mass. This happens, for instance, when the charge on a spinning bar is positive at one end of the bar and negative at the other. Radiation produced in this way is called *dipole radiation.* But everything exerts a gravitational force exactly proportional to its mass or inertia—this is Einstein's famous "equivalence principle," according to which the gravitational acceleration, at any point, is the same for all bodies. This means that the "center of gravitational charge" is *identical* to the center of mass. The former cannot, even in principle, wobble with respect to the latter. Thus, there is no such thing as dipole gravitational radiation. Gravitational radiation resembles the case when there is a positive charge at each end of the bar: the "center of charge" remains fixed at the center, and much less radiation is then emitted because all that changes is the so-called *quadrupole moment,* which measures the extent and shape of the charge distribution. Quadrupole radiation is intrinsically weaker than dipole radiation.

As a consequence of gravitational radiation, a binary system loses energy and its orbit shrinks. This makes the stars speed up and the orbital period decrease. The effect is extremely small when the members of the binary are widely separated and orbiting each other slowly. But as the orbit shrinks, the rate of energy loss accelerates, eventually driving the binary stars toward coalescence.

*The Hulse-Taylor binary pulsar follows a highly elliptical orbit around its companion star. As a result of gravitational radiation, its orbit is becoming smaller and more circular. In 300 million years, the stars will spiral together and merge. Shrinkage of the orbit can be measured with high precision by timing the closest approach of the two stars, using the pulsar as a clock. If the orbital period were fixed, the points would lie along the horizontal line in the graph. Instead, the orbital period is decreasing exactly as predicted by general relativity (red curve).*

The predicted rate of orbital shrinkage caused by gravitational radiation has been verified observationally, with high precision. A binary pulsar was discovered in 1974 by Russell Hulse and Joseph Taylor, using the giant Arecibo radio telescope, while they were conducting a general survey for pulsars. They immediately saw that the pulsar was peculiar in that its pulse period seemed unsteady. Hulse and Taylor quickly concluded, however, that it wasn't the pulsar itself that was unusual, but rather the pulsar's environment—it was orbiting around a close companion. Its unsteady period was a consequence of the changing Doppler effect, as the pulsar first swung around its companion toward us, and then away from us. The pulsar has an 8-hour orbit with a high eccentricity—that is, the distance from its companion changes significantly during each orbit.

Over the last twenty years Taylor and a succession of collaborators have pursued a sustained program of measurements, recording the arrival times of pulses with a precision of around a millionth of a second. These cumulative observations have revealed that the orbital period is gradually

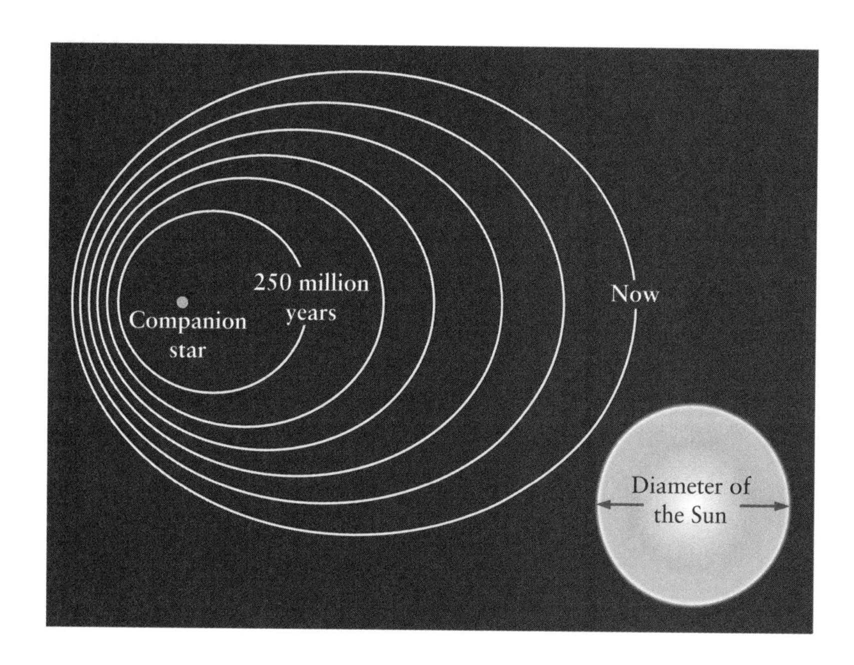


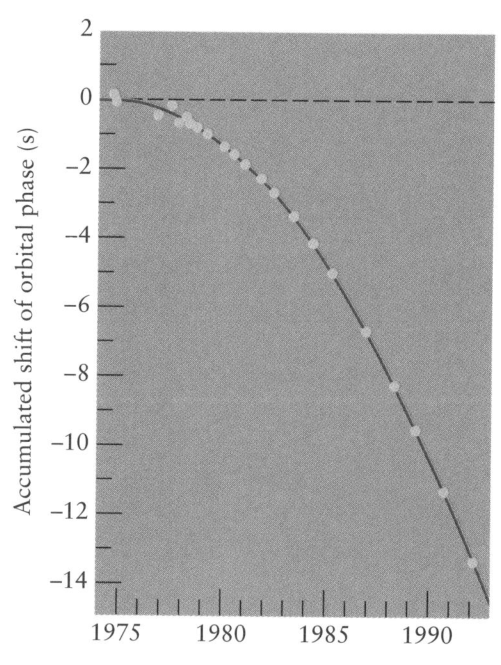

decreasing, by 2.7 parts in a billion (0.27 millionths of a percent) per year. The decreasing period tells us that the orbit is shrinking, although by only a few centimeters per year. Such is the precision of the observations, however, that this tiny effect has been measured to better than 1 percent. Moreover, it agrees, to that same precision, with what should happen if Einstein's theory is correct. This is in fact the only empirical corroboration that gravitational radiation is a real phenomenon. And it is the only observation that probes a distinctive prediction of Einstein's theory, one that does not appear as a correction to Newtonian theory in Solar System tests.

The most dramatic bursts of gravitational radiation should occur when two compact stars in a binary—neutron stars, black holes, or one of each—coalesce. The coalescence of two neutron stars, lasting only a fraction of a second, would be a violent event that triggers a release of neutrinos (as in a supernova) and perhaps also a flash of gamma rays. If two black holes spiral together, the outcome is, in a sense, simpler than the merging of two neutron stars. The only relevant force is gravity, or "curved empty space" in the language of general relativity. No "messy" physics involving nuclear matter or magnetic fields plays any role. Everything that happens is simply a consequence of Einstein's equations. When two black holes come together, their separate "horizons" merge into a single, larger one, and the original binary ends up as a solitary black hole. After it has settled down, this hole must, according to the "no hair" theorems, revert to the simple, standardized form described completely by Kerr's solution of Einstein's equations. The hole ends up with a mass and a certain amount of angular momentum (or spin), but once these two numbers are fixed we can be sure that the hole is a clone of any other hole having the same mass and spin.

It is during the merger of the two holes that gravitational waves are produced in abundance. The gravity field is then fluctuating wildly; one can imagine the fabric of space itself being violently shaken about. By the emission of gravitational waves, the merging structure shakes off its "hair," the asymmetries and inhomogeneities that depend on exactly how the merger took place. It thereby erases most traces of its history, and settles toward a new equilibrium as a Kerr black hole. The gravitational waves produced in this "settling down" process would vary with time in a complicated way—they would certainly be more complicated than the nearly periodic waves emanating from the premerger binary. Not even theorists with gigaflops of computer power at their disposal have yet had much success in solving Einstein's equations in such "dynamical" situations, when the curvature of space is changing so violently. But computational techniques are advancing rapidly, and such calculations should be accomplished during the coming decade. Until then, we cannot be sure what pattern of gravitational waves would be generated, nor even how

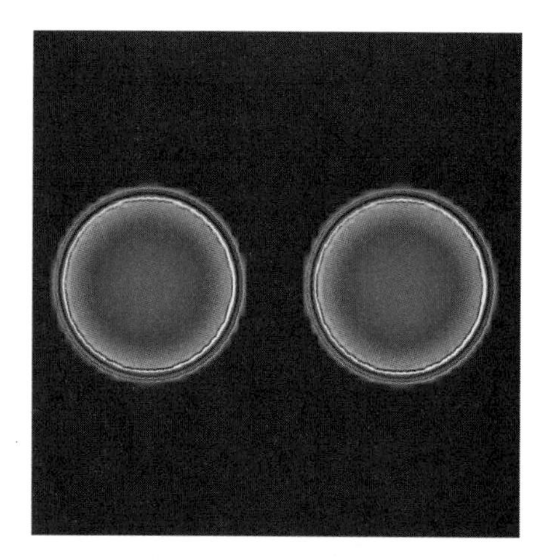

much (10 percent? 20 percent?) of the original mass would be radiated away during the settling down process.

Even without detailed calculations we can, however, place general limits on the amount of gravitational wave energy that can be radiated during the merger of two black holes. Stephen Hawking proved, very generally, that the sum of the areas of the horizons cannot decrease. This means that when two black holes merge, the horizon around the final hole must have an area equal to or bigger than the sum of the original two. The area of a black hole's horizon depends on both mass and spin; for a nonspinning (Schwarzschild) hole, whose radius is proportional to mass, the area scales as the square of the mass. So, by placing a lower limit on the area of the final hole's horizon, Hawking also set a minimum to the final hole's mass. Suppose, for example, that two identical Schwarzschild holes, of 1 solar mass each, collided head-on to produce a bigger nonrotating hole. According to Hawking's law, the mass of the final hole must then be equal to *at least* the square root of 2 (that is, $2^{1/2}$) solar masses. Gravitational waves couldn't, therefore, carry away more than $(2 - 2^{1/2})$ solar masses, equivalent to about 29 percent of the mass of the two original holes. For a Kerr hole, with spin as well as mass, the arithmetic is slightly more complicated, but the results are similar. There are some indications that the amount of gravitational wave energy is often far lower than the theoretical limit. In fact, the head-on collision of two Schwarzschild holes, a relatively simple problem, has already been computed in enough detail to show that the amount radiated is quite small.

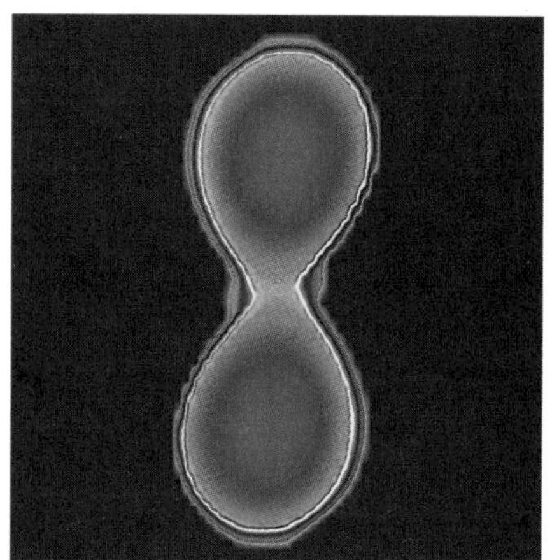

*The merger of a neutron star binary, in a computer simulation, by Max Ruffert, to a single object with a disk around it; it will collapse to a black hole. Gravitational radiation is emitted during the merger.*

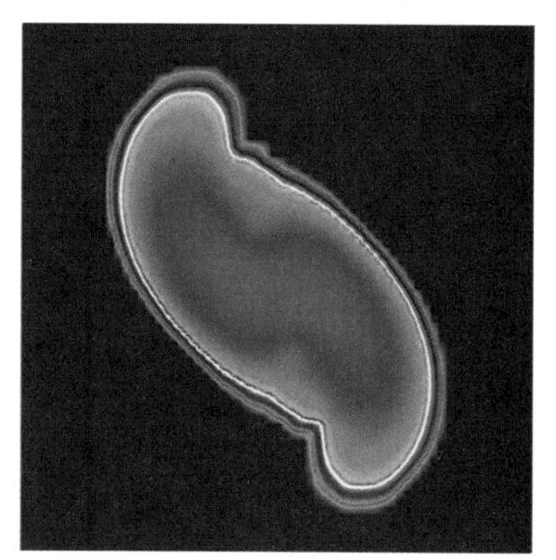

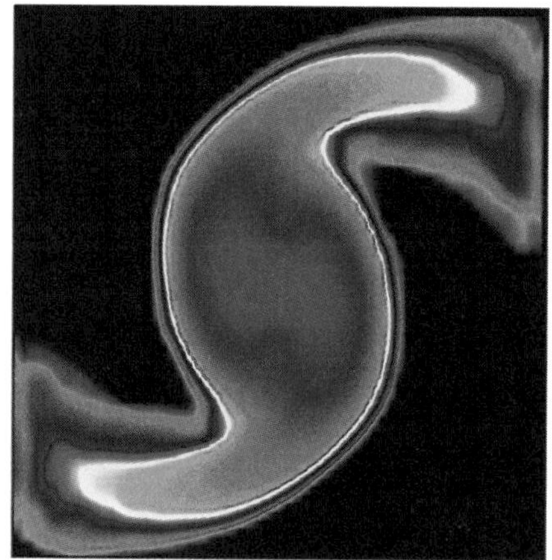

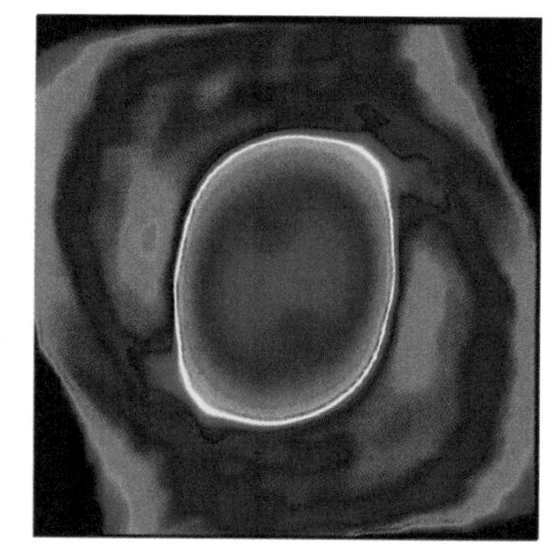

## Can Gravitational Waves Be Detected Directly?

To assure ourselves that gravitational radiation is real, it would obviously be more convincing if the waves could be detected directly. Right now, the sole indication that gravitational radiation truly exists is the rather indirect evidence that orbital energy in one binary pulsar has drained away at a rate consistent with that expected. How can such waves be observed? The gravitational wave is a signal, moving at the speed of light, telling us that the shape (and hence gravitational field) of its source has changed—it would be, in effect, a slight distortion of space. The relative positions of freely moving masses, placed in the path of the wave, would jitter when the wave passed them, as though they were exposed to a varying force acting perpendicular to the direction in which the wave propagates.

Imagine a circular ring or hoop in a plane at right angles to the direction of the source. As the waves pass, the hoop is distorted into an ellipse. It contracts in one direction and is stretched in the other. The previously shorter axis then stretches, and the larger one gets squeezed. This pattern of alternate squeezing and stretching, in two directions at right angles to each other, is the characteristic signature of gravitational radiation. It is as though the ring of matter is feeling a tidal gravitational force that reverses during every cycle of the wave. However, an ant crawling on the hoop during one cycle would erroneously conclude that the source of the tidal force was in a direction perpendicular to the direction of the true source of the waves.

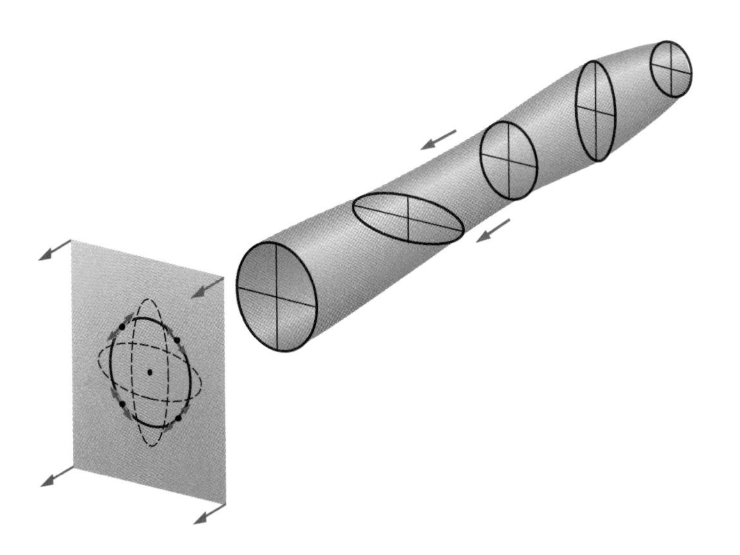

*A gravitational wave causes a "distortion" of space that travels at the speed of light. Any body in the path of the wave would feel a tidal gravitational force that acts perpendicular to the wave's direction of propagation. Thus a circular hoop, intercepting the wave face-on, would be distorted into an ellipse. Waves could be detected by monitoring the relative positions of four masses, one in each quadrant (or indeed of any three of these, forming an* L*).*

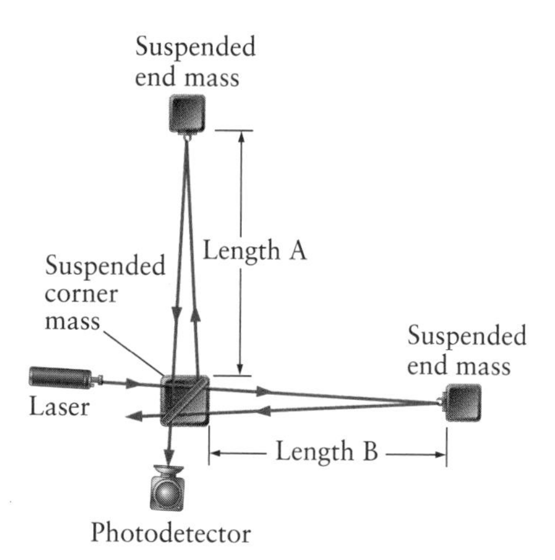


*In an interferometric gravitational wave detector, laser beams are reflected off mirrors attached to two end masses lying several kilometers away, in two perpendicular directions. A gravitational wave causes the two lengths A and B to change: A shortens relative to B, and then B shortens relative to A. There will be a consequent change in the interference pattern between the light reflected from the two masses. The detection of gravitational waves is a daunting experimental challenge because the expected change in path length amounts to only about ten trillionths of the wavelength of light!*

To detect such a wave requires equipment that can sensitively record tiny changes in distance—and the effects are indeed tiny. When the two neutron stars in the Hulse-Taylor binary pulsar finally merge, the intense pulse of gravitational waves, lasting a few seconds, will cause fractional changes in distance of only one part in $10^{17}$ as measured at the Earth. The distance between two particles 100 meters apart would change by less than the diameter of an atomic nucleus. But, assuming that we can measure such tiny changes in distance, we still have to wait 300 million years before the system ends its life in a burst of gravitational radiation. There may be other similar binaries in our Galaxy, but probably not enough to give more than one coalescence every hundred thousand years. Few experimenters derive satisfaction from building ultrasensitive equipment that detects nothing in an entire lifetime—one event per year is a minimum incentive. To achieve this goal they need equipment sensitive enough to record a coalescence not just in our own Galaxy, but in any of the nearest 100,000 galaxies. The typical such event would be ten thousand times farther away than the binary pulsar: the amplitude of the motions induced by the waves would then be ten thousand times lower. A gravity wave detector must therefore be able to measure fractional distance changes of one part in $10^{21}$!

This is a daunting experimental challenge. Early efforts concentrated on designing "bar detectors"— large metal bars, to which "strain gauges" are attached that can detect the motions if the bar is set vibrating by a passing gravitational wave. There would clearly be spurious "events" caused by earthquakes and other local disturbances, so two or more detectors, at different locations, would be needed to pick out genuine gravitational waves. Although some experimenters are developing bars, it is hard to make them sensitive enough.

A more promising technique uses laser beams to measure the distance between two masses. In the United States, the Laser Interferometric Gravitational Observatory (LIGO) experiment uses laser interferometers with a 3-kilometer baseline; one instrument is in Washington State, the other in Louisiana. LIGO should (at least in its second stage of development, shortly after the turn of the century) achieve enough sensitivity to detect the final stages of coalescing neutron star binaries out to 300 million light-years—far enough to have a good chance of detecting one event per year. If there were also a population of binaries consisting of black holes that were, say, 10 times heavier than a neutron star, LIGO could "see" their mergers at even larger distances. LIGO enthusiasts emphasize that other types of as yet unknown sources might be discovered. But there is a fundamental limitation on the kind of "gravitational wave astronomy" that can be done from the ground. Terrestrial detectors are sensitive only to high-

*An artist's impression of one of the LIGO (Laser Interferometer Gravitational-Wave Observatory) detectors being built in the United States. One detector will be near Hanford, in Washington State; the other near Livingstone, Louisiana. It is hoped that these instruments may detect the coalescence of neutron star binaries. (More speculative possibilities include coalescence of stellar-mass black holes, or even supernova explosions that are sufficiently asymmetrical.)*

frequency waves, those oscillating 10 to 100 times per second. The frequency declines as the mass of the colliding objects rises, so only coalescing bodies with a total mass below a few hundred suns would trigger a response from these detectors. Even if a low-frequency detector were built on the ground, seismic background noise would mask the signal. Thus, LIGO will not be able to detect mergers of very massive black holes, because the waves would vary too slowly. Such mergers are expected among the supermassive holes in the nuclei of galaxies. And, fortunately, their gravitational wave signals are detectable—but only from space.

## Supermassive Black Hole Binaries

Our discussion so far may have given the impression that single massive black holes at the centers of galaxies are always "home grown," either introduced as the seed around which the galaxy formed, produced in the collapse of a dense star cluster, or grown by accretion during the first episode of quasar activity. But such black holes can also be acquired all at once by hole-less galaxies, and can even be lost by galaxies. Many, perhaps even most, galaxies have experienced a merger with another galaxy since the first generation of massive holes formed—indeed, merging galaxies are

*Kip Thorne, of the California Institute of Technology, was a student of John Wheeler at Princeton. Thorne and his "school" have been responsible for many of the key developments in relativistic astrophysics. Over the last decade, he has been the prime promoter of the LIGO project.*

quite commonly observed. Whenever two galaxies merge, the orbits of their constituent stars get mixed up, and the resultant "star pile" resembles an elliptical galaxy. If each of the original galaxies contained a massive black hole, these holes would spiral together into the center of the merged galaxy, forming a supermassive black hole binary.

The holes gradually approach each other, as they transfer kinetic energy to stars and perhaps suffer drag on gas. At some point they are close enough that each is influenced more by the gravitational pull of the other than by the collective effect of the surrounding stars. The black hole binary continues to shrink, until the two holes are sufficiently close for gravitational radiation to carry energy away faster than the stars can. During these latter stages of binary evolution, the rotation axis of each black hole can precess in the gravitational field of the other. If a pair of jets is produced by one of the holes, the jets should also precess—this may be the explanation for the "inversion symmetric" radio sources discussed in Chapter 6. (A similar explanation was tried for the jet-producing X-ray binary SS 433, but ran into difficulties. It is more likely that the precession in SS 433 arises mainly from properties of the ordinary star and accretion disk, rather than the compact object.) Could both members of a black hole binary produce pairs of jets? The answer is not known, but images of at least one radio galaxy—3C 75—give the impression that this might be happening.

A powerful burst of gravitational radiation is emitted during the final coalescence. Since that burst is likely to be focused toward one direction, at the moment of final coalescence the merged black hole may experience a recoil—a strong-field gravitational effect that depends on a lack of symmetry in the system. It won't be possible to calculate the speed and direction of recoil until computers can cope adequately with fully three-dimensional general relativistic calculations; as already mentioned, this ability may still be a decade away. Very crude approximations suggest that the recoil speed may attain several hundred kilometers per second, which could be enough to propel the merged hole clear out of the galaxy. At the very least, the recoil would eject it from the nucleus. The hole would then gradually regain its central location by losing energy to the stars in the galaxy.

Binary black holes can be ejected from their host galaxies even before the final coalescence occurs. If a third hole drifts in before the binary has merged, it will form a triple system with the existing pair of black holes. "Three-body" systems are notoriously unstable according to Newtonian physics, and a "slingshot" effect can lead to the ejection of *all three* holes, with a "kick" comparable to the orbital velocity of the binary, several thousand kilometers per second. The disconcerting possibility exists that some massive black holes could now be hurtling through intergalactic space.

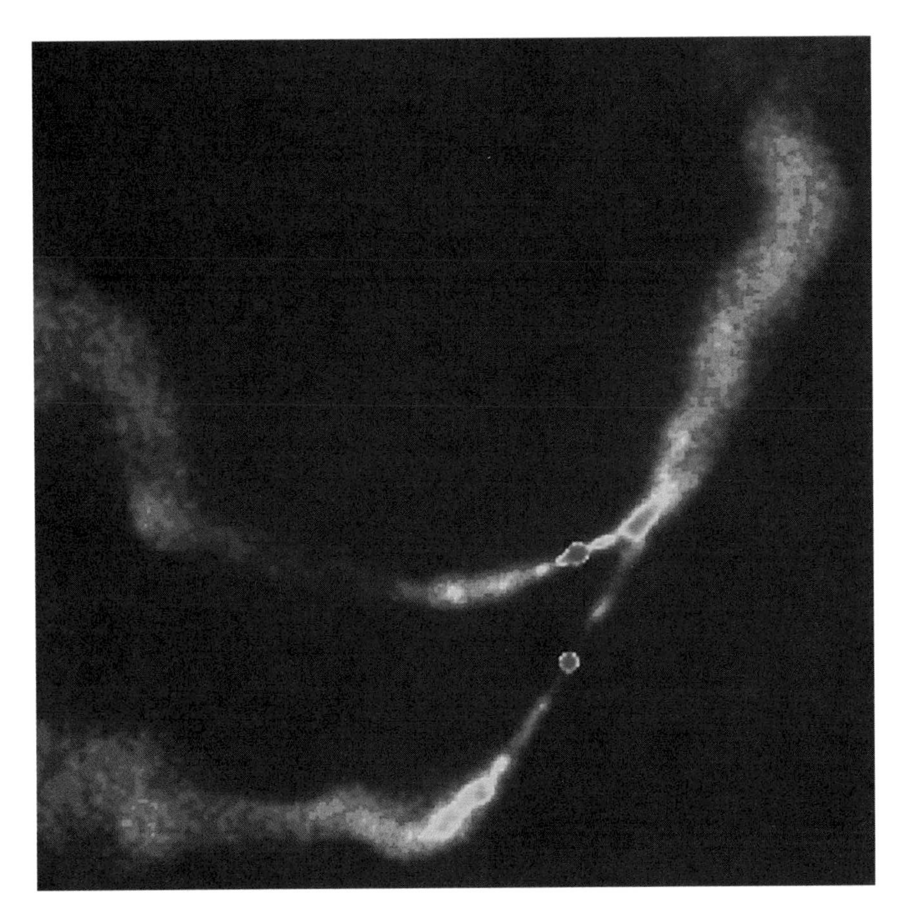

*The radio galaxy 3C 75 appears to contain* two *active cores, each of which produces a pair of jets. The projected separation is about 10,000 light-years—small compared to the size of the optical galaxy. Has this galaxy captured a second massive black hole, which will eventually form a close binary with the hole in the nucleus, leading to a merger? Or are we witnessing the chance alignment of two quite distinct active galaxies on the sky? It is also unknown whether the apparent intertwining of the jets on one side is real or an illusion. Galaxy mergers are common: if each galaxy harbors a supermassive black hole, then the holes will spiral toward coalescence. What is unusual about this system is that both holes manifest the rare form of activity that leads to double radio jets.*

Suppose that two merging galaxies each harbor a supermassive black hole. As the holes spiral together toward coalescence, energy equivalent to millions of solar masses could be emitted in gravitational waves. How could we detect this colossal gravitational wave energy? Since no force other than gravity is at work, nothing in the physics differentiates a merger between two black holes of, say, 10 solar masses each from a merger between two black holes of a billion solar masses each. One merger is simply a scaled-up version of the other—radii, wavelengths of gravitational radiation, and timescales are all just proportional to the masses. The size of a black hole scales with its mass; so also does the minimum orbital period just before coalescence, which determines the frequency of the waves radiated. When supermassive black holes coalesce, the periods are hours rather than milliseconds, and thus the wave frequencies are low rather than high.

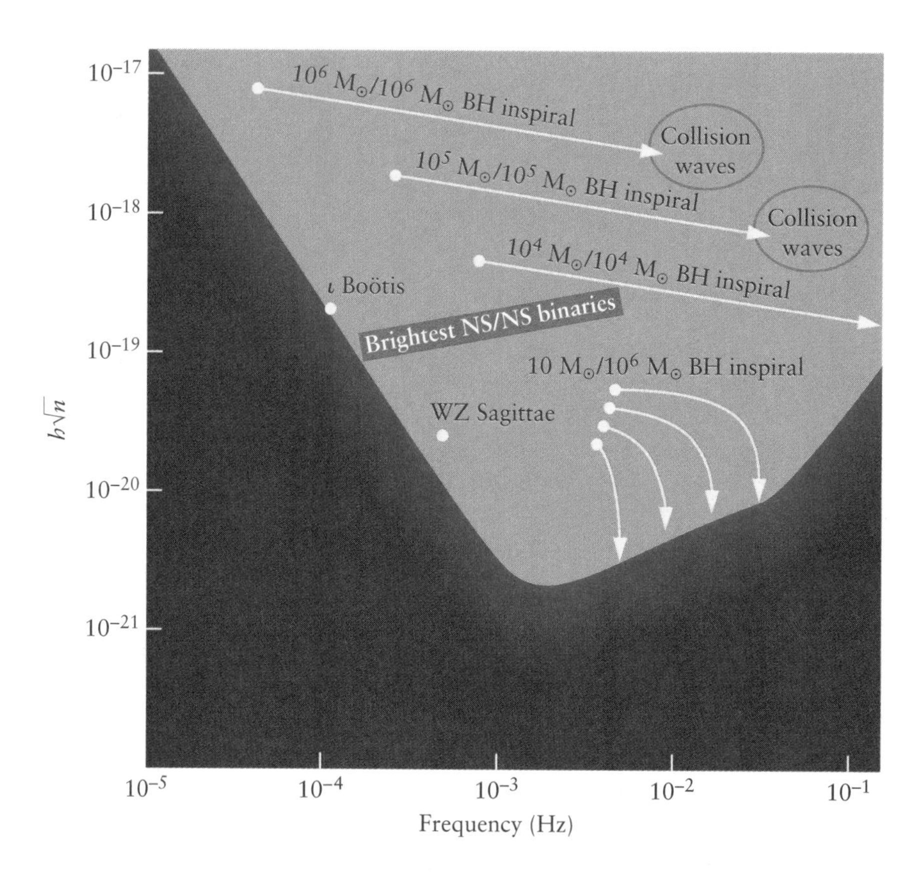


*Predicted sensitivity of LISA, the European Space Agency's planned space interferometer, to various gravitational-wave sources. The red band shows the minimum signals that should be detectable at different wave frequencies. The vertical scale is the wave amplitude multiplied by the square root of the number of wave periods collected during the observation—a weak signal is easier to distinguish from background noise if it can be summed over many periods. Arrows show the evolution of signal strength and frequency from the spiraling together, and eventual coalescence, of black hole (BH) binaries 10 billion light-years away (at "cosmological" distance). Even the inspiral of a 10 $M_\odot$ hole into a $10^6$ $M_\odot$ hole is detectable (the multiple arrows plotted for this case show the range of signals, depending on the spin of the holes). Detectable sources within our Galaxy include neutron star (NS) binaries and certain nearby binary stars. Two examples of the latter, ι Boötis and WZ Sagittae, which have known periods and would be detectable, are plotted.*

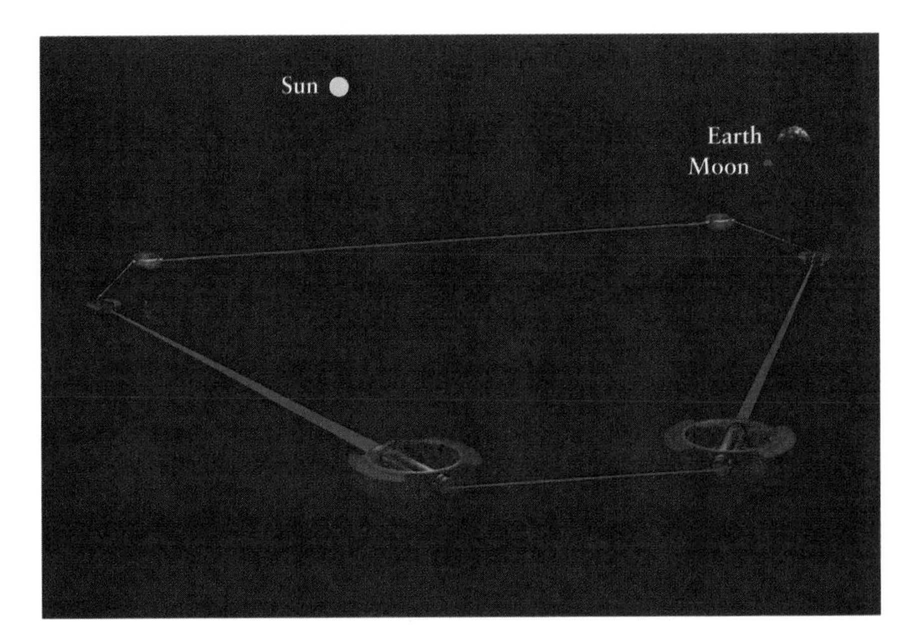


*An artist's impression of the LISA gravitational-wave detector in space. The array of six spacecraft would go around the Sun in the same orbit as the Earth, but about 20 degrees away. The separations between the spacecraft (about 5 million kilometers) would be monitored by laser beams. This instrument, free of the seismic interference that bedevils terrestrial instruments like LIGO, would be sensitive to gravitational waves of low frequency, such as those that arise from orbits around supermassive black holes.*

As we discussed earlier, waves of such low frequency leave no sign as they pass through ground-based gravitational wave detectors. But they can be detected by a system of spacecraft, such as the Laser Interferometer Space Antenna (LISA) project being studied by the European Space Agency.

LISA would consist of six separate spacecraft deployed in a huge hexagonal pattern, at separations of 5 million kilometers, all orbiting around the Sun. Small changes in the relative positions of the spacecraft, caused by passing gravitational waves, would be monitored by transmitting laser beams between them. This array could easily detect the merger of two black holes of a million solar masses each, even if it occurred in a very distant galaxy. LISA's sensitivity would be so great that it could follow in detail how the waves gradually strengthen and the orbital motion speeds up as the black holes spiral together.

LISA will not be in orbit until 2015 at the earliest. Well before then, computational techniques will surely have advanced far enough for theorists to have calculated the precise wave patterns expected from coalescing black holes, and how these depend on the masses, spins, and orientations of the two holes. Astrophysicists will be able to compare these calculations with LISA data to probe phenomena that are dominated by strong-field gravity, and thereby obtain "clean" tests of the crucial predictions of Einstein's theory. The discouraging news for experimenters is that on average a merger between two supermassive black holes takes place less than once per decade over the entire observable Universe.

Fortunately, LISA need not sit idle between the rare mergers of quasar-size holes. If there is a large population of intermediate-size black holes, of ten thousand to a million solar masses, then these should merge with greater frequency. Moreover, because of its great sensitivity, LISA can detect the fainter signals from objects that might be much more common than supermassive black hole binaries. There are binary stars in our Galaxy with orbital periods of about an hour; the nearest of these should be detectable. A more interesting prospect is that LISA might detect the waves from low-mass compact stars orbiting supermassive black holes. The waves emitted in this case are, of course, much weaker than those emitted by a supermassive binary—the amplitude is less by the ratio of the star's mass to the hole's mass. But LISA may be sensitive enough to detect the almost-periodic signals from such stars.

## Is General Relativity the Right Theory?

Einstein's theory dates from 1915, but even fifty years later empirical support for the theory was meager. Two of its predictions—the bending of light near the Sun and the precession of Mercury's orbit—had been tested, but only with about 10 percent accuracy. Several rival theories were still being taken seriously in 1965. But the evidence in support of general relativity is now much more compelling. In our Solar System, there have been amazingly precise measurements of how radio waves are bent near the Sun, of the small extra time delay of radar echoes from planets, and of deviations in the orbits of space probes. These have confirmed that the Sun slightly "curves" the surrounding space exactly as predicted—theory and observation now match to one part in 1000. Moreover, the gradual shrinkage of the binary pulsar's orbit due to gravitational radiation confirms a more subtle and distinctive prediction of Einstein's theory.

We are now more confident that Einstein's theory can be extrapolated into the "strong-gravity" regime relevant to black holes. But we have seen in this chapter that, despite our confidence, there is yet no direct confirmation that general relativity applies under the most extreme conditions—nor that black holes are described exactly by Kerr's equations.

Astrophysicists believe that active galaxies are powered by black holes, just as stars are powered by nuclear fusion. Einstein's theory stands in the same relation to quasars and the centers of galaxies as nuclear physics does to ordinary stars. When astrophysicists model stars, they feed in the relevant nuclear physics as they understand it to be, then calculate the stars' properties and check them against observations. If our stellar models were detailed enough, astronomical data might allow us to check whether the nuclear physics is right, or even to learn some new physics. This hasn't happened yet, though the dearth of neutrinos detected from the Sun (using the same equipment that detected the neutrino burst from Supernova 1987A), compared to theoretical predictions, may be telling us something important about neutrino physics. Likewise, those who try to model galactic nuclei (and their small-scale counterparts, the X-ray sources described in Chapter 3) are hoping to probe the fundamental physics of strong-field gravity. They may eventually be able to do so, but "relativistic astrophysics" is at a much earlier stage of development than stellar structure theory, so the goal is still remote.

The renaissance in gravitational research that began in the late 1960s was stimulated partly by the use of new and more powerful mathematical techniques (which led to the proof of the "no hair" theorems), but also by the realization that astrophysical phenomena may actually exist where the effects predicted by the general theory of relativity are large. Theories of gravity can be tested fully only by astronomical observations and by experiments in a cosmic environment, rather than by laboratory work. In the short run, the most useful tests of rival gravitational theories may well come from further high-precision experiments in the Solar System. But in order to study situations where gravity is overwhelmingly strong, one must look farther afield. The highly suggestive evidence already obtained surely justifies an intensive quest for black holes using all promising techniques of space- and ground-based astronomy.

General relativity is, of course, crucial for understanding the Universe itself. The universal expansion from a "big bang" resembles the collapse inside a black hole, but with time's arrow reversed. In our final chapter, we discuss the role of black holes in cosmology, and how they might influence the final fate of the Universe.

*Chapter* 9

# A Universe of Black Holes

*An artist's impression of a flight through a supercluster of galaxies.*

The black holes we have discussed in this book—mainly the supermassive ones in galactic nuclei, but also the stellar-mass black holes that are the evolutionary end points of some stars—are *faits accomplis.* We clearly see their manifestations, and with a few diversions our main task has been to describe what they are like. There is nothing mysterious about the origins of the stellar-mass holes—they are the inescapable fates of stars too massive to settle into retirement as white dwarfs or neutron stars. The origins of the hugely massive holes in galactic centers are less certain, but we can be confident that much of their growth, at least, is attended by the spectacular luminosity of quasars or the vast jets of radio galaxies, which are easily visible.

Black holes tend to be thought of in apocalyptic terms that make it difficult to consider them as being "commonplace," at least as much so as stars and galaxies. The title of a Disney Studios film of a few years ago—*The Black Hole*—suggested (at least to us) that the screenwriter hoped to convey the sense of some unique place of doom, a *terra incognita* at the edge of the Universe. We have tried to dispel such over-romanticized notions by concentrating on the concrete ways in which black holes have been sought and found.

Perhaps, even among our colleagues, we are near an extreme in viewing black holes with equanimity. But we also recognize that black holes *are* extraordinary features of the Universe. And while general relativity gives us a nearly perfect description of their interactions with the outside world, their interiors—where all paths must converge toward a point at which physics as we know it breaks down—fill even jaded astrophysicists with awe. So (to make a physicist's pun), while black holes are not unique, they are *singular.*

To say that the Universe is sprinkled with black hole singularities is quite a statement indeed. A useful concept for characterizing the trajectory of a particle is to describe its "worldline" through "hyperspace," that is, its trajectory through space and time. There is a lot of hyperspace in the Universe, and the chance of any two worldlines intersecting by design is small. For example, suppose you were playing interstellar billiards and you wished to hit the eight ball with the cue ball from a distance of 30,000 light-years. It would not be easy. But if the cue ball and the eight ball were both falling inside the same black hole, their worldlines would be *guaranteed* to meet (albeit not under conditions conducive to the continuation of the game). Black holes are places where many worldlines converge and are, figuratively speaking, tied together into a knot from which they can never escape. Black holes are therefore catchment basins where matter and energy are "taken out of circulation" and can no longer interact with the rest of the Universe *except* gravitationally. (Trajectories in a contracting Universe would behave very similarly to the worldlines of matter and radiation falling into a black hole.)

On a less poetic level, we can ask whether the gravitational collapse of matter to form black holes has broader implications than "merely" the creation of quasars, radio jets, and X-ray binaries. This final chapter considers black holes from the viewpoint of cosmology—we discuss how they could, collectively, affect the structure and evolution of the entire Universe. Did they form in other ways than through the collapse of stars or the coalescence of material in the centers of galaxies, as described in Chapter 2 and Chapter 4? Could they compose much of the dark matter, which makes up 90 percent of a typical galaxy's mass? And could they even determine the fate of the Universe—whether it will expand forever, or eventually recollapse to a "big crunch"?

The answers to these questions depend on whether black holes could have formed early in the history of the Universe, and whether these holes could have quite different masses from those discussed so far.

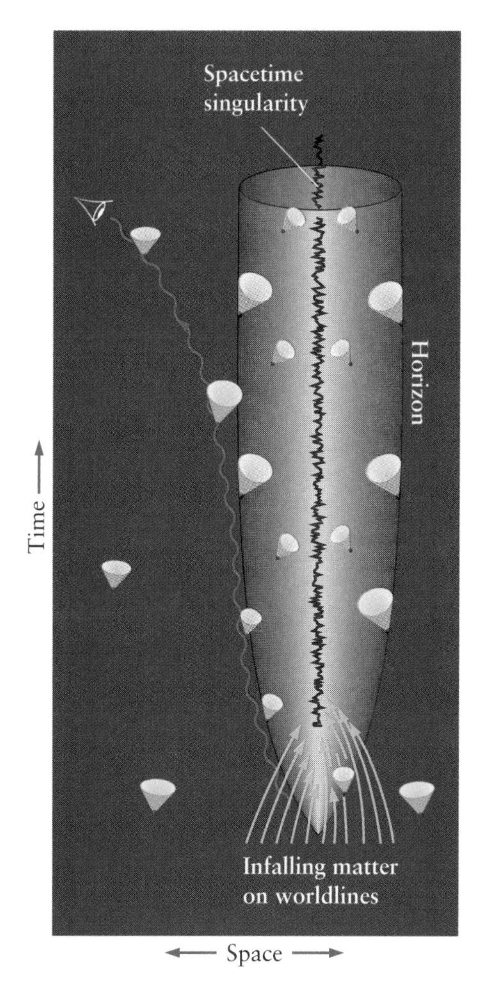


*In this "spacetime diagram," the trajectories represent the "worldlines" discussed in the text; worldlines passing each red dot must lie within the attached "light cone." Near a collapsing mass, the light cones are tilted: light can more readily move inward than outward. Material that collapses to form a black hole is crushed in a central singularity, but this singularity is shrouded from view by a horizon. Light emitted just outside the horizon takes a very long time to reach a distant observer. Inside the horizon the light cones are so strongly tilted that even outwardly directed radiation is pulled in toward the singularity.*

## The Cosmological Context

Edwin Hubble, in the 1930s, discovered that our Universe is expanding; because of this expansion, the light from quasars and distant galaxies is redshifted. In 1965, Arno Penzias and Robert Wilson made an equally momentous discovery while scanning the sky with a radio antenna at the Bell Telephone Labs at Holmdel, New Jersey. (This was the second major serendipitous discovery from that laboratory—Carl Jansky's pioneering studies in radio astronomy, discussed in Chapter 6, were carried out there as well.) Penzias and Wilson found that the Universe is pervaded by dilute "heat radiation" having no identifiable source. Cosmologists quickly interpreted this radiation as a relic of the so-called "hot big bang," when the very young Universe was filled with intense radiation and squeezed to immense densities and temperatures.

The "primordial" radiation can never disappear—it fills the Universe and has nowhere else to go. But it does become cooled and diluted during the expansion, so that now, 10 to 15 billion years after the big bang, its temperature is −270 degrees Celsius (less than 3 degrees above the "absolute zero" temperature). Over the thirty years since its discovery, the "cosmic background radiation" has been probed with increasing precision, most notably by the experiments on board NASA's Cosmic Background Explorer (COBE) satellite. COBE's measurements showed that the radiation has exactly the form expected if it is indeed an afterglow of the big bang.

During the first second of its expansion, the Universe would have been hotter than 10 billion degrees—far hotter even than the center of a supernova explosion. How did this primordial fireball evolve into our present Universe of galaxies? If the Universe had started off absolutely uniform, it would still be composed mainly of dilute cold hydrogen, with a healthy admixture of helium and trace amounts of other light atomic nuclei, synthesized during the first few minutes following the big bang. There would be no stars, no galaxies, and, worst of all, no cosmologists. Its present-day complexities couldn't have emerged unless the Universe had been slightly irregular right from the start. Galaxies and other structures could then have evolved from patches that started off slightly denser than average, or were expanding slightly more slowly than average. The expansion of these patches would have lagged further and further behind their surroundings, and eventually their excess gravity would have halted their expansion, allowing them to condense.

The COBE satellite was actually able to observe this condensation process in its early stages, by comparing the temperature of the background radiation on different parts of the sky to a precision of one part in

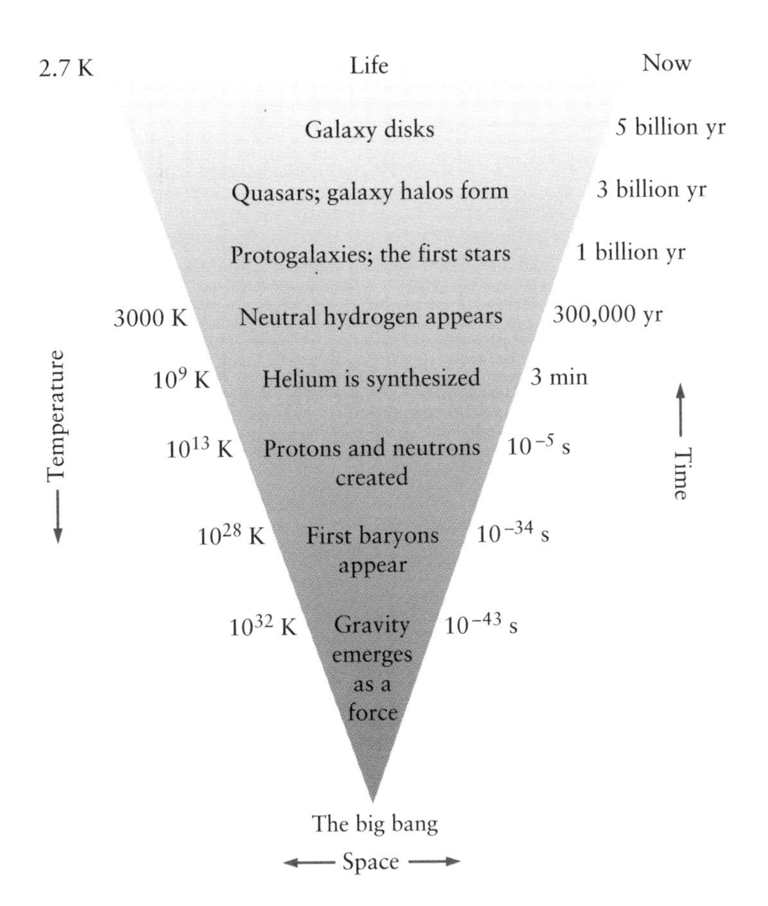


*Key stages in the evolution of the Universe according to the standard "hot big bang" theory. The earlier stages are more speculative than the later ones, because the basic physics is uncertain—at early epochs, densities and particle energies were higher than can be achieved in any laboratory.*

100,000. These measurements revealed slight nonuniformities. Since the radiation is a direct messenger from early pregalactic epochs, these temperature fluctuations—the much-hyped "ripples in space"— directly confirm that the early Universe indeed possessed the kind of irregularities from which cosmic structures could have evolved.

But did the galaxies themselves form directly from the primordial structures, or were there intermediate steps? Were the sizes and masses of galaxies specially "imprinted" in the early Universe, or did galaxies aggregate from smaller objects? Could the early irregularities in the Universe have been pronounced enough to produce black holes that became the seeds of new galaxies? These questions are still being hotly debated. And there is certainly no consensus about the role of black holes in the early Universe, nor about the total number that might exist today. Were the supermassive black holes in galactic nuclei entirely grown from the debris of galaxy formation and nourished on quasar activity, or do they predate the galaxies?

## Pregalactic Black Holes?

Astrophysicists use computers to simulate how structure emerges in the expanding Universe. The calculations do not attempt to represent the entire observable Universe, but follow what happens in a volume large enough to represent a "fair sample"— what Carlos Frenk of Durham University has dubbed a "Universe-in-a-Box." The biggest "virtual Universe" to date has represented about 300 million light-years on a side. Some assumptions are made about the initial fluctuations—how large they are, and how much their density deviates from the average—and the calculations start at some early time when the fluctuations are just small fractional changes in the density. The computer then follows how the density contrasts grow, and structures condense, as the expansion proceeds.

The larger the initial "overdensity," the sooner gravity overcomes the expansion. The early Universe would have been "roughest" on small scales; that is, on these scales the density fluctuations would have been most marked. In other words, if you had averaged the density over larger and larger volumes, the early Universe would have appeared increasingly smooth. This is not a surprising characteristic of the fluctuations—you would get a similar effect if you measured a coastline on a very fine scale, then "averaged" it over larger distances in order to draw a large-scale map. This graininess has an important consequence for the development of structure in the Universe. It means that, were gravity the only relevant force, small scales would condense first. A cloud of matter with the mass of a star, for example, would condense before a cloud with the mass of a galaxy. The resultant objects would then cluster on progressively larger scales, stars gathering to form galaxies, galaxies gathering to form clusters of galaxies, and so on. This sequence of events is called a "hierarchical clustering" scenario.

This is close to but not exactly what happens. Gravity doesn't act alone—the gas exerts a pressure as well. This pressure prevents the collapse of very small clouds but is overwhelmed by gravity for scales above a million solar masses. We therefore have good reason to believe that the first cosmic structures to form were clouds weighing about a million solar masses: pressure stopped smaller clouds from condensing, and on larger scales the early Universe was smoother, so that it took longer for the excess gravity to reverse the initial expansion.

It is much harder to compute what happens inside a cloud after it starts to collapse. It could fragment into a million separate stars. Alternatively, the cloud (or most of it) could collapse as a single entity, forming a supermassive star that quickly evolves into a black hole. Computers even more powerful than those we now have will be needed to settle this question. We cannot yet rule out the possibility that a large fraction of the gas in the Universe collapsed into black holes, at an early epoch before galaxies formed. The holes would have had masses characteristic of the first clouds likely to condense in the expanding Universe. They could, individually, have been as heavy as a million suns and might have formed when the Universe was a few million years old.

After forming, the black holes would have clustered together in a hierarchical fashion, creating progressively larger gravitationally bound clusters of holes. After a billion years, these structures would have been built up to the scale of galaxies. Whereas most material that had been in overdense patches in the early Universe could now be locked up in black holes, the material that was in underdense patches would still exist as gas. If the remaining gas fell into these systems of black holes and turned into stars, we would then end up with galaxies embedded in dark "halos" of black holes. As we discussed in Chapter 4, each galaxy is indeed embedded in an extensive halo of dark matter. It is possible that the dark matter could consist of black holes much heavier than stars—each perhaps as heavy as a million suns.

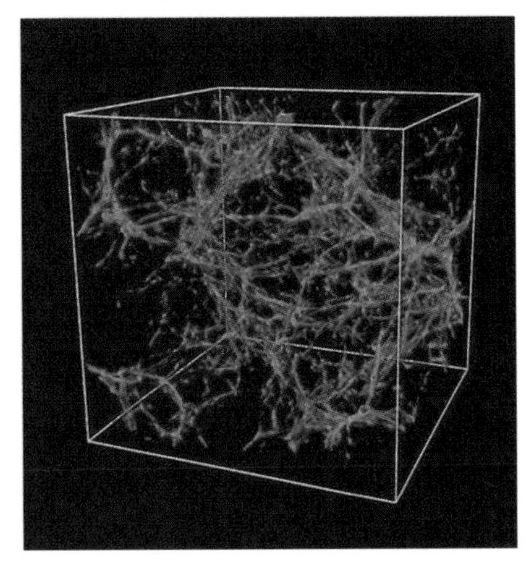

*A "snapshot" from a three-dimensional computer simulation by Greg Bryan and Michael Norman, showing the formation of large-scale structures in a synthetic "Universe." This image shows how (initially much smoother) gas has coagulated into a complex network of filaments, with density increasing from blue to yellow to red. This simulation simultaneously followed the growth of structures made of "stars" and "dark matter," but their distributions are not shown in this image. By analyzing the statistical properties of these structures and comparing them with observations, researchers can test cosmological theories.*

Whenever one of these black holes passed through the central parts of the galaxy, where the stars are packed relatively close together, it would transfer some of its energy to the (much lighter) stars. This energy transfer would act like a friction that slows down the hole's motion. Black holes that spend too much time in this star-laden environment may suffer so much "drag" that they spiral into the center. The first two to reach the center would form a binary; they may then spiral together and coalesce as a consequence of their release of gravitational radiation. When a third hole falls in, the process repeats, and so on. In this way, a single very massive hole may form. But the outcome is very different if a third hole drifts in toward the center before the binary has coalesced. The resultant three-body system is unstable and disrupts very quickly: a binary shoots off one way, and the third hole recoils in the opposite direction, leaving nothing behind in the galactic center at all. If black hole binaries were always disrupted by a third body before they had been driven to coalescence, then a galaxy could have zero, one, or two holes in the center at any given time, but these would never accumulate into a heavier one.

If black holes exist before galaxies do, and can settle toward the centers of galaxies forming among them, they could serve as seeds for the formation of AGNs. Once present, a central black hole could grow through repeated mergers or the accretion of gas during the quasar epoch. Mergers and accretion represent two alternative routes to the buildup of large central black hole masses.

Black holes of thousands (rather than millions) of solar masses could also exist. This mass scale does not arise "naturally" from theories of how the early Universe evolved, as does the scale of a million solar masses. Such holes could, however, exist as remnants of an early generation of very massive stars. Black holes may form from ordinary massive stars, although it is hard to calculate which types of supernovae give rise to neutron stars and which leave behind black holes. However, any "superstar" weighing more than 200 solar masses would certainly become a black hole—indeed, it would end its life by collapsing without any spectacular explosion. We know that stars as heavy as this are not found now. However, there could have been a bigger proportion of such objects among the first generation of stars.

Many astrophysicists, especially Bernard Carr of London University, have explored how many of these "very massive objects" there might be. It turns out that they could be important contributors to the dark matter. In contrast, we can assert quite confidently that black holes in the ordinary stellar-mass range (from a few solar masses up to 100) can't be major contributors to the dark matter. They have been ruled out because the formation of each lower-mass hole would have been accompanied by a supernova explosion. Every time a supernova occurs, several solar masses of "processed" material—containing carbon, oxygen, and iron—are ejected into space. If these holes contributed even 1 percent of the dark matter in our halo, their precursors would have produced more of these chemical elements than we observe. So our halo cannot consist of black holes that are remnants of the kinds of star that exploded as supernovae. On the other hand, there is no such constraint on holes of more than 200 solar masses, because they swallow, rather than expel, the material processed by their precursor "superstars."

Thus there could be enough black holes, with masses substantially greater than 200 suns, to provide most of the dark matter in the halos of galaxies. How could we detect such a population? One of the most promising techniques makes use of gravitational lensing, already discussed in Chapter 4. A black hole's gravitational field would deflect light rays passing close to it, and a distant quasar could appear magnified and distorted if one of these holes lay in an intervening galaxy along the line of sight; the lensing could also produce multiple images of the distant source. For black holes at the upper end of the allowed mass range, around a million solar masses, the images would be separated by a milliarcsecond—a tiny angle, but one that radio astronomers can resolve using very-long-baseline interferometry.

Black holes of more modest mass can also disclose themselves by lensing distant objects; but it is easiest to detect these holes if they are in our own Galactic halo. If a black hole in the Milky Way's halo moves across the line of sight to a background star, its gravity focuses the light so that the star appears brighter. The magnification rises to a peak (when the alignment is closest) and then declines again; so the background star

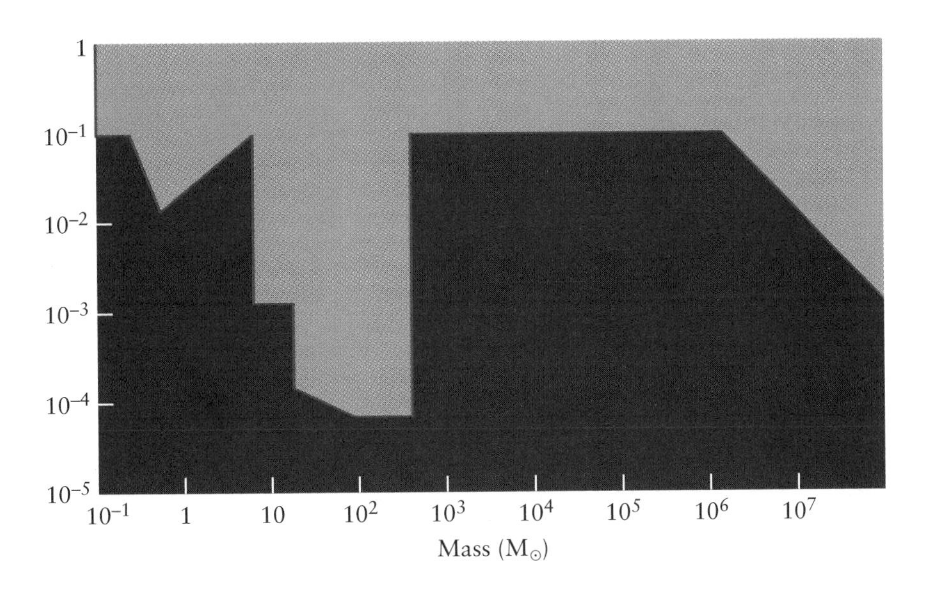


*In this diagram (adapted from Bernard Carr), the height of the purple region shows the maximum amount of matter permitted in stars or stellar remnants of different masses. The limits are shown (vertical scale) as a fraction of the "critical density" required to bring the cosmological expansion to a halt. If there were too many ordinary stars—that is, stars of up to a few solar masses—they would produce too much background light. The limits on higher-mass stars are more stringent, hence the purple region is lower here. Before producing dark remnants (neutron stars or black holes), these stars would have synthesized chemical elements such as oxygen and iron and expelled them into the environment; the limits on their number come from the observed abundances of these elements. This part of the diagram has a complicated shape because the way heavy stars evolve is sensitive to their masses. Very massive stars (VMOs), heavier than a few hundred solar masses, end their lives by imploding and do not eject material that has undergone nuclear processing. Constraints on the black-hole relics of VMOs are therefore weak unless they are heavier than $10^6$ solar masses, in which case their gravitational effects within our Galaxy would have been noticed. The best candidates for dark matter are therefore "brown dwarfs" below 0.07 solar mass, or black holes of at least a few hundred solar masses.*

brightens and fades in a predictable way. Unfortunately, such an event shouldn't happen very often—alignment must be very close to get the effect. Suppose we look at a background star, and ask what the chance is that a black hole lies along the line of sight to it, sufficiently well lined up to cause significant magnification. Even if there were enough black holes to make up all the dark matter in our Galaxy, the chance is only about one in 10 million.

To stand a chance of detecting the effect, one must be prepared either to wait for a very long time indeed or (more optimistically) to observe not one but millions of background stars. Until recently, merely handling the reams of data generated by the latter approach would have been too daunting. But the project has now become feasible, and two independent groups have been monitoring, every clear night, several million stars in the small nearby galaxy known as the Large Magellanic Cloud.

We would expect these programs to pick up thousands of variable stars of many types. And they have. Many astronomers make a living out of studying pulsating stars, flare stars, and binaries; but for this purpose the intrinsic "variables" are a nuisance. The challenge is to pick out the rare instances of variability that manifest the characteristic symmetrical rise and fall of a lensing event, and which are also "achromatic," in the sense that the amplitude of variability is the same in blue and red light. Lensing events have already given us some very impressive candidates for "brown dwarfs" of around 0.1 solar mass, but such objects seem to make up no more than 20 percent of the halo. Black holes are harder to find because the "size" of the lens, and the time it takes for the brightness of a lensed object to rise and fall, goes up as the square root of the lens's mass. So any events caused by the kinds of black hole that might make up the rest of the dark matter in our Galactic halo would be at least 30 times slower than those caused by brown dwarfs. It may therefore be several years before this technique can yield useful results.

We close this section with a sociological digression. The scientists who embarked on these searches for lensing had backgrounds in particle physics—even though the techniques and instruments being used are primarily of an astronomical kind. Particle physicists had already faced up to the challenge of sifting through enormous amounts of data to find rare interactions occurring in particle accelerators. Traditional astronomers were more easily discouraged by the difficulties of monitoring millions of stars and distinguishing lensing events from all the different kinds of intrinsically variable stars. Whatever the outcome of the search, this cross-fertilization between fields has paid off: gravitational lensing has now proved itself to be an effective technique for discovering dark compact objects.

## Miniholes and the Hawking Effect

We have seen that there are many circumstances under which black holes of a star's mass or higher could form, but what about black holes of much less mass than a star? These could have formed, if at all, only in the ultra-dense initial instants of the big bang—the era when temperatures and densities were far higher than can ever be attained in a laboratory, even inside a particle accelerator. These "miniholes" are especially interesting to physicists because they may yield fundamental insights into how gravity links to the other forces of nature.

Finding a "unified theory" that joins all types of forces is perhaps the most ambitious aim of physicists. Gravity is well described by Einstein's general theory of relativity—undoubtedly one of the great achievements of twentieth-century physics. But Einstein's theory, like Newton's, is a "classical" theory. It doesn't incorporate the concepts of quantum mechanics, which revolutionized and deepened our understanding of atoms and forms the basis for modern descriptions of electromagnetism and the forces within atomic nuclei. A key tenet of quantum theory is Werner Heisenberg's famous "uncertainty principle." The speed, or the energy, of a particle cannot be pinned down instantaneously: the more precisely we wish to know these quantities, the longer the measurement takes.

The lack of a fully "quantized" version of Einstein's theory hasn't impeded progress in astronomy. Gravity is so weak that it can be ignored on the scales of atoms and molecules, where quantum effects are important. Conversely, when gravity is important—in planets, stars, galaxies—the scales are so large that we can ignore quantum effects. Thus, in ordinary circumstances there is no real overlap between the domains in which these two great pillars of physics, general relativity and quantum mechanics, are relevant. But developing a theory of "quantum gravity" is conceptually of great importance for physicists, and it is of immediate relevance for astrophysicists and cosmologists. Such a theory is needed before physicists can develop a fully "unified" theory of all the forces. In the absence of such a theory we will never have a consistent picture of the very first instants of the big bang (when the density would have been so high that quantum effects were crucial for the entire Universe), or of what happens near the "singularity" deep inside black holes, where conditions become just as extreme.

There is no accepted theory of quantum gravity—indeed, there is not even a consensus among specialists about the most promising path toward it. An important first step, however, has been taken. Relativists have developed an approach that takes account of quantum effects on the particles and radiation in space, but still treats gravity, and space itself, by Einstein's

*Stephen Hawking's greatest contribution was his realization, in 1974, that black holes are not completely black but radiate with a well-defined temperature. This finding has no significance for the holes that have actually been discovered, but it would be astrophysically important if "miniholes" exist. Hawking's discovery revealed deep conceptual links between gravity, quantum theory, and thermodynamics. Earlier, he had been responsible, with Roger Penrose, for showing that singularities form inside black holes, even when the collapse happens in a nonsymmetrical way. His more recent researches have dealt mainly with the consequences of quantum theory for the entire Universe.*

theory. This "hybrid" approach—quantizing the particles and radiation in a "classical" gravitational field—led Stephen Hawking to a remarkable new insight that linked Einstein's relativity with quantum concepts; as a bonus, he discovered unexpectedly deep connections between gravity and thermodynamics. Freeman Dyson rated Hawking's conceptual breakthrough as "one of the great unifying ideas in physics."

Hawking's results apply in principle to any black hole, but their consequences would be significant only for very tiny ones—naturally, given the link to quantum theory. To introduce the "Hawking effect," let us consider the properties of holes with different masses. A black hole's radius is proportional to its mass, but according to Einstein's theory there is no qualitative difference between holes of low and high mass—the former are just miniature versions of the latter. Some other object, or some other physical process apart from gravity, must come into the picture to set a scale. For instance, when a star orbiting a massive black hole feels a tidal force, the outcome depends sensitively on the size of the black hole relative to the star: black holes exceeding a billion solar masses are so big that they could swallow a star whole; a star falling toward a somewhat smaller hole would be torn apart first, creating a more conspicuous display.

A black hole weighing as much as the Sun would be 6 kilometers across; one with the mass of the Earth would have a diameter of only 9 millimeters. What would happen if there were black holes the size of atoms? Such holes would still be enormously heavier than atoms. A hole as big as an atomic nucleus would weigh as much as a small asteroid. To make such a hole, about $10^{36}$ protons would have to be packed into the dimensions normally occupied by one. This particular large number should come as no surprise. We note in Chapter 1 (and also in the Appendix) that the gravitational force attracting two protons is only $10^{-36}$ of the electric force repulsing them. As many as $10^{36}$ protons must therefore be packed together before gravity can compete with the vastly stronger forces that hold protons apart.

We might expect quantum effects to be important around such a small hole. But how do these effects arise, and what is their connection to gravity? The key lies in the nature of the tidal force near the black hole. Supermassive holes in the centers of galaxies pull more strongly on the near side of a passing star than on its far side; this differential (or "tidal") force tears apart any star that comes too close to the hole. The tidal forces are fiercer around miniature, atomic-scale black holes—so much so that even individual electrons and protons are perturbed by them.

Empty space, what physicists call "the vacuum," is seething with activity on the microscopic quantum scale. When high-energy particles crash together, new particles are created, along with their antiparticles. Such collisions, observed routinely in particle accelerators, may create a shower of electrons and positrons, for instance. These electrons and positrons, and indeed all other possible kinds of particles and antiparticles, can be thought of as latent everywhere in space, as "virtual" particles. Normally, virtual particles can exist only very transiently, because the Heisenberg uncertainty relation allows them to "borrow" their rest-mass energy for only a very short time. A virtual proton–antiproton pair, for example, can exist for a mere $10^{-24}$ seconds before disappearing back into the vacuum. However, these pairs can be made "real"— that is, long-lived—if they can acquire enough energy during the time of their brief appearance. For instance, an electron and positron can be permanently "created" if an electric field pulls them apart strongly enough to supply their rest-mass energy before the Heisenberg relation demands that it must be paid back.

The strong gravity near a black hole can create particles in a similar way, except that the force that pulls the particles apart is the tidal gravitational force—a miniature version of the effect that tears a star apart around a supermassive hole. By chance, one virtual particle will appear closer to the black hole than the other. If the tide is strong enough, it can pull these particles apart so violently that they acquire their entire rest-

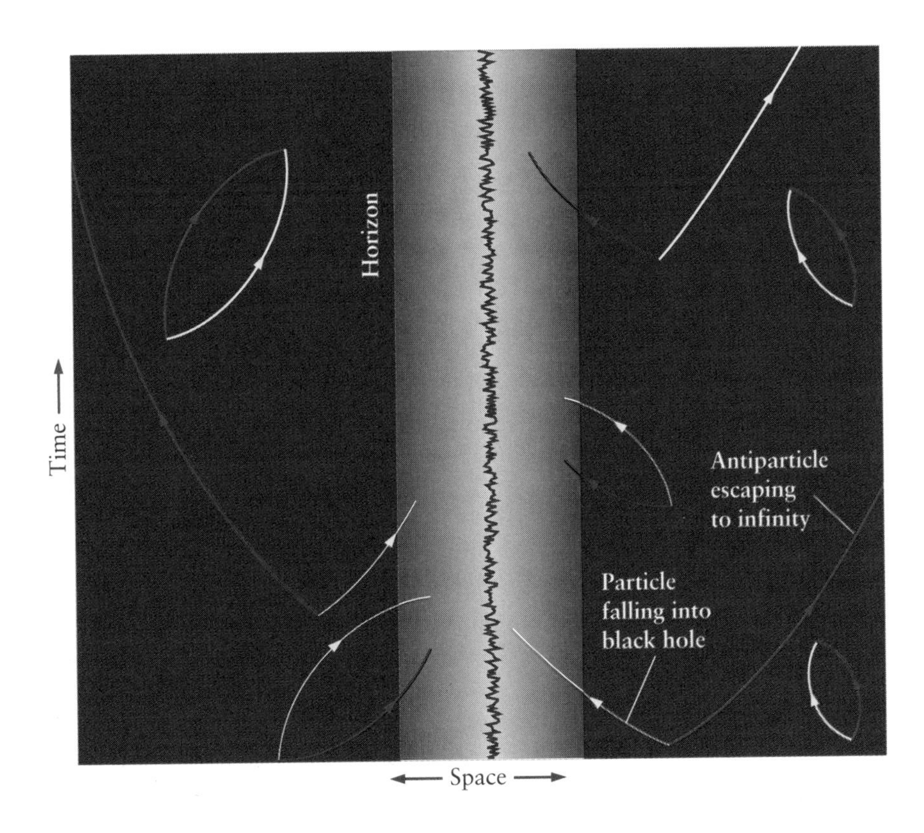


*The worldlines of virtual particles close to a small black hole. Particle pairs, consisting of every possible particle and its antiparticle, are constantly appearing and disappearing at each point in space. Under normal conditions, Heisenberg's uncertainty principle allows these particles to exist only very briefly. Close to a black hole's horizon, however, tides can pull them apart with such force that they acquire enough energy to last indefinitely. When this happens, one particle of the pair is always swallowed by the hole, having lost so much energy that it decreases the hole's mass; the other escapes to infinity. Through this analysis, Stephen Hawking discovered in 1974 that black holes would emit heat radiation and could ultimately "evaporate."*

mass energy $mc^2$ within the brief span of time for which the uncertainty relation allows this energy to be "borrowed." The particles then need not annihilate to repay the energy debt: they instead transform into a real particle and a real antiparticle. But there is a catch—they can't both escape from the hole. One escapes; the other is trapped inside the hole with, in effect, *negative energy*. What do we mean by the expression "negative energy"? When a black hole swallows a particle that comes in from far outside, it grows in mass. Such particles always have positive energy. But absorbing a particle with negative energy actually diminishes the hole's mass: the other particle, carrying positive energy away from the hole, may escape the hole's clutches. Hawking's remarkable inference was that black holes actually emit something. They glow, rather than being completely "black," and in doing so they gradually lose mass. Thus, a black hole isolated in space will actually "evaporate"!

Hawking's mechanism permits the creation only of particles that are as large or larger than the hole itself. According to quantum mechanics, the effective size of a particle is inversely proportional to its mass. Only very tiny holes, therefore, can emit electrons and positrons. Holes of any mass, however, can emit photons: the only requirement is that the wavelength of the light must be larger than the size of the hole's horizon. Depending on the hole's mass, the dominant radiation could be X-rays, visible light, or microwaves. (If neutrinos have zero rest mass, these particles would be emitted as well.) Just as a hot ingot of steel glows red, orange, or blue-white, according to its temperature, so Hawking reasoned that black holes must glow with a well-defined temperature, proportional to the gravitational force just outside the hole.

Before Hawking derived his result, several people had applied the concepts of thermodynamics to black holes. It was known that the surface area of the horizon around a black hole could increase if, for instance, things fell into the hole and it gained mass. But that same surface area could never decrease; if two holes merged, the horizon around the resultant hole would have a bigger area than the two original holes added together. To some physicists this sounded a lot like the second law of thermodynamics, which states that the entropy of a system—a measure of its disorder—can never decrease. They speculated that, somehow, the area of a black hole's horizon was analogous to entropy. They could hardly have imagined how profound this analogy would prove to be.

When we say that the disorder, or entropy, of a system is increasing, we mean that information is being lost about how the system was set up. That lost information is irretrievable. When equal volumes of hot and cold gas are mixed together, for example, the increase in entropy is related to the amount of information that is lost about the microscopic properties of the gas. Before mixing, all the slow atoms might be confined to one half of the volume, and all the fast ones to the other half; we are able to say which atoms are slow and which fast. Afterward, they are inextricably mixed, and we can no longer make that distinction. The formation of a black hole entails a similar loss of information. When a black hole forms or grows, all trace disappears of what fell into it. We can think of not just the objects themselves disappearing into the black hole, but also all information about those objects. A 10-solar-mass black hole created by the collapse of a burnt-out star's iron core is *identical* to a black hole that has grown to the same mass by accreting hydrogen or swallowing pineapples. The "entropy" of a black hole can thus be characterized by the amount of information lost in making it; amazingly, the area of its horizon is a precise measure of this loss.

The Israeli physicist Jacob Bekenstein (then one of John Wheeler's students at Princeton) pushed this analogy further. Still reasoning from the principles of thermodynamics, he pointed out that if the area of a black hole resembled entropy, then the strength of gravity on its surface must resemble temperature. But he was uneasy about whether this insight could be more than just a metaphor, because black holes were then presumed to be truly "black," in the sense of absorbing radiation but emitting none. Bekenstein's insight was vindicated when Hawking showed that the surface isn't completely black but radiates with a well-defined temperature that is indeed proportional to the strength of gravity at its surface.

Hawking's ideas on "black hole evaporation," proposed in 1974, are now generally accepted. They were, however, so contrary to earlier intuitions that they proved hard to assimilate at the time. When Hawking presented his work for the first time, at a conference in Oxford, the session chairman was overtly disparaging, and soon afterward published a paper "refuting" the idea. It took more than a year for some leading experts—people such as Yakov Zel'dovich and his colleagues in Moscow—to become convinced of Hawking's claims.

## Evaporating Black Holes?

The theory of black hole evaporation, sometimes called the "quantum radiance" of black holes, has been "battle tested" in the sense that it has been rederived in many different ways by different people. But that is no substitute for actually observing the predicted radiation, as we perhaps could if very small black holes actually existed. Whether or not the predicted effects can be observed, the Hawking effect ranks as a pinnacle in our understanding of gravity. But is there a chance of actually observing the "quantum radiance" of a black hole?

The predicted temperature is proportional to the gravitational force just outside the hole. This force depends on the mass of the hole divided by the square of its radius. The radius of a hole scales with its mass $M$, so the "Hawking temperature" will be large for small holes and small for large ones. The black holes that form when massive stars die would be only a millionth of a degree above absolute zero. For these (and for the even bigger black holes in galactic nuclei), Hawking's process is utterly negligible: big holes radiate energy far more slowly than they soak it up from the primordial radiation that pervades even intergalactic space. Only if the Universe were to continue to expand for $10^{66}$ years (enormously longer than the current age of $10^{10}$ years) would stellar-mass holes evaporate.

But the temperature of a "minihole" the size of a proton is a billion degrees, high enough for it to radiate X-rays and gamma rays. This radiation would be detectable if such an object came anywhere near our Solar System. As a minihole radiates, it shrinks, and its radiation gets hotter and still more intense, until the minihole finally disappears in a flash of gamma rays. (It is controversial among theorists whether it disappears completely, or leaves some kind of particle as a relic.)

The fabric of space is punctured by black holes, each marking the death of a star or the outcome of some runaway catastrophe in the center of a galaxy. But no astronomical processes operating today could create a black hole smaller than 2 or 3 solar masses. Any star of lower mass, even if it had exhausted all nuclear sources, could survive indefinitely, quite stably, as a neutron star or white dwarf. It could transform into a black hole only if a huge external pressure squeezed or imploded it to even higher densities than those of a neutron star. Anything smaller than a star—say, a planet or asteroid—would need to be squeezed to a higher density still.

John Wheeler, in the 1960s, envisaged how miniholes could be created by the artificial "implosion" of a suitable mass to colossal densities. Wheeler's idea has lately generated renewed interest, though still in the spirit of "science fiction." Physicists have speculated that space inside a black hole, rather than being crushed indefinitely toward a "singularity," could instead sprout into an entire new universe, expanding almost from nothing into a space of its own, quite distinct from and inaccessible to our Universe. According to the popular "inflationary" cosmology, our own Universe may have actually formed in a similar way. Alan Guth of the Massachusetts Institute of Technology, the leading early proponent of "inflation," has written semifacetiously about "creating a universe in the laboratory."

If we were to find a minihole, however, we need not interpret it as the artifact of a supercivilization: it could be a relic of the early Universe. In the first microsecond of its expansion, our Universe was far denser than a neutron star. The tremendous pressures could have created much smaller black holes than are produced astrophysically at current epochs. Whether this would have actually happened depends on how chaotic or irregular conditions were.

It is not sufficient that the pressure be high: pressure *differences* between one place and another are necessary to drive an implosion. The best current guess about the early Universe suggests that the pressure variations would have been too small. Minihole formation would have required the early Universe to have been very much choppier on small scales than we would estimate by extrapolating from the larger scales at which galaxies form.

The gravitational effect of one of these miniholes would be undiscernible unless it came within the inner Solar System. This would be an unlikely event even if there were enough miniholes to make up all the dark matter in the halo. The Hawking radiation itself offers the best hope of detecting miniholes. Lower-mass holes are hotter and radiate more brightly, and we can conclusively exclude the possibility that the dark matter is made up of miniholes with masses below $10^{18}$ grams (corresponding to a size of $10^{-10}$ centimeters) simply because they wouldn't be "dark" enough: their combined radiation would make the sky brighter in X-rays and gamma rays than it is observed to be. But miniholes of this tiny size could still of course exist in smaller numbers, without violating observational limits.

Black holes that formed in the big bang with masses far below $10^{15}$ grams would not have survived to the present epoch: they would have "evaporated" long ago. Those that started off with masses of $5 \times 10^{14}$ grams would just now be in their final stages of evaporation. The lifetime goes roughly as the cube of the hole's mass; those that started off with twice the mass of the holes now disappearing would be only an eighth of the way through their lives. Each of these miniholes—with a mass of $10^{15}$ grams, but smaller than a single atom—could radiate 10 gigawatts of gamma rays for 10 times the lifetime of the Solar System. Even though we can rule out there being enough of these holes to contribute to the dark matter, it would be extraordinarily interesting to find even one of them. These entities are a truly remarkable creation of pure thought—a rigorous deduction from the two best-established theories in twentieth-century physics: quantum electrodynamics, combined with general relativity.

Miniholes are possible "fossils" of the ultra-early Universe, "missing links" that might illuminate the quest for a theory unifying particle physics and gravitation theory. And they are eminently detectable, if enough of them exist. In their death throes, they would emit gamma rays intense enough to be detected many light-years away. (However, the gamma rays from evaporating holes don't have the right properties to account for the mysterious gamma-ray bursts described in Chapter 3. Ironically, we seem to have a fantastic phenomenon predicted by theorists that has never been observed, and some remarkable observations requiring a quite different explanation that still eludes the theorists.) The details of the final explosions are sensitive to uncertainties in the particle physics. According to some theories, the final explosion may eject an intense fireball of electrons and positrons. There would then be an even more sensitive way of seeking evaporating miniholes. Sweeping up the ambient magnetic field as it plowed into the surrounding matter, the fireball would emit a flash of radio waves. Radio telescopes are enormously more sensitive than gamma-ray telescopes and could detect such an event, caused by an entity smaller than an atomic nucleus, even if it happened in the Andromeda Galaxy—2 million light-years away.

## How Should You Bet?

Miniholes would be a missing link between the cosmos and the microworld; the quest for these extraordinary objects is worthwhile even if the chance of finding them is small. We would ourselves, however, bet long odds against their existence. They seem unlikely ever to have been created, because their formation requires rather contrived initial conditions, which are hard to reconcile with other constraints on what the early Universe could have been like. They are the least likely category of black holes to exist.

Much less improbable are the black holes that might be remnants of an early generation of very massive stars. Such a population could form at any stage when the Universe was between a million and a billion years old—after it has cooled down sufficiently for objects to condense, but before galaxies are themselves fully formed. We are open-minded about these black holes: whether they exist depends on uncertain details of how the first generation of stars formed. It also depends (just as the formation of miniholes does) on the nature of the irregularities in the early Universe. There is, however, nothing unnatural about scenarios that would allow most of the material in the early Universe to condense into massive black holes. An important population of black holes—perhaps even the overwhelming fraction of those that now exist—could therefore predate the first galaxy. If present in sufficient numbers, however, these holes should soon reveal their presence to astronomers via gravitational lensing.

But a main theme of this book has been the black holes that form in galactic centers. The existence of remote quasars tells us that some of these black holes have existed for 90 percent of cosmic history. Today these same objects—starved black holes in nearby galaxies all around us—make their presence known by their gravitational effects. The case for these supermassive holes has strengthened greatly in recent years and seems to us to be almost incontrovertible.

There is also overwhelming evidence for black holes in our own Galaxy, formed when ordinary massive stars die, each weighing a few times as much as the Sun. When these objects are supplied with fuel (from, for instance, a close binary companion star), they manifest intense X-ray emission; sometimes they expel jets. Their intense radiation and jets of gas are miniature speeded-up counterparts of the features observed in active galaxies and quasars that are energized by holes millions of times more massive. These luminous objects are, however, an atypical minority of the stellar-mass holes in our Galaxy. Most black holes in this mass range would be solitary wanderers through interstellar space, where the ambient diffuse gas cannot supply enough fuel to energize them conspicuously. Like neutron stars, these holes would have been formed in supernovae. It is hard to estimate theoretically how many there should be—supernova explosions aren't well enough understood for us to be sure what kinds of precursor stars leave neutron stars and what kinds leave black holes—but the number is likely to be around 10 million. At this number, of course, these holes cannot amount to a significant fraction of the Galaxy's mass. If there is a population of black holes that contributes importantly to the dark matter, they must weigh at least 200 solar masses apiece and have formed from a hypothetical population of very massive stars that collapsed without an explosion.

The challenge for astrophysicists is to understand in detail the complex ways in which all of these populations of black holes manifest themselves, and to find new ways of testing whether the holes actually have the properties that Einstein's theory predicts.

## Looking Forward

The entire history of the Universe is the story of how gravity gradually overwhelms all other forces. First, the original cosmic expansion must be defeated: because some parts of the Universe start off slightly denser than average, or expand slightly slower than average, they condense into structures held together by their own gravity. If the early Universe had been very irregular on small scales, this condensation process could have happened very early, giving rise to primordial black holes.

Systems condensing later would become protogalaxies, in which gas is gradually transformed into successive generations of stars. In individual stars, gravity is balanced by pressure and nuclear energy; low-mass stars, even when their fuel runs out, can survive as white dwarfs or neutron stars. But the cinders of very massive stars collapse all the way to black holes. These holes are hard to detect, except when they are orbited by a companion star that provides a supply of fuel—they are then revealed by their emission of X-rays. In the centers of galaxies, runaway processes lead to the formation of supermassive holes that sometimes manifest themselves by generating the tremendous luminosity of quasars, or by squirting out the jets of strong radio sources.

The amount of matter locked up in black holes will continue to grow into the long-range cosmic future: more and more black holes will form; existing holes will grow by capturing gas, radiation, and, in the case of supermassive holes, even entire stars. Popular books sometimes overdramatize this process by asserting that everything in our Galaxy will eventually be engulfed by the central black hole. This is not realistic. Our Sun, and most of the material in the Galaxy, has so much angular momentum that it will never approach close enough to the center to be endangered. The Sun

*In the early 1960s, Roger Penrose stimulated a "renaissance" in relativity theory by introducing new mathematical techniques that could deal with situations where symmetry was lacking and Einstein's equations couldn't be solved exactly. With Hawking, he used these methods to prove theorems about when "singularities" would occur. Penrose has contributed to many other topics, including gravitational wave theory; he has also developed a fundamentally new approach (using the so-called twistor concept) to the study of space and time. More recently, he has speculated about new connections between gravity, quantum theory, and consciousness.*

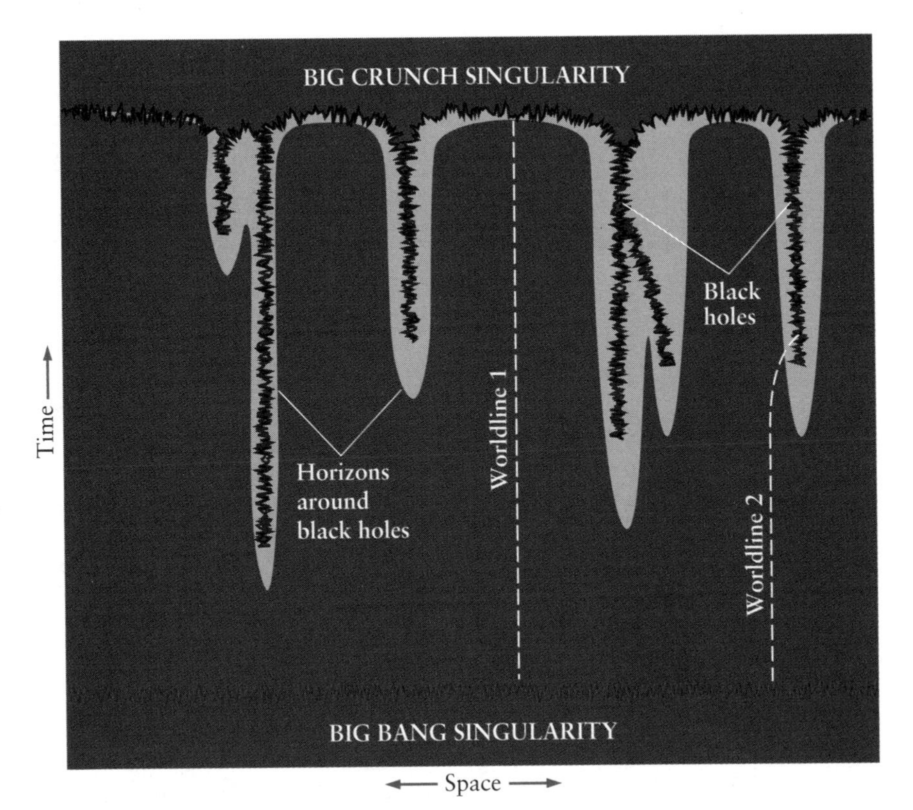


*Our Universe expanded smoothly after the big bang. But if it ends in a "big crunch," the final collapse will be irregular. The processes that led to the emergence of cosmic structure will continue to enhance density contrasts, even during the collapse phase. Regions that collapse or fall into black holes in effect reach a final singularity sooner than those that survive until the overall crunch. A clock on worldline 2, which falls into a black hole, would have recorded less time before it gets crushed than another clock, also synchronized to the big bang, that follows worldline 1. Black holes are represented by "stalactites" in this kind of spacetime diagram, and the final singularity is jagged rather than smooth.*

is far more likely to be kicked out of the Galaxy by close encounters with other stars.

What about the fate of our entire Universe? Each piece of material in the Universe exerts a gravitational pull on everything else and decelerates the cosmic expansion. The long-range forecast depends on what the average density of the Universe is—whether there is enough dark matter to supply the critical density needed to stop the expansion entirely and even reverse it. This dark matter may be composed of particles left over from the big bang; some may be in black holes. It is easy to calculate that if the average density exceeds about five atoms per cubic meter (the exact value depending on the Hubble constant), gravity will eventually bring the expansion to a halt. The Universe would then recollapse. The light from distant galaxies, instead of being weakened by the redshift, would be brightened as they fall toward us and display increasing blueshifts. Everything—even stars that have escaped into intergalactic space—would eventually be destroyed by the heat from blueshifted radiation, and shredded or compressed by the overwhelming gravity of the "big crunch." All our

*These results of computer simulations show how "gravitational instability" enhances density contrasts as the Universe expands. Starting at very early epochs (corresponding to very high redshifts) in the upper left-hand corner, the evolution of stars, gas, and dark matter is shown in projection within a "box" large enough to contain several hundred galaxies. The densest regions are brightest. This model Universe becomes increasingly inhomogeneous, as the most prominent density enhancements cluster and eventually merge to form still larger "lumps." Powerful telescopes are probing the Universe at the later frames, when galaxies are believed to have formed.*

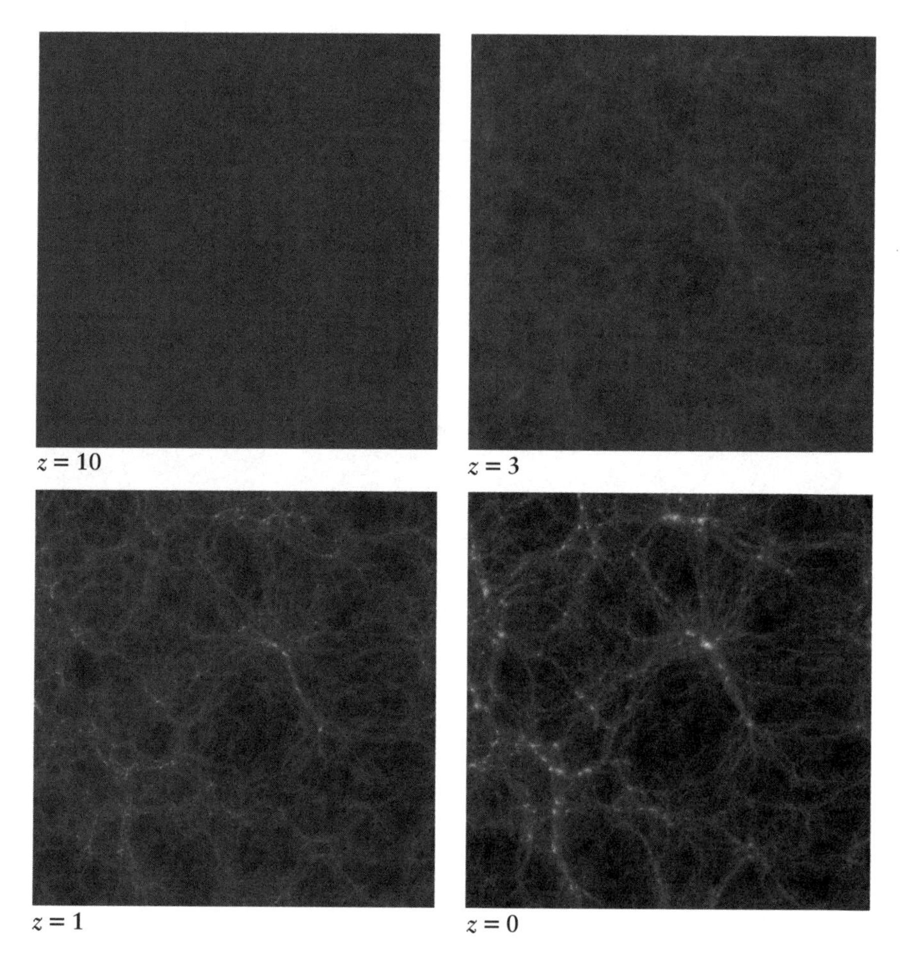


remote descendants would experience the fate of an astronaut who falls into a black hole.

If you're of an apocalyptic temperament, then head for a black hole—there you'll encounter a foretaste of the cosmic "big crunch," created by a local gravitational collapse. You should aim, preferably, for one of the largest black holes. The tidal forces are more gentle, and these black holes are so capacious that, even after falling inside, you would have several hours for leisurely observation before being shredded near the central singularity. A more cautious course would be to remain in orbit just outside the hole. From that vantage point, if the hole were spinning fast, you'd be safe and would have a blueshifted and speeded-up preview of the future of the external Universe.

That future could be quite different from the "big crunch." Public opinion among cosmologists is fickle, but most now believe that there isn't enough dark stuff ever to halt the expansion. Instead, the expansion will inexorably slow down but will never quite stop. In an ever-expanding Universe, stellar-mass black holes would continue to form, until all the gas was tied up in dead stellar remnants. The galaxies would not evolve in isolation, however, but rather would merge into larger systems. After a trillion years, the typical "galaxy" would be an amorphous system of stellar remnants (including stellar-mass black holes) as massive as a supercluster of galaxies is today.

Nowadays, Hawking's evaporation process is important only for microscopic black holes, but in a perpetually expanding Universe even the largest holes would be affected eventually. Given enough time, they would release, in the form of radiation, all the matter and energy they had ever swallowed. Stellar-mass black holes would take $10^{66}$ years to evaporate. Supermassive holes, the remnants of long-dead quasars, would take even longer—more than $10^{90}$ years for the largest so far discovered in galactic nuclei. But black holes might never finally disappear—indeed they could become ever larger. Long before they could be eroded by evaporation, black holes in the nuclei of merging galaxies would spiral inward and coalesce into still larger ones at the centers of the merged systems. This process could happen repeatedly as the scale of structure grew. The time it takes a hole to evaporate depends on the cube of its mass. Black holes could grow, by repeated coalescence, always remaining big enough to survive however long the Universe continued to expand.

So black holes, the causes of so many remarkable cosmic phenomena in our present Universe, may become ever more dominant even if the Universe continues forever. Hidden from view inside their "horizons," they hold secrets that transcend the physics we understand. The central "singularity" involves the same physics that occurred at the initial instants of the big bang and will recur again if the Universe recollapses. When we really understand black holes, we will understand the origin of the Universe itself.

# Appendix

## Gravity and Cosmic Dimensions

The force of gravity is described by an inverse square law, just like the electrical attraction between positive and negative charges. We can get a "feel" for the relative strength of gravity by considering its effect in a single hydrogen molecule, containing two protons. The gravitational attraction between two protons is only $10^{-36}$ as strong as the electrical attraction, which is of course why atomic physicists can safely ignore it. But every gravitating body has the same "sign" of gravitational charge: there is no near-cancellation of positive and negative, as there is for the electrical forces within any macroscopic body. Gravity can therefore become significant when huge numbers of atoms are packed together.

We can measure gravity's importance by working out the "gravitational binding energy" per hydrogen atom. This quantity, discussed in Chapter 1 in the cases of globular clusters and planets in the Solar System, measures the amount of energy needed to move a particle away from a mass to a large distance. The gravitational binding energy of a particle orbiting a mass $M$ at a distance $r$ depends on $M/r$. When $N$ atoms are packed together in a sphere (at a given density), the mass $M$ will be proportional to $N$ (and thus $M/r$ to $N/r$), and the radius $r$ will be proportional to $N^{1/3}$. The gravitational binding energy per atom (proportional to $N/r$) therefore goes as $N/N^{1/3}$, equal to $N^{2/3}$. Gravity starts off with a "handicap" compared to the electrical forces that hold atoms together. But as we consider larger and larger aggregations of atoms, it gradually "catches up": specifically, it gains by a factor of 100 for each thousandfold rise in $N$, and becomes dominant when $N$ is as large as $10^{36 \times 3/2} = 10^{54}$. This simple argument encapsulates the basic physical reason why stars are as massive as they are: a cold body containing more than $10^{54}$ hydrogen atoms (or protons) would be compressed, because gravity is too strong to be balanced by the same interatomic forces as in ordinary solids. A cold body with $N$ larger than $10^{57}$ is so completely crushed that it forms a black hole. (This is the famous "Chandrasekhar limit.") A hot body of this mass can become a hydrogen-burning star.

These and other "numerological" relationships can be instructively displayed on a single diagram such as the one on the facing page. In this diagram, mass is plotted vertically and radius horizontally. The plots are

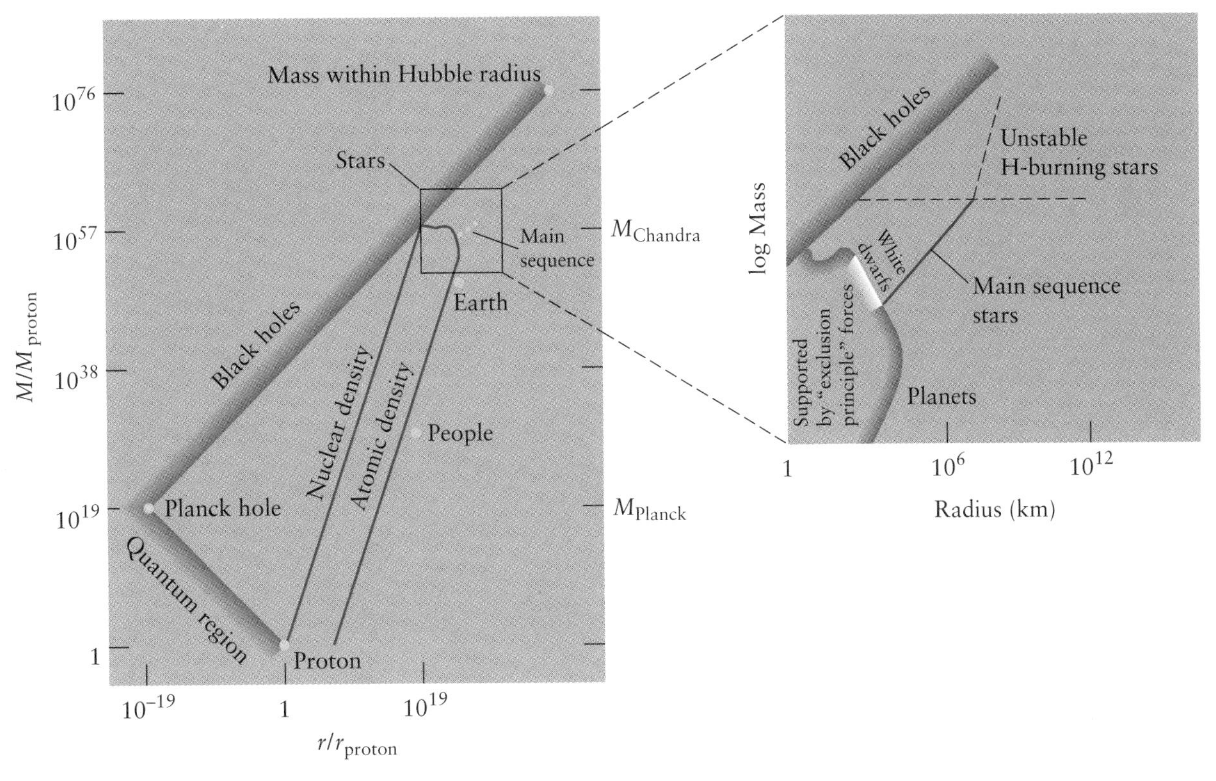


*This diagram summarizes the physics of stars, planets, and other bodies in a plot of mass against radius. Ordinary solid lumps lie on the line marked "atomic density": the mass, plotted in units of the proton mass, scales as the cube of the radius. That line would eventually cross the "black hole" line at a mass of about 100 million suns, except that it is curtailed before it can do so. The reason is that any object containing more than about* $10^{54}$ *atoms (about the mass of Jupiter) would be crushed by gravity to a higher density than an ordinary solid. The smaller "inset" diagram shows in more detail the limited range of scales in which stars can exist. Along the thin red diagonal line, the central temperatures (related to* M/r *by the virial theorem) are high enough for hydrogen fusion to occur. Main sequence stars lie along this line. Lower-mass stars need to become denser than higher-mass stars to achieve a given central temperature. No object below about 0.07 solar masses can burn hydrogen because the degeneracy pressure resulting from the exclusion principle would prevent it getting hot and dense enough. Stars of more than about 100 solar masses are supported mainly by radiation pressure, which makes them vulnerable to instabilities.*

given on a log scale: an increase by a factor of 10 in mass or radius is the same fixed length everywhere on each axis. Masses are given as multiples of the proton mass and lengths as multiples of the proton radius. The other distinctive feature of the diagram is that it spans an enormous range: the vertical (mass) axis goes from a single proton up to a mass that is larger by nearly 80 powers of 10.

The relation between the mass and radius of a black hole is shown in the diagram by a line with slope 1. Nothing can exist to the left of this line; anything very far to the right of it has such a large radius, in relation to its mass, that gravity is unimportant. Note that a black hole the size of a proton weighs as much as $10^{38}$ protons. Miniholes have to be so heavy simply because, for a single proton, gravity is weaker than the microscopic forces by just the inverse of that number. (Chapter 9 explains why holes with this particular mass are especially interesting.)

We can also plot on the same graph the mass–radius relation for objects of fixed density. These lines have slope 3 (mass being proportional to the cube of radius). Two such lines are shown. The lefthand one corresponds to the density of an atomic nucleus. It meets the "black hole" line at a mass of $10^{57}$ protons: neutron stars, confined by gravity to dimensions not much larger than their Schwarzschild radii, are indeed around this mass.

The righthand line corresponds to objects of "ordinary" density, where the atoms, rather than their central nuclei, are "touching." It is parallel to the nuclear density line, but displaced to the right by about 100,000—this is the factor by which an atom is larger than its nucleus. This straight line extends only up to about $10^{54}$ protons, which is roughly the size of Jupiter. In bodies of larger mass, gravity starts to overwhelm the ordinary atomic forces that hold solids together, crushing such bodies to higher densities. The behavior becomes especially interesting as the mass approaches $10^{57}$ protons. This part of the diagram is shown magnified at the right, and the legend explains the physical processes further. We can readily see, for instance, why stable stars (gravitationally bound fusion reactors) can exist only for a range of masses around $10^{57}$ protons.

Galaxies and AGNs are also plotted in the diagram; unfortunately there is no neat physical argument that determines their location in this diagram, as there is for stars.

The diagram extends up as far as the mass of the observable Universe (i.e., the mass within the Hubble radius). This mass is of course very poorly known, but it is about $10^{76}$ protons—the square of $10^{38}$. This is actually not a coincidence. It can be shown that the lifetime of a star is about $10^{38}$ times longer than the time light would take to cross an atom. So by the time the Universe is old enough for stars to have evolved in it, its "size" (the Hubble radius, or the distance light has traveled since the big bang) must be $10^{38}$ times larger than an atom. The Universe had to be old by the time we came on the scene; it therefore (if it is homogeneous) has to be large as well.

Also plotted in the diagram is a line of slope $-1$, which represents the uncertainty in the location of a mass (or, in a sense, its minimum size) set by quantum theory. This line meets the black hole line for a mass of $10^{-8}$ kilogram ($10^{19}$ protons) and a radius of $10^{-35}$ meter ($10^{-19}$ proton radius). Known respectively as the Planck mass and the Planck length, these are the dimensions at which quantum and gravitational effects overlap. The Planck scales lie many powers of 10 away from the "ordinary" scales confronted in the laboratory, or even in astronomy. We can also define a basic timescale, the Planck time, by dividing the Planck length by the speed of light. A big bang cosmological model could not be validly extrapolated back to the Planck time, whose value is $5 \times 10^{-44}$ second, until it is understood how quantum effects operate over the entire Universe.

The important message of the figure is that the characteristic masses and radii all involve simple powers of a single large number, whose vastness is a consequence of the weakness of gravity in the microworld of atoms. It is because gravity is so weak that structures dominated by its effects have to be so massive. If gravity were not quite so weak, the basic shape of the diagram would be unchanged, but the horizontal and vertical scales would be compressed: they wouldn't stretch over so many powers of 10. In a hypothetical Universe where gravity was stronger, cosmic structures could be neither so large nor so long lived. But, if the gravitational force were zero, no cosmic structures could exist at all. Paradoxically, the weaker gravity is (provided it isn't exactly zero), the grander are its consequences.

# Further Readings


Audouze, Jean, and Guy Israël, eds. *The Cambridge Atlas of Astronomy.* 3d ed. New York: Cambridge University Press, 1994.

Friedman, Herbert. *The Astronomer's Universe.* New York: W. W. Norton, 1990.

Hawking, Stephen W. *A Brief History of Time.* New York: Bantam Books, 1988.

Hawking, S. W., and W. Israel, eds. *Three Hundred Years of Gravitation.* New York: Cambridge University Press, 1987.

Kaler, James B. *Stars.* Scientific American Library. New York: W. H. Freeman, 1992.

Luminet, Jean-Pierre. *Black Holes.* New York: Cambridge University Press, 1992.

Novikov, Igor. *Black Holes and the Universe.* New York: Cambridge University Press, 1990.

Pagels, Heinz R. *Perfect Symmetry.* New York: Simon and Schuster, 1985.

Rees, Martin. *Before the Beginning: Our Universe and Others.* (Helix Books) Addison-Wesley, 1997.

Rees, Martin. *Perspectives in Astrophysical Cosmology.* New York: Cambridge University Press, 1995.

Silk, Joseph. *A Short History of the Universe.* Scientific American Library. W. H. Freeman, 1997.

Thorne, Kip S. *Black Holes and Time Warps: Einstein's Outrageous Legacy.* New York: W. W. Norton, 1994.

Verschuur, Gerrit L. *The Invisible Universe Revealed: The Story of Radio Astronomy.* New York: Springer-Verlag, 1987.

Wheeler, John A. *A Journey into Gravity and Spacetime.* Scientific American Library. New York: W. H. Freeman, 1990.

Will, Clifford M. *Was Einstein Right? Putting General Relativity to the Test.* 2d ed. New York: Basic Books, 1993.

# Sources of Illustrations

The depiction of stellar evolution on page 28 is by Tomo Narashima; all other diagrams and graphs are by Fine Line Illustrations.

*Cover image:* Christopher A. Perez and Robert V. Wagoner, Stanford University.
*viii, 179:* Drawing prepared by Joh Kagaya for Makoto Inoue, National Astonomical Observatory of Japan, © Hoshi No Techou.

**Chapter 1** *5:* AIP Niels Bohr Library, The Hebrew University of Jerusalem. *8:* Robert Bein, AIP Emilio Segrè Visual Archives. *13:* University of Canterbury, New Zealand. *15:* Isaac Newton, *A Treatise of the System of the World* (London, 1728), University Library, Cambridge. *17:* European Southern Observatory. *20:* Sidney Harris, *"You Want Proof? I'll Give You Proof!"* (New York: W. H. Freeman, 1991.)

**Chapter 2** *22:* J. Bally, D. Devine, and R. Sutherland. *27, left:* Anglo-Australian Observatory; *right,* C. R. O'Dell (Rice University) and John Bally and Ralph Sutherland (University of Colorado). *30:* Subrahmanyan Chandrasekhar. *33:* Modified from *Cambridge Atlas of Astronomy.* *34, 35:* Anglo-Australian Observatory. *36:* J. Marcaide. *40:* "On the nature of core-collapse supernova explosions," Ap. J., 1994, Adam Burrows, John Hayes, and Bruce Fryxell. *41:* S. J. Bell Burnell. *42:* F. R. Harnden, Jr., Smithsonian Astrophysical Observatory. *45:* From Jacob Shaham, "The oldest pulsar in the Universe," *Scientific American 256* (1987), 50.

**Chapter 3** *48:* Dana Berry and Keith Horne, Space Telescope Science Institute, NASA, 1990. *51:* Igor Novikov. *52:* A. C. Fabian. *53:* Ira Wyman. *56:* Adapted from Eric Chaisson and Steve McMillan, *Astronomy Today* (Englewood Cliffs: Prentice Hall, 1993), 132. *58, 59:* Modified from *Cambridge Atlas of Astronomy.* *62:* Edgar Allan Poe, *Tales of Mystery and Imagination,* illustrated by Arthur Rackham (London: George G. Harrap, 1935). *64:* Raymundo Baptista and Keith Horne, University of St. Andrews, 1995. *65:* Adapted from Philip Charles, Oxford University. *67:* G. Fishman et al., Burst and Transient Source Experiment, NASA. *68:* J. R. Eyerman, Life Magazine © Time Inc. *69: left,* "Cubic Space Division," 1952, M. C. Escher/Cordon Art B.V.; *right,* "Circle Limit IV," 1960, M. C. Escher/Cordon Art B.V. *71:* Modified from Jean-Pierre Luminet, *Black Holes* (New York: Cambridge University Press, 1992), 233.

**Chapter 4** *74:* K. Y. Lo, University of Illinois, and N. E. B. Killen, Australian Telescope National Facility. *76:* Thomas Wright, "Original Theory and New Hypothesis of the Universe" (London: Dawson, 1750). *77:* Richard Griffiths (Johns Hopkins University), the Medium Deep Survey Team, and NASA. *78:* European Southern Observatory. *79:* J. Bally, D. Devine, and R. Sutherland. *81:* Modified from Martin J. Rees, "Black holes in galactic centers," illustrated by George Retseck, *Scientific American 263* (1990), 56. *82: top,* European Southern Observatory. *82–83:* Joshua Barnes and Lars Hernquist. *84:* Richard Griffiths (Johns Hopkins University), the Medium Deep Survey Team, and NASA. *85:* Hubble Space Telescope WFPC 2 Team and NASA. *87:* Photograph by Tony Hallas. *88:* Hubble Space Telescope photograph by J.-P. Kneib, R. S. Ellis, and I. Smail. *90, 91:* Modified from Charles Alcock et al., "Possible gravitational microlensing of a star in the Large Magellanic Cloud," *Nature 365* (1993), 621. *96: left,* J. Trauger, JPL and NASA; right, Hubble Space Telescope WFPC 2 Team and NASA. *100:* John Bally et al. *101: left,* F. Yusef-Zadeh, M. R. Morris, and D. R. Chance, NRAO/AUI; *right,* Naval Research Laboratory.

**Chapter 5** *102:* Space Telescope Science Institute, NASA. 105: The Mary Lea Shane Archives of the Lick Observatory. *106: top,* John Biretta, Space Telescope Science Institute, NASA; *bottom,* Patrick Shopbell (California Institute of Technology), Joss Bland-Hawthorn (Anglo-Australian Telescope Board), and Brent Tully (University of Hawaii). *108:* Palomar Observatory. *109:* Chris Carilli and Robert Perley, NRAO. *110:* California Institute of Technology. *118:* E. E. Salpeter Collection, AIP Emilio Segrè Visual

Archives. *119:* AIP Emilio Segrè Visual Archives. *123:* D. B. Sanders et al., "Continuum energy distributions of quasars: shapes and origins," *The Astrophysical Journal 347* (1989), 29.

**Chapter 6** *126:* NRAO/AUI. *130:* University of Cambridge. *132:* Roger Ressmeyer, © Corbis. *135: left,* Anglo-Australian Observatory; *center,* NRAO/AUI; *right,* High Energy Astrophysics Division, Harvard-Smithsonian Center for Astrophysics. *136:* NRAO. *137:* Adapted from Bruce Margon, "The bizarre spectrum of SS 433," *Scientific American 243* (1980), 54. *138: left,* David Parker, SPL/Photo Researchers. *139:* S. A. Baum and R. Elston, NRAO. *140:* M. Norman, J. O. Burns, et al. *141:* left, NRAO; *right,* C. P. O'Dea and F. N. Owen. *142:* Dinshaw Balsara and Michael Norman, National Center for Supercomputer Applications. *143: top,* R. A. Perley and A. G. Willis, NRAO; *bottom,* E. B. Fomalont, C. Lari, P. Parma, R. Fanti, and R. D. Ekers, NRAO/AUI. *145:* From Kenneth I. Kellermann and A. Richard Thompson, "The very-long-baseline array," illustrated by George Retseck, *Scientific American 258* (1988), 54. *146:* R. C. Walker, J. M. Benson, and S. C. Unwin, NRAO. *147:* Luigina Feretti (Bologna, Italy), Robert Laing (Cambridge), and Alan Bridle and Robert Perley (NRAO). *148:* Photographs by Anglo-Australian Observatory. *150:* Ann Wehrle, Infrared Processing and Analysis Center, Jet Propulsion Laboratory and California Institute of Technology. *151:* From Roger D. Blandford, Mitchell C. Begelman, and Martin J. Rees, "Cosmic jets," *Scientific American 246* (1982), 124. *155:* From Richard H. Price and Kip S. Thorne, "The membrane paradigm for black holes, *Scientific American 258* (1988), 69. *156:* Roger Blandford. *157:* J. Hester (Arizona State University), the WFPC 2 Investigation Definition Team, and NASA. *158:* R. M. Hjellming and M. P. Rupen, NRAO.

**Chapter 7** *160:* Alan Dressler (Carnegie Institution) and NASA. *165:* Data from D. P. Schneider et al., "PC 1247 + 3406: An optically selected quasar with a redshift of 4.897," *The Astronomical Journal 102,* 837. *166, 167:* Data from M. Schmidt, *Highlights of Astronomy 8* (1989), 31. *172:* Space Telescope Science Institute, NASA. *173:* J. Kormendy and R. Bender, *The Astronomical Journal* (in press). *174:* T. R. Lauer (NOAO), NASA. *180, 181:* M. Miyoshi et al. "Evidence for a massive black hole from high rotation velocities in a sub-parsec region of NGC4258," *Nature 373* (1995), 127. *183:* Adapted from C. R. Evans and C. S. Kochanek, "The tidal disruption of a star by a massive black hole, *The Astrophysical Journal 346* (1989), L31.

**Chapter 8** *186, 191:* Christopher A. Perez and Robert V. Wagoner, Stanford University. *190:* Diagram modified from Jean-Pierre Luminet, *Black Holes* (New York: Cambridge University Press, 1992), 143; computer simulation by Jun Fukue. *192:* Data from Y. Tanaka et al., "Gravitationally redshifted emission implying an accretion disk and massive black hole in the active galaxy MCG-6-30-15," *Nature 375* (1995), 659. *193:* Adapted from Kip S. Thorne, *Black Holes and Time Warps: Einstein's Outrageous Legacy* (New York: W. W. Norton, 1994), 292. *196:* Kevin Rauch, Canadian Institute for Theoretical Astrophysics (CITA), University of Toronto. *199:* From R. Ruthen, "Catching the wave," illustrated by George Retseck, *Scientific American 266* (1992), 90. *200: left,* J. M. Weisberg et al., "Gravitational waves from an orbiting pulsar," *Scientific American 245* (1981), 74; *right,* J. H. Taylor, Jr., "Binary pulsars and relativistic gravity," *Reviews of Modern Physics 66* (1994), 717. *202:* Max Ruffert, Max-Planck-Institut für Physik und Astrophysik, Munich. 203: Adapted from R. Ruthen, "Catching the wave," illustrated by George Retseck, *Scientific American 266* (1992), 90. *204:* Adapted from Kip S. Thorne, *Black Holes and Time Warps: Einstein's Outrageous Legacy* (New York: W. W. Norton, 1994), 383. *205:* LIGO. *206:* Kip S. Thorne. *207:* NRAO. *208, 209:* European Space Agency.

**Chapter 9** *212:* "Over the Great Wall," digital image by Margaret Geller and enhanced by Roger Ressmeyer, © 1994 Smithsonian Institution/Corbis. *219:* Greg Bryan and Michael Norman, National Center for Supercomputer Applications. *221:* Bernard Carr, *Annual Review of Astronomy and Astrophysics 32* (1994), 531. *224:* Stephen Shames/Matrix. *225:* Modified from S. W. Hawking, "The quantum mechanics of black holes," *Scientific American 236* (1977), 34. *232:* Anthony Howarth/Photo Researchers. *234:* Carlos Frenk, University of Durham, and the Virgo Consortium.

# Index

*Note: Page numbers in italics indicate illustrations.*

Selected hardcover books in the Scientific American Library series:

GENES AND THE BIOLOGY OF CANCER
by Harold Varmus and Robert A. Weinberg

EARTHQUAKES AND GEOLOGICAL DISCOVERY
by Bruce A. Bolt

THE ORIGIN OF MODERN HUMANS
by Roger Lewin

THE LIFE PROCESSES OF PLANTS
by Arthur W. Galston

THE ANIMAL MIND
by James L. Gould and Carol Grant Gould

THE EMERGENCE OF AGRICULTURE
by Bruce D. Smith

AGING: A NATURAL HISTORY
by Robert E. Ricklefs and Caleb Finch

INVESTIGATING DISEASE PATTERNS: THE SCIENCE OF EPIDEMIOLOGY
by Paul D. Stolley and Tamar Lasky

LIFE AT SMALL SCALE: THE BEHAVIOR OF MICROBES
by David B. Dusenbery

CYCLES OF LIFE: CIVILIZATION AND THE BIOSPHERE
by Vaclav Smil

CONSERVATION AND BIODIVERSITY
by Andrew P. Dobson

COSMIC CLOUDS: BIRTH, DEATH, AND RECYCLING IN THE GALAXY
by James B. Kaler

THE ELUSIVE NEUTRINO: A SUBATOMIC DETECTIVE STORY
by Nickolas Solomey

LASERS: HARNESSING THE ATOM'S LIGHT
by James P. Harbison and Robert E. Nahory

Other Scientific American Library books now available in paperback:

POWERS OF TEN
by Philip and Phylis Morrison and the office of Charles and Ray Eames

THE EVOLVING COAST
by Righard A. Davis

THE NEW ARCHAEOLOGY AND THE ANCIENT MAYA
by Jeremy A. Sabloff

THE HONEY BEE
by James L. Gould and Carol Grant Gould

EYE, BRAIN, AND VISION
by David H. Hubel

PERCEPTION
by Irvin Rock

HUMAN DIVERSITY
by Richard Lewontin

SLEEP
by J. Allan Hobson

THE SCIENCE OF WORDS
by George A. Miller

DRUGS AND THE BRAIN
by Solomon H. Snyder

BEYOND THE THIRD DIMENSION
by Thomas F. Banchoff

SEXUAL SELECTION
by James L. Gould and Carol Grant Gould

PLANTS, PEOPLE, AND CULTURE
by Michael J. Balick and Paul Alan Cox

ATMOSPHERE, CLIMATE, AND CHANGE
by Thomas E. Graedel and Paul J. Crutzen

If you would like to purchase additional volumes in the Scientific American Library, please send your order to:

Scientific American Library
41 Madison Avenue
New York, NY 10010